MW01487171

Multi-Component Treatment Manual for Post-Traumatic Stress Disorder

Including Strategies from Clinical Psycho-Physiology and Applied Neuroscience

Dr. John A. Carmichael R Psych

The ISNR Research Foundation
International Society for Neurofeedback and Research
email: cynthia@isnr.org
2011

2011 ISNR Research Foundation
1925 Francisco Blvd. E. #12
San Rafael, CA 94901
cynthia@isnr.org

All rights reserved

"Multi-Component Treatment Manual for Post-Traumatic Stress Disorder: Including Strategies from Clinical Psycho-Physiology and Applied Neuroscience"" is a publication of the ISNR Research Foundation. Opinions expressed herein are those of the author and do not necessarily reflect the official view of ISNR-RF.

No part of this book may be reproduced, translated, stored in a retrieval system, or transmitted, in any form or by any means, electronic, mechanical, photocopying, microfilming, or otherwise, without written permission from the Publisher.

Printed in the United States of America

ISBN: 978-0-9846085-2-2

Cover photograph: Jeevs Sinclair http://77percent.com

Contact the Author:
Dr. John A. Carmichael R Psych
Clinical, Police, and Military Psychology
Practice Restricted to Reactions to Traumatic Experiences
215 Arrowstone, Kamloops, British Columbia, Canada, V2C 1P8
Telephone: (250) 374-3215
email: dr.john@telus.net

TABLE OF CONTENTS

Preface

My solo private practice is restricted to Post-Traumatic Stress Disorder (PTSD), a psychological condition that can result from exposure to traumatic stress.   have been interested in this area of service-provision since about 1985. To date, I have seen well over 200 clients with this disorder most of whom have come from police or military backgrounds. Also, I have seen other adults with symptoms of PTSD resulting from such traumatic incidents as life-threatening illness, physical and/or sexual assault, motor vehicle accidents, and the tragic death of a loved one.

At the clinical level, most clients with PTSD have any of a number of associated issues, including depression, panic disorder, sleep disturbance, psycho-physiological conditions, chronic pain, and cognitive impairment.   Such conditions need to be addressed in addition to the symptoms of PTSD.

In addition to revising this manual continuously since beginning it in 1985 and providing it to clients, I have published on this topic and given workshops to professionals at international meetings. Also, I am qualified as an expert witness in regards to both PTSD and depressive conditions in the Supreme Court of British Columbia.

In terms of post-secondary education, in 1964 I obtained my B.A. from Mount Allison University with a double major in philosophy and religious studies. From 1966 to 1968 I completed undergraduate courses in psychology at the University of British Columbia preparatory to applying to graduate school. I obtained my M.A. in clinical psychology from the University of Victoria in 1970 and completed requirements for my Ph.D. as well as the certificate in clinical competence from that university in 1975.   From 1972 to 1995, I worked first as a clinical psychologist with the Provincial Mental Health Service before managing a portion of this service for five years.   By 1980 I had established a part-time private practice that became full-time in 1995 after I retired from government service.

I am registered by the College of Psychologists of British Columbia to practice clinical psychology and am registered also with the Canadian Health Services Providers in Psychology. Additionally, I am a member in good standing of the Canadian Psychological Association, the American Psychological Association, the Psychopharmacology Division of the APA, the Psychological Trauma Division of the APA, the Military/Emergency Trauma Special Interest Group (SIG), the Military Psychology Division of the APA, the International Society for Traumatic Stress, the Association for Applied Psychophysiology and Biofeedback, the International Society for Neurofeedback and Research, the Society for Applied Neuroscience, the International Association for the Study of Pain, and the Clinical and Neuroscience Society.I have long been committed to providing competent services that are up to date with the continuing developments in such professional disciplines as psychology, psychiatry, psycho-physiology, and affective neuroscience in relation to my area of practice. Since it is one of my interests as well as how I earn my living, I spend an average of 500 hours each year reading relevant material published in professional journals/books, attending post-doctoral institutes/workshops, attending meetings of the professional associations to which I belong, and in supervision/consultation arrangements with recognized experts whenever I begin a new area of treatment. Such experts have included Dr. Tony Hughes, Mr. John Anderson, the late Dr. Jeff Cram, the late Dr. Joseph Horvat, and currently Dr. Bob Thatcher.

Simply put, my overall objective is to help clients reduce their suffering and maximize their well-being. Thus, relief from distressing or dysfunctional thoughts, feelings, bodily sensations, and actions is emphasized along with their replacement with rational thoughts, physical and mental calmness, positive feelings, effective actions, and successful stress management strategies.

The methods I use during the phases known as assessment, diagnosis, treatment/intervention, and follow-up/relapse-minimization are referred to in the professional literature as the *biopsychosocial* approach, since I take into account the multiple determinants of human thoughts, bodily sensations, feelings, and actions. Such "causes" can include ones that are biological, cognitive, spiritual, historical, environmental (physical and social/cultural),psycho- physiological, as well as related to brain functioning. In this regard, each client is a unique human being who has an active part to play in the healing process that is facilitated by the therapeutic relationship we are able to establish. One of the client's roles is to implement consistently interventions we agree to try with the objective of taking advantage of the client's own natural healing power.

In regards to intervention, I am committed to implementing only those aspects with clinical or research evidence of effectiveness. This means that my practice is *informed by evidence*. Given my extensive continuing education activity, I remain current in the research, clinical base, and implementation details of common approaches to the individualized treatment of trauma conditions. Additionally, I utilize new developments with a good clinical, research, or theoretical base, thereby remaining on the "cutting edge."

A careful assessment including structured clinical interviews, an evaluation of psycho-physiological (mind-body) interactions, administration of standardized psychological tests, a screening assessment of cognitive status, and the gathering of other relevant information is the foundation upon which I build a formulation-driven, evidence-informed, and effectively sequenced treatment plan. For some clients, direct measurement of brain processes is completed also. Thus, depending on their presentations, clients receive a specific range of interventions from the menu in this treatment manual; not every client needs every intervention. Furthermore, I work closely with physicians, since some clients within my speciality area require medical management as well as psychological treatment. Sometimes crisis management is required before beginning an assessment since police and military clients often wait until the very last minute before calling me. Consequently, when these clients are seen, (a) their psychological condition is usually towards the severe range of intensity and (b) other aspects of their lives are out of control.

Once the assessment is completed, I discuss the results with the clients client and provide them with a copy of the assessment report. Also, the report is forwarded to the client's physician and often includes suggestions about medical conditions/he might want to investigateand/or medications that the physician might want to consider. Psychologists in Canada are not authorized to prescribe medication, although I stay current field in-so-far as PTSD and related conditions are concerned.

Thereafter, the psychological treatment plan is implemented one step at a time. Clinical observations, repeat testing, information from the client's significant others, as well as client feedback are the basis for moving to the next step or for modifying the treatment plan.

As I become aware of new information, I insert it into the treatment manual. Then, according to where we are in the treatment sequence, the relevant portion of this treatment manual is given to the clients. This serves many functions such as keeping clients well- informed and aiding memory, which often is less than optimal among those with PTSD. Clients are encouraged to retain the various sections in a binder and to write down any questions, implementation problems, and insights that arise between sessions. It is recommended that they bring the binder to each session. May this treatment manual prove helpful to clients as they learn how best to manage their conditions.

Then, late in 2009 I offered the most current version of the manual to the International Society for Neurofeedback and Research.   Their Research Foundation would publish the manual, advertise it, and sell it to interested professionals. The Foundation would retain all monies from sales and direct it in furtherance of their research objectives.   The current edition of the manual became available in 2010. May this treatment manual be helpful to colleagues as they provide psychological services to their clients.

While I developed the steps and details of the treatment method, many others contributed.
First, I am indebted to all of those who published thoughts and studies on PTSD and related matters. Some are noted in the reference section in Appendix VIII. My belief is that clinical practice should be guided by research findings and that clinical experiences should play a role in informing the research enterprise.
Second, I am thankful for the discussions I have had with other clinical practitioners and to those mentioned in the manual who have had a special role to play as my mentors.
Third, I owe a great deal to my clients with PTSD who have shared their struggles and insights as well as taking the chance to learn and implement interventions that were not typical treatment methods at that time.
Fourth, a special mention should be made about the support of the ISNR Research Foundation and in particular its founding president, Dr. David Trudeau, who agreed to publish this manual.
Fifth, a thank-you to Ms. Marsha Calhoun for her copy-editing expertise and to Dr. Cynthia Kerson for shepherding the book through the publication process.
Last but certainly not least, this manual and everything leading up to it would not

have been completed without the continuing support of my wife, Lynne; marrying her in 1965 was the single smartest decision I ever made.

The Assessment of PTSD

In summary, although the symptoms of PTSD have been known for centuries (but under different names), the intensive study of PTSD is fairly recent. Nevertheless, good research that continues to emerge demonstrates clearly that many psycho-physiological and other process are involved in this condition. Consequently, my clinical experience is that such processes need to be assessed so that therapeutic attention can be directed to as many relevant factors as possible if relief from PTSD is to be achieved.

The complexity of PTSD is well-noted, as found for example in a chapter by psychologist Dr. John P. Wilson in the book *Assessing Psychologial Trauma and PTSD* (Wilson & Keane 2004), who described PTSD as a

> syndrome of dynamically related psycho-biological processes that include the brain, the nervous and hormonal systems, psychological systems of memory, cognition, emotion, motivation, perception, and behavioral expression of the organismic changes caused by the trauma. (p. 8)

Further, Clark and colleagues conclude as follows from their 2009 extensive review of research studies on the role of various electro-physiological brain processes in PTSD:

> Thus, electrophysiological measures provide objective evidence for PTSD as a disorder of information processing. Studies indicate abnormal asymmetry and activation in basal levels of cerebral arousal, a loss in inhibitory processing mechanisms relevant to the normal filtering of stimulus information, and concomitant difficulties with evaluating and remembering the significance of everyday information. There are also aberrations with the processing of threat-related information, with an orienting bias toward threatening or trauma-related stimuli.
>
> Electrophysiological measures also provide some insight into the role of fear and related structures in the pathophysiology of PTSD. A recent review by Southwick clearly demonstrates the involvement of the hypothalamic-pituitary-adrenal axis in the pathophysiology of PTSD, noting that long-standing neurobiological responses to fear and stress are likely to be maladaptive and contribute to the development of the disorder. Neuro-biological markers of the fear responses identified in the present review include a hypersensitivity to traumatic stimuli, as reflected in enhanced autonomic response (e.g., HR, BP, SCR, EMG, startle response), and enlargement of both the P3, LPC, and CNV components of the ERP to traumatic stimuli. (p. 97)

During the early years of the railway in England, when accidents were very common, the symptoms of what we now call PTSD were variously referred to as tunnel disease, railway spine, and railway brain without appreciating that train accidents were the cause of the symptoms.

Also, it should be noted that the symptoms of PTSD among military personnel have been identified for thousands of years (as early as in *The Iliad* by Homer) and described variously since then as irritable heart, soldier's heart, Da Costa's syndrome, traumatic neurosis, fright neurosis, shell shock, battle fatigue, war sailor syndrome, Vietnam war syndrome, and Gulf War syndrome.

However, the intensive study of PTSD did not really begin until the 1980s, spurred in part by returning U.S. military personnel who saw action in the Vietnam War and by its clear inclusion into the diagnostic systems of the day. Thus, there is much we still do not know about PTSD. On the positive side, there is much we do know and helpful research is being published at a breath-taking pace. Thus, there is much we can do to reduce symptoms as noted below. Also, as discussed later, the combination of approaches I use as described in this manual result in remission of PTSD among police and military veterans over 90% of the time. This is in contrast to most published studies whose measure of treatment impact is a percentage decrease in symptoms; such results, while helpful to clients, usually are the preferred result of an actual remission of the condition.

Even though PTSD has been part of the diagnostic system since the 1980s, efforts to use other terms continue partly because as individuals and as a society, we do not wish to believe that traumatic incidents can lead to psychological symptoms. We accept that environmental things (germs and toxins) can affect our bodies but not that they can affect our mind. A recent example of this tendency is that after Iraq invaded Kuwait, returning western service personnel pushed for legitimization of the Gulf War syndrome as a result of their belief that toxic chemicals must have been used by Iraq and these chemicals were causing their symptoms. Then a study of personnel with Gulf War syndrome concluded that only one unique symptom remained once characteristics associated with both depression and PTSD were removed.

Due to the numerous and complex combinations of factors that can lead to PTSD, over the years I have developed the following assessment strategy.

Once a client first gets to my office, s/he is given information in writing and, after discussion of the contents, signs an *informed consent* allowing us to proceed. A copy of this is found in Appendix I of this treatment manual. Also at this time and based on decades of experience with police and military folks suffering from PTSD, I inform the client that I need only brief and general information about the traumatic incidents. *I do not delve deeply into traumatic incidents* at this time since (a) doing so can make things worse because most clients do not yet have in place effective strategies for adjusting to the trauma and symptoms (that's why they are seeing me), (b) speaking of one trauma seems to open the door to being flooded by the memories of multiple additional traumatic experiences, (c) many studies have demonstrated a 30% drop-out rate when a major focus has been on the traumatic experiences, and (d) it is not necessary for initial planning of the treatment methods I have found to be the most useful. Later on in treatment when other strategies are well in place, clients often will describe details spontaneously thereby giving us a chance to develop reality-based views of the trauma(s) as required. Admittedly, this is a significant departure from traditional approaches in the treatment of PTSD but it is justified by my results below as well as by outcomes reported by other professionals. For example, discontinuation rates are below 5% and remission rates are above 90%.

Typically, the next step is to ask the client what brings/him her to this meeting. Often, the client will mention symptoms, note briefly the type of trauma experienced, and mention other issues of importance. Since s/he knew this question was coming, usually the information is presented rapid fire, and not always in any structured order. If the client apologizes for this, I note that it is my job to make sense of what s/he says and there is no need to worry about a logical sequence.

After having their say, most often clients become more settled than when they entered and we can proceed with my requirements for the assessment. In doing so, a number of assessment tools are used.

First, especially if the assessment relates to Veterans Affairs Canada, I use the Clinician Administered PTSD Scale (CAPS), a structured interview about the type and frequency of symptoms. Many, including Veterans Affairs in both Canada and the United States, consider the CAPS as the gold standard for the assessment of PTSD. CAPS is available from the US Veterans Administration on the internet.

Second, clients complete the Trauma Symptom Inventory (TSI), a standardized psychological test developed by Dr. Briere, a recognized expert in the field of PTSD. The TSI has good psychometric properties and I use it also from time to time throughout treatment as a quick way to determine progress. This test is available through PAR in Lutz, Florida.

Third, I obtain basic information using the structured clinical interview I have developed over the past 30 years. A copy of the most recent revision is found in Appendix II. This is supplemented by client completion of the Personal History Checklist for Adults (PHC) and the LISRES-A, a checklist survey of stressors. The LISRES is standardized while the PHC is not. Both are available from PAR in Lutz, Florida.

Fourth, clients complete a number of additional standardized psychological tests such as the Personality Assessment Inventory (PAI), the MultiScale Dissociation Inventory (MDI), the Beck Depression Inventory (BDI-II), the Beck Hopelessness Scale (BHS), the Beck Scale of Anxiety (BAS), and the State-Trait Anger Expression Inventory (STAXI-2). All these tests are available through PAR in Lutz, Florida.

Fifth, as described in detail later in the manual, I use a psycho-physiological protocol that records both a client's autonomic nervous system functioning and level of muscle tension directly without being either invasive or embarrassing. Typically, I clip a lead onto one finger to measure blood oxygen levels, place a strain gauge around the middle to measure aspects of respiration, attach a cuff to measure blood pressure, and place sensors on the surface of the skin to measure skin conductance, heart rate, finger temperature, and muscle activity. This is not quite the simple assessment it appears to be since the effects of medications need to be factored in, use of appropriate norms are required (knowing, for example, that a 15% to 20% difference in sEMG between right and left paraspinals is normal), clients should not be speaking during the evaluation since it will alter levels, and it is important to remain current with the new studies that are published.

Sixth, clients complete the MicroCog, a computer-administered and standardized screening test for cognitive functioning. This test is available through Pearson. Also, some clients complete one or more sections of the Cognitive Symptom Checklists that are available through PAR in Lutz, Florida.

Seventh, if my structured clinical interview and/or the sleep observation sheet completed by the client's bed partner indicates a sleep disorder, then specialized assessment methods are recommended such as polysomnography (PSG).

From the above, my basic multi-component treatment plan is generated. Thereafter, the resulting steps of that plan are implemented, usually in the sequence noted in the remainder of the book. This sequence is the one I have discovered to be the most effective and there is a certain logic to it.

If my basic multi-component treatment approach described below does not result in a complete remission of symptoms, then I complete a quantitative EEG (qEEG). Results of the client qEEG are compared with the NeuroGuide normative sample and processed by LORETA for precise localization of identified anomalies. From these, protocols for the biofeedback training of the central nervous system (BFT-C) are developed. All of this is discussed later in the book.

Treatment Step A: Preliminary Information

After the assessment is completed and prior to our first treatment session, clients have the opportunity to read the information in Section A below so that they have some basic information before proceeding. During the first fact-to-face treatment session, this information is discussed as needed.

Contents

A-01: Purpose
The primary purpose of this treatment manual is to provide clients with current information about PTSD. Also, the continuous revising of this manual is good discipline for me as I try to make sense of all the information that continues to emerge at conferences and in publications, as I integrate the valuable insights of clients as they struggle with their conditions, and as I reflect on my own clinical experiences. Thus, the manual represents my understanding to date about PTSD.

Much is known about this condition and helpful new information continues to be circulated at a breath-taking pace if you know where to look for it. On the other hand, we still have much to learn about this nasty condition. Thus, this manual is a *work in progress* and will continue to be so for the foreseeable future.

A-02: Client Groups I Have Seen
Since the mid 1980s, many civilians who came to my attention were victims of a single episode of trauma that led to the symptoms of PTSD. This psychological condition is one of the most common among civilians. The results of many studies indicate that an average of 10% of females and 5% of males will meet the diagnostic criteria for PTSD at least once in their lifetimes. While not everyone who experiences a single traumatic event develops PTSD, research indicates that about 17% do. Thus, some refer to PTSD as a failure of natural recovery since about 83% of civilians recover from single-episode trauma pretty much on their own.

At about the same time, I began also to see many police officers, both active and retired, suffering from PTSD and associated symptoms. I was functioning as a police psychologist with responsibilities that included assessments of fitness for duty (for example, for deployment to isolated postings or UN duty and for assignment to specialized units such as the emergency response team, the underwater recovery team, and undercover operations), group interventions immediately after a traumatic incident, and consultation to the hostage negotiation team.

Then, starting in the 1990s veterans of the Canadian Forces with PTSD began to be referred to me by Veterans Affairs Canada.

Since 2000, 80% to 90% of my solo private practice has been focused on PTSD among active police and military personnel as well as veterans of these two groups. Most of the remaining 10% to 20% of clinical time has been spent working with adult civilians suffering from PTSD as a result of a motor vehicle accident, the traumatic death of a loved one, a life-threatening illness, or subsequent to a physical attack.

A-03: PTSD Due to Duty Related Accumulated Trauma (DRAT)
Rather than being associated with a single traumatic episode, symptoms among the police and military folk seemed more related to multiple traumatic events they experienced in their careers. For example, patheits who were police officers could easily refer to many traumatic incidents and could count hundreds of them by referring to their notebooks; most of my clients had 15 to 20 years of front-line policing before coming to my attention. The military veterans I saw had been involved in long periods of combat (WWII and Korea), shorter periods of combat (the first Gulf War), or peace-keeping / peace support operations (for example, in Gaza, Cyprus, Bosnia, and Rwanda) where their lives had been in constant danger and where they witnessed much death and serious injury. Others had experienced or witnessed traumatic incidents during military training.

My clinical experience later confirmed by research evidence, indicates that for those exposed to repeated or accumulative trauma, such as police officers and military personnel, the symptoms are more pronounced, the consequences are more far-reaching, and the time required for successful treatment is much longer than than for those exposed to a single incident. Additionally, the trajectory is different. For example, the 2006 book by Calhoun and Tedeschi cites a 2002 study by Johensen and colleagues demonstrating that some PTSD symptoms tend to decrease during the first 12 months ost-trauma among those exposed to a single trauma but tend to increase during the first 12 months among those exposed to multiple incidents.

Until the number of traumatic incidents reaches a threshold or other factors are present such that significant symptoms emerge, generally police and military folk continue to function during any particular incident by going into "automatic mode." Doing so allows them to function and survive under difficult circumstances as well as to remain members of the team. One analogy I use to help police and military clients understand what has happened to them is the balloon. It stretches as more and more air is pumped into it. But it reaches a point where it can no longer expand and when one more unit of air is added, it bursts. Multiple trauma is like that; the last traumatic event or even a non-traumatic stressor can finally exceed our capacity to cope or adjust or contain it.

Then a number of physical and psychological systems break down (as discussed later) and the person can no longer function as before.

Also, whether or not they receive appropriate treatment, as long as they remain operational, their duties are such that they continue to be exposed to trauma both in terms of what they see and the actions against them that have the potential of being life-threatening. As well, they observe what has happened to colleagues and civilians. Moreover, coffee breaks and other informal gatherings of peers usually includes trauma-related conversations that serve as reminders of their own experiences. Furthermore, other reminders of trauma are ever-present, such as driving by crime scenes, anniversary dates such as Remembrance Day, and what is presented by the media as news and "entertainment". Often, it does not occur to clients to avoid these triggers. All of these experiences can increase the difficulty of putting the pieces together again.

In addition, a number of studies indicate that the type and seriousness of stress is markedly worse for police officers than for providers of other emergency services such as those associated with fire/rescue or ambulance work. No doubt, the same is true for military personnel.

Unfortunately, generally police officers and military personnel wait and suffer until they are in really bad shape. They may be unable to concentrate or remember important things, stay closed up in the house for days on end, have increasing and persistent thoughts of suicide, and be given ultimatums from their partners such as "Get help or I and the children will leave" before thinking that they may need help for escalating symptoms. One study of police officers suggested stated that they waited an average of 7 years after the onset of symptoms before seeking help. Even when they do seek help, too many of them never meet with a psychologist who is competent in treating PTSD due to DRAT (see definition below). For example, in one study, only 7.3% of police officers with PTSD following the 1992 riots in Los Angeles indicated that they would see a psychologist. The same is true among military personnel and veterans. Most of the WWII veterans I have seen have struggled with PTSD off and on for 60+ years without ever telling family or health care professionals until symptoms increased dramatically when the patients were in their late seventies and into their eighties.

As research in this area grew and as I gained experience and knowledge, it became clear that there were some significant additional differences between police and military folks and those suffering from a single traumatic incident, Thus, some modifications in traditional methods of treatment were necessary in treating police and military patients in order to get their symptoms under control. Because of this and since the conditions were the direct result of their work, I named this PTSD Due to Duty-Related Accumulated Trauma (DRAT). Since then, my perceptions have been confirmed by others and by research studies indicating that the lifetime number of different traumas predicts both the number and severity of symptoms.

A-04: PTSD as a Democratic Condition

Also, in the research studies and my own experience, it became clear that whether the client was a civilian or not, PTSD occurred independent of such variables as race, religion, economic status, gender, educational attainment, sexual orientation, and the historical era in which individual lived. Thus, PTSD is very much a democratic condition.

Canadian Army Lt. General Romeo Dallaire was commanding officer of the UN Forces in Rwanda. While there, he observed many examples of "ethnic cleansing." He and his troops were under fire from one side or the other frequently. Also, he had to follow his UN mandate in spite of a public radio station that openly advocated killing members of the ethnic minority. Moreover, he had to deal with lies, games, counter-strategies, and so on while trying to achieve agreements between opposing forces after each skirmish. Furthermore, he had to manage in spite of an inept and unresponsive UN, which effectively placed his troops under extreme difficulties. All of these experiences are described in the book he wrote, *Shake Hands with the Devil: The Failure of Humanity in Rwanda* (published by Random House in 2003). Towards the end of his tour and for years afterwards, he suffered from severe PTSD that included suicidal ideation. Courageously, he made his condition known publicly. If a person of his stature, military training, and expertise can develop PTSD, then one can conclude that no one is immune.

A-05: Co-occurring Conditions
Among the folks I have seen, both civilian and police/military, often their PTSD was accompanied by one or more *associated conditions* such as depression, day and/or night-time panic attacks, sleep disturbances/disorders, chronic pain, and psycho-physiological problems (for example, headache, functional cardiovascular conditions, fibromyalgia syndrome, and irritable bowel syndrome). In fact, most of my clients with PTSD have 2 to 4 additional conditions. Others have reported that this is typical among clinical populations. Moreover, many with PTSD reported that pre-existing medical conditions such as arthritis and diabetes worsened following exposure to trauma. Furthermore, many victims demonstrated impairment in one or more aspects of cognitive functions including attention, memory, information processing, and the capacity to multi-task; there was reason to believe that some had experienced a brain injury. So over time, I gained the capability to assess and treat both PTSD and these accompanying conditions. For example, I developed and revised my structured clinical assessment interview and I included an increasing number of standardized psychological and psycho-physiological tests in the assessment. Also, in some instances I included a qEEG, as discussed later on.

The facts above raise at least two questions. *One* question is why other conditions co-occur with PTSD. The answer is not known fully at this time. Accidental co-occurrence due to the individual base rates of various conditions, other conditions occurring because of the effects of PTSD, multiple conditions arising from a common set of vulnerabilities, and trauma having an impact on multiple symptoms are among the possible explanations. Only one good study of these possibilities has come to my attention. Wittman and associates reported in a 2008 post-accident assessment of 225 survivors that 82% of the co-occurring symptoms could be explained by the combination of PTSD and the consequences of trauma-type mechanisms.

A *second* question is how best to proceed once it is known that the person meets the diagnostic criteria of both PTSD and other conditions. Sadly, there is no reason to believe that PTSD symptoms will decrease if a co-occurring condition is treated. For example, study by Teng and colleagues (2008) treated panic disorder only among veterans who suffered from PTSD as well. Over 60% of those treated for panic disorder were free of panic after specific treatment compared to the 19% who were simply given information. Neither group demonstrated any positive changes in PTSD symptoms afterwards. On the plus side, my experience is that two of the treatment components I use for PTSD, clinical psycho-physiology and applied affective neuroscience, not only reduce the symptoms of PTSD but decrease also the symptoms associated with many co-occurring conditions. Also, both of these components are successful in treating conditions among those who do not have PTSD. Colleagues in the professional associations in which I hold membership have reported similar results.

A-06: Some Common Findings
Related to the above, it was not uncommon early on in my career to see one of two patterns.

First, clients would see me because of reported symptoms of depression, sleep problems, stress/anxiety, or panic attacks. For many, during the assessment it would become evident that they met all of the diagnostic criteria for PTSD as well.

Second, I would see clients, treat them for the presenting conditions that did not include PTSD, and terminate sessions, only to have them return one to three years later with symptoms of PTSD that had not been present initially. For example, we had been able to get the symptoms of depression under control for three clients except that depression breakthroughs kept happening a number of times each year without any identifiable cause. Re-assessment indicated symptoms of PTSD that had not been present previously. Following treatment of PTSD including the use of BFT-C (discussed later), the PTSD symptoms decreased to remission and depression breakthroughs stopped.

Recent studies among the elderly are instructive in regards treatment of WWII and Korean veterans. A 2008 community study by Creamer and Parslow of a little more than 1800 Australians over the age of 65 reported a linear increase in lifetime exposure to trauma across the life span for men but not for women. Combat exposure seemed to explain the difference. Also, rates of PTSD were low among those over age 65. However, 10% still reported re-experiencing symptoms, one of the diagnostic criteria of PTSD. A 2008 study by Spitzer and colleagues of 850 citizens over age 65 living in communities in Germany reported lifetime PTSD rates of 3.1% for both men and women, that the rates of trauma exposure were higher for men than for women but no different than exposure among other age groups, and that those over age 65 with PTSD were more likely than those without to have additional psychological conditions, in particular depression and anxiety. The latter finding may suggest that those over age 65 who present with symptoms of depression or anxiety may be victims of trauma. We do know that trauma can result in any one of or a combination of conditions including PTSD, depression, panic attacks, sleep disturbances and disorders, and psycho-physiological reactions.

A-07: Not Recognizing Symptoms, Under-reporting, and Denying

For a number of reasons, many with PTSD (especially police officers and military folk) often do not recognize or admit that they have the condition. Dr. Rudofossi (2007), a well-known police psychologist, discovered that both combat veterans and police officers are prone to under-report, minimize, and deny.

First, typically, they ignore or are otherwise unaware of their symptoms or they work hard maintaining ways to remain in control of symptoms. This is functional in the performance of duties; for example, if they succumb to pain or fear during a physical altercation with a suspect or enemy combatant, they put their lives in danger, as well as the lives of their mates.

Second, symptoms develop slowly at first and increase in number and intensity a bit at a time so that changes are difficult to notice.

Third, when noticed, symptoms often are mistakenly attributed to other influences; for example, an emerging illness, children going through a ratty phase, financial concerns, sleep disruption, shift work, stress-producing managers, and administrative policies and procedures that are unenlightened or frankly dysfunctional. Often these are relevant but are not usually the key factors in the individual's condition.

Fourth, many males and some females believe that it is a major sign of weakness to admit that trauma is bothering them. For men, the tendency is to act like the movie characters played by John Wayne for whom to be a man means that nothing bothers them. Relatedly, many females believe that in order to be accepted in a male-dominated occupation they have to show less emotional reactivity than their male colleagues.

Fifth, typically police and military cultures do not tolerate discussion among personnel or from administration regarding the possibility of suffering from PTSD; men do not talk about such things, administration charged with providing the community with sufficient staff do not want more employees to go off duty since they have no replacement options, and the effectiveness of military and other units can be compromised when personnel are absent.

Finally, in part because the costs of effective treatment are significant, some organizations do not inform their employees of the professional assistance available to them if they suffer from PTSD; my own clinical experience indicated that this was so with respect to RCM police officers and for those in reserve military units who returned from tours of peace-keeping duty.

As a consequence of these and other reasons, many with PTSD wait until things reach crisis or near-crisis proportions before seeking professional attention. Being completely unable to function, experiencing increasing thoughts about suicide as the only method to decrease pain and suffering, developing serious physical symptoms, or receiving ultimatums from marital partners are often required to spur the individual to seek treatment. The unfortunate outcome of delay is that treatment is more difficult and takes longer than when symptoms are acted upon early in the developmental progression of PTSD. Therefore, clients and those around them suffer for much longer than is necessary.

On the other hand, the failure to disclose symptoms among those who have faced traumatic incident(s) does not mean necessarily that unhealthy denial or other processes are at work. For example, depending on such things as their pre-training characteristics, the type of basic and subsequent training they receive, their stress-management practices, and the support available from various sources, not every police officer or military person experiences symptoms of PTSD. This was very clear to me when, as a police psychologist, I conducted extensive interviews and psychological testing of those officers applying for specialized assignments. In spite of their prior exposure to trauma, most had no sign of PTSD symptoms. Also, there is reason to believe that offering such persons psychological aid or encouraging them to delve deeply into nasty experiences is counter-therapeutic. For example, some studies of those forced to attend a particular form of psychological debriefing indicated that about 33% of attendees who were not symptomatic before developed symptoms after the debriefing.

A-08: Reasons for Assessing and Treating PTSD
In addition to the obvious reason of wanting to decrease the symptoms of PTSD and the suffering associated with them, there are many reasons for assessing and treating clients with PTSD.

First, studies demonstrate that, as compared with others, those with PTSD are less satisfied with life, have higher unemployment rates, have suicide rates greater than with any other type of anxiety disorder, have decreased educational attainment, and often struggle with marital, family, and other interpersonal difficulties. Such quality of life aspects are likely to be more severe and pervasive among those with PTSD than among folks with any other anxiety disorder.

Second, for a variety of reasons, many with PTSD fail to engage in preventative health strategies including exercise, diet, safe sex, adequate sleep, regular health care, sensible risk management, and so on, thereby giving rise eventually to other conditions.

Third, physical health can be affected (for example, see the 2009 review by Kendall-Tackett). The evidence is that PTSD is associated with a wide range of serious and life-threatening medical conditions such as diabetes, cardiovascular disease, gastrointestinal disorders,, and cancer. This is so even when risk factors such as smoking are controlled. For example, in 2008 Dzubur and colleagues reported that chronic PTSD among combat veterans was associated with higher plasma lipids leading to increased risk of coronary artery disease; this was not so among combat veterans without PTSD. A 2008 community study by Sledjeski and co-investigators comparing those with and without PTSD found that persons with PTSD had a higher likelihood of having one or more of 15 chronic medical conditions than non-traumatized individuals. They found also that those with the greatest number of medical conditions and number of traumas were most likely to have PTSD. In their 2009 study, Calhoun and associates reported that women who had PTSD, with or without major depressive disorder, experienced poorer health than those without either diagnosis. At this time it is difficult to determine whether or not medical conditions are caused by the PTSD but they certainly are worse among many with PTSD. One of the inter-related mechanisms may be that (1) trauma activates the chemistry of the stress response, which then (2) promotes the immune system to release cytokines or other inflammatory markers (for example, C-reactive proteins and fibrogens), thereby ultimately (3) increasing levels of inflammation while at the same time (4) normal checks and balances fail. Thus, it is not surprising that those with PTSD are high users of health care services.

Fourth, conditions known to be influenced by stress, such as chronic fatigue syndrome, irritable bowel syndrome, multiple chemical sensitivity, fibromyalgia syndrome, other pain conditions, and psoriasis, are exacerbated after traumatic exposure and co-occur frequently with PTSD. Moreover, most sufferers see their physicians for their medical symptoms and neither discuss the PTSD with the physician nor consult a mental health professional. In their 2008 study, Sledjeski and co-investigators examined the relationship between the number of lifetime traumas and 15 chronic medical conditions. With the exception of headaches, the presence of chronic medical conditions was correlated with the number of lifetime traumas experienced. Those with PTSD had the highest likelihood of chronic medical conditions. With respect to headache, Peterlin and colleagues (2009) surveyed roughly 600 headache patients from six headache clinics in the U.S.A. Results indicated that about 30% of those with chronic daily headache had PTSD, as did approximately 22% of those with episodic migraine. In addition, patients with PTSD reported greater headache-related disability than those without PTSD.

Fifth, mortality rates are higher among those with PTSD as compared to those without symptoms; sometimes medical conditions are the cause (such as diseases like cancer) and sometimes death is the result of preventable accidents or suicide.

Sixth, as is discussed later, traumatic brain injury can result from various types of experiences involving blows to the head (for example, physical assault) and significant movement of the brain within the skull (for example, motor vehicle accidents and percussion waves from the firing of big guns or explosives). Although not detectable by conventional structural medical imaging (like CAT and MRI scans), electrophysiological evaluations (such as qEEG) and other instruments that also measure functional aspects of brain processes (including PET and fMRI) can indicate disturbances in brain processes. Evidence of such brain dysfunction can be associated with various kinds of cognitive impairment (for example, problems with concentration and memory) that contribute additional symptoms and frustrations to the client.

Seventh, several investigators have suggested that severe and prolonged trauma may accelerate cognitive decline such as the onset of dementia among aging individuals as well increasing the symptoms of depression in later life.

Eighth, particularly among military veterans with PTSD, rates of physical aggression and violence towards others are high. For example, one study reported that about one-third were violent with their partners, a rate 2 to 3 times higher than among veterans without PTSD.

Finally, PTSD is costly from a societal point of view. One estimate placed the productivity loss in the U.S.A from PTSD at three billion dollars per year and this does not include the costs of providing medical services for the associated health problems.

A-09: Posttraumatic Growth
However, it is true also that a percentage of persons exposed to a traumatic event or events will experience post-traumatic growth (PTG) in addition to the symptoms of PTSD. For example, Solomon and Dekel (2007) reported that some Israeli combat personnel from the Yom Kippur War and some of those held as POWs gave responses to a questionnaire indicating a greater understanding of self and a clear concept of priorities in life in addition to symptoms of PTSD as a consequence of their traumatic experiences. A process of assessing and re-balancing of life priorities as well as an increased appreciation of what life has to offer are among the outcomes common among the folks I have seen after PTSD treatment is completed successfully. After the U.S.A experience of 9/11, Poulin and co-investigators asked a national sample of adults if they perceived any positive societal results from the terrorist attacks. About 58% believed there had been an increase in positive behaviour to others, religiousness, and political engagement.

The two psychologists most associated with post-traumatic growth (PTG) are Calhoun and Tedeschi, who also edited a 2006 book on this subject. They write in the first chapter:
> As we and others . . . have indicated, the assumption that, at least for some people, an encounter with trauma which may contain elements of great suffering and loss can lead to highly positive changes in the individual is ancient and widespread. The possibilities for growth from the struggle with suffering and crises is a theme that is present in ancient literature and philosophy and, at least in some ways, the problem of human suffering is central to both ancient and contemporary religious thinking. (p. 3)

Research studies by the pair as well as by other authors in the book indicate that *some* survivors of traumatic incidents experience some of the following at various times after the trauma is over:

1. the sense that one has been tested and survived, suggesting that one is indeed quite strong;
2. the emergence of new possibilities, new interests, new activities, and new paths in life;
3. human relationships that are highly positive compared to pre-trauma relationships;
4. increased connection with and compassion for others who suffer;
5. increased freedom to be oneself, less inhibited by unnecessary constraints;
6. an enhanced appreciation for life; and
7. a changed sense of what is truly important in life, resulting in the development of new priorities.

Studies have reported such PTG in respect to cancer (chapter 8), bereavement (chapter 9), war (chapter 10), HIV/AIDS (chapter 11), disaster/emergency work (chapter 12), and among some survivors of the Holocaust (chapter 13). Some studies have included military personnel, police officers, and staff of other emergency services.

However, authors in the book do not deny that trauma does involve pain. As the editors state: "But the richer life may come at the price of the discomfort that tragedy and loss almost always produce. . . . The encounter with trauma may indeed produce growth, but it also tends to produce significant pain" (p. 4).

Also, the editors are clear that PTG describes a process that probably unfolds gradually, showing that the early phases of the struggle to cope do not define trauma survivors, but are simply part of a longer term process. Following the process over time can lead to some surprising discoveries in many people, particularly if the survivors are fortunate enough to find the kind of support that encourages continuing psychological work on the trauma and its aftermath. . . . We have pointed out repeatedly that in no way are we suggesting that trauma is, in itself, good. (p. 6)

Furthermore, they emphasize that many people do not experience PTG, and such an outcome is not necessarily negative.

Finally, they acknowledge that

> survivors often suffer at the hands of others who expect them to be recovered from their trauma and loss rather quickly. If they show distress, they are often regarded as poor copers who are wallowing in their pain. We honor people by acknowledging what they are up against following a trauma, not by holding out false hope that if they have the right
> personality characteristics, if they process the event in the right way, and if they adopt the right coping strategies, they will be able to grow from their experience. If outsiders believe that growth is prevalent, this can become the new standard that survivors' progress is measured against. Such a standard may lead to negative judgements towards those who do not show personal growth, making them feel like coping failures. (p. 293)

As a result of my reading and clinical experiences with hundreds of trauma victims, I focus our treatment on reducing the symptoms until PTSD and associated conditions are in full remission. Only after this is achieved do we devote some time in the treatment or follow-up process to determining how the trauma has altered beliefs, objectives, and priorities. Changes will take place and some may be viewed as post-traumatic growth.

Also, I have changed as a result of intimate clinical work with victims of trauma. This includes realizing and accepting that trauma can happen to anyone, including me, recognizing the importance of my own Health practices with a capital "H", gaining an enhanced view of life's possibilities, receiving constant reminders to insure that my priorities are in order, developing greater compassion for those suffering from trauma by knowing the processes involved in both the development and successful treatment of PTSD, and celebrating the individual's spirit that strives for a meaningful adjustment to the pain that is part of our human condition.

A-10: Genetics

While an exposure to trauma is the direct cause of the symptoms, many factors determine whether or not an exposed person will develop PTSD. For instance, mention will be made later of risk factors. Also, there is some evidence of a genetic vulnerability to PTSD. For example, studies of Vietnam War veterans have shown that identical twins (monozygotic) who saw combat action have higher PTSD rates than fraternal twins (dizygotic) who participated in combat. The best guess at this time is that what is inherited is dysfunction in the biological response to traumatic stress, which is discussed later. For example, research demonstrates that between 13% and 34% of the propensity to increased arousal level and avoidance strategies to a stressor is inherited. Perhaps relatedly, in the 2007 book edited by Lehrer and associates, it was reported that compared to full-term babies, neonates born prematurely have lower heart rate variability, which increases their vulnerability to later stressors. The 2008 report by Goenjian and others of Armenian families exposed to the 1988 earthquake documented heritabilities of 41% for those who developed PTSD, 61% for other anxiety symptoms, and 66% for depression; they concluded that for some individuals, their genetic make-up increases their vulnerability to develop symptoms of PTSD, anxiety, and depression. Thakar and associates published their study in 2009 of 41 people who had PTSD as a consequence of motor vehicle accidents; higher chronic PTSD was found among those without the ss allele gene than those with the ss and sl genotypes (55% versus 20% respectively). Fortunately, regardless of any genetic component, the symptoms of PTSD can still be decreased successfully.

Treatment Step B: Structure of the First Session

Contents

B-01: Discussion of Preliminary Information

During our first treatment session, clients and I have the chance to discuss any matters arising from the readings I provided in *Components of Treatment: Step A Preliminary Information*.

B-02: The Assessment Report

Also, in the first treatment session, I review results of the assessment, provide a diagnosis, and inform the client of the specific components of psychological treatment I recommend. Most of this information is contained in the written report provided to clients following completion of the assessment. Normally, the client's physician receives received a copy also so that s/he is fully informed when the client and physician meet.

B-03: Interim Matters

In addition, in the first treatment session, as needed, conversations take place on how best to increase both security and quality of life. For example:

_____we pay careful attention to managing any safety issues including (a) any accident-proneness due to cognitive impairment and (b) suicidal thoughts;

_____we plan for how best to control anger and to provide safety for family members;

_____we review how best to handle emergencies;

_____we consider how best to compensate for any problems with memory;

_____we determine whatever practical assistance is needed;

_____we address work issues including personal, collegial, and public safety as well as likely responses by managers

_____we consider what help is needed and available from relevant agencies;

_____ we talk openly and honestly on how best we can work together; and

_____we decide what (if anything) to tell relatives, friends, and employers.

B-04: Cognitive Therapy
During the first and other sessions and when the timing is right, attention is directed towards aspects of cognitive behavioural therapy (CBT), as discussed later. Sometimes clients are unable to take full advantage of CBT until autonomic and/or central nervous systems become properly regulated, as described later.

B-05: Early Strategies For Emotion Regulation
Many folks with PTSD are not aware of effective strategies for the regulation of emotion. But the strategies are effective and I recommend them strongly. They are taken from the *Handbook of Emotional Regulation* edited by Dr. James Gross (2007). The strategies below are given to the client to read and subsequently we discuss them as required.

> *Attentional Deployment.* When symptoms arise within us, we can focus on them and allow the "monkey-mind" to take over, bringing up a lot of other negative experiences and feelings. Or, we can modify or redirect the focus of conscious attention to positives such as what is going well in life at this moment, a fun task to complete, or a hockey game to watch.

> *Cognitive Transformation or Appraisaltransformation or re-appraisal.* The explanatory style we use, that is, the way we understand things or explain them to ourselves, can be helpful or distressing. For example, an unhelpful response when symptoms arise can be constant worry that this means either that something is really wrong with us, that we are getting worse, that we are incurable, or that we do not deserve happiness; these become the explanation for why the symptoms are there. This of course leads to a worsening of symptoms and more stress. Cognitive transformation or reappraisal can be effective in that the meaning or emotional significance of the symptoms can be changed. "I am just having a bad moment", "It is just that my amygdala is acting up for some reason" or "Here comes another brain fart" or "Here is an opportunity to see if tactical breathing will help", or "I am determined to beat this" are effective and realistic ways of dealing with the symptoms. Studies have indicated that cognitive transformation or re-appraisal in fact activates those parts of the brain (frontal cortex) so as to regulate those pathways involved in generating emotional reactions (for example, the amygdala, which is discussed later). Following this with a positive activity (see mood repair below) helps even more.

> *Situation Selection* can be employed, whereby the client avoids situations that produce symptoms of PTSD consistently. After returning from WWII, many of my veteran clients noticed that specific activities such as watching TV news and war movies activated PTSD symptoms. So quite rightly in my professional opinion, they decided to avoid these

stimuli, thereby decreasing the chances of re-triggering the symptoms. During the historical era following that war, there really were no reasonable alternatives, and who in their right mind would want to re-trigger symptoms of PTSD? Accordingly, until clients have mastered ways to reduce reactivity to triggers, I recommend strongly that they avoid such triggers.

Situation Modification is similar to situation selection and helps by reducing negative emotional impact. For example, no matter where they are gathered, police and military folk tend to "talk shop" and share stories of traumatic experiences and/or stressors caused by management. Thus, members of both groups with PTSD may experience an increase in symptoms when these conversations begin but believe they must stay to the end regardless of personal cost. An effective alternative is to change aspects of the situation rather than avoiding it altogether by re-directing the topic from stories of traumatic events to last night's sport event, by telling friends beforehand that "shop talk" should be kept to a minimum, or simply by moving to a different group of people at the gathering.

Mood Repair. When symptoms arise, clients often just try to wait them out without doing anything specific to help themselves, in part due to low levels of motivation and energy. Ultimately this leads to reactions such as anxiety, agitation, or feelings of depression. An alternative to waiting is moving right away to doing something that is engaging and enjoyable or to begin listing all of the positive things in one's life. Studies show that doing so leads to a positive change in mood.

Response Modulation. Stabilization strategies, medication, strategic breathing, and decreasing muscle tension, which we cover later in the basic approach, are among the methods of response modulation. Through these, anxiety symptoms can be reduced. Without them, the nervous systems of those with PTSD remain in a state of hyper-arousal.

B-06: Preparation for the Next Session
At the close of the first treatment session, clients are given relevant readings from Section C below.

Treatment Step C: Psycho-Education of PTSD and Related Conditions

Generally, clients leave the first treatment session with basic information about PTSD relevant to their situation and any related psychological conditions they may have. The goal is for clients to understand the nature of their condition(s) as fully as possible. Usually this serves also to help them to adopt accurate beliefs about their condition, to decrease any false beliefs, and to increase the chances that they will find additional ways to decrease symptoms.

The first treatment session (Step B) concludes with providing clients with information below that is relevant to their situations. These are reviewed with recommendations to be considered during the next session.

Contents
C-01:   Trauma Defined
C-02:   Types of Experiences Considered to be
                Traumatic
C-03:   Risk Factors for PTSD
C-04:   Prior Trauma
C-05:   Factors That Promote Coping
C-06:   Diagnostic Criteria B, C, and D for PTSD
C-07:   More Information about Specific PTSD
                Symptoms
C-08:   Complexity of PTSD
C-09:   Timing of Symptoms
C-10:   Associated Psychological Symptoms
C-11:   Traumatic Brain Injury
C-12:   Pain and Its Management

C-01: Trauma Defined
Some events/situations/stressors are so significant that they are considered to be traumatic. As defined by the diagnostic classification system presented in the *DSM-IV*, in respect to PTSD a trauma is one in which
(A-1)
the person experienced, witnessed, or was confronted with an event or events that involved actual or threatened death or serious injury or a threat to the physical integrity of self or others and

(A-2)
the person's response involved intense fear, helplessness, or horror.

All of the types of trauma noted in C-2 below certainly meet the A-1 criterion for PTSD and may meet criterion A-2, depending on the person's response to the traumatic incident. Courtois and Gold (2009) cite a well-designed study that found that 80% of adult respondents had experienced at least one event conforming to criterion A-1 above.

I rely on what the client tells me when considering whether s/he meets the A-1 and A-2 criteria I. As indicated in the *Handbook of PTSD: Science and Practice* (Friedman, Keane, & Resnick, 2007), a study by Dohrenwend and colleagues (2006) found a high reliability between self-report data of Vietnam vets and data obtained from personnel files, military archival sources, and historical accounts. Also, the psychological tests, the individual's psycho-physiology, and other sources of data in my assessment provide information about validity of self-reported information.

The ICD-10 diagnostic system of the World Health Organization followed the *DSM-III-R* definition by defining PTSD as

> a delayed and/or protracted response to a stressful event or situation (either short or long-lasting) of an exceptionally threatening or catastrophic nature which is likely to cause pervasive distress in almost anyone (e.g., natural or man-made disaster, combat, serious accident, witnessing the violent death of others, or being the victim of torture, terrorism, rape, or other crime). (World Health Organization, 2007)

C-02: Types of Experiences Considered to be Traumatic

The intent of this sub-section is to provide a list of types of trauma and details of the trauma most relevant to each client. From this clients come to realize that they are not alone; others have suffered from similar types of trauma.

The bottom line is that PTSD is the most common psychological condition that can result from trauma; although the rates vary depending on the study. A recent publication (Darves-Bornoz and associates, 2008), after surveying 8,797 civilians from 6 European countries (Spain, Italy, Germany, Holland, Belgium, and France), indicated a 12-month PTSD rate of 1.1%. Those with the condition had experienced an average of 3.2 potentially traumatic events that included rape, having a child with a serious illness, and having been beaten by a partner or a caregiver. The 2009 population study of civilians in the Netherlands by Bronner and co-investigators found the PTSD rate to be 3.8% with rape and physical assault the most common traumas.

Both research studies and clinical experience indicate that usually clients do not provide information about traumatic experiences following general questions by the clinician (for example, "Have you had any traumatic experiences?") but will discuss matters openly if specific questions are asked (for example, "Have you ever been in a car accident?" "Have you witnessed any ethnic cleansing?" or "Has anyone ever touched your genitals without your consent?"). Relatedly, one cannot assume that the traumas for police or military personnel are limited to experiences while on duty; these people too can be victims of such experiences as family violence and life-threatening illness. Information about the questions I ask are found in the structured clinical interview in Appendix II. However, even with a careful approach to discovering trauma incidents, it is not unusual for clients to reveal other traumatic incidents during the treatment process. Sometimes they are not disclosed initially because the client is embarrassed about then and sometimes they were not in conscious memory earlier in the treatment process.

Sections of C-02 are:
a. Military Combat
b. Prisoners of War
c. Peace-keeping/ Peace-support Deployments
d. Police Services
e. War Experiences of Civilians
f. Terrorist Attacks
g. Physical Assault
h. Spousal Assault/Domestic Violence
i. Rape/Sexual Assault
j. Torture, Political Oppression, Displacement, Ethnic Cleansing, Refugees
k. Motor Vehicle Accidents
l. Airplane Crashes
m. Other Human-caused Disasters
n. Natural Disasters
o. Life-Threatening Illness and Other Medical Conditions or Procedures
p. Tragic Death Of a Loved One
q. Severe Psychiatric Illness
r. Human Service Work

a. Military Combat
In summary, to 1995 someone has calculated that since WWII, there have been 127 wars and over 20 million people have died because of them. Depending on the study, statistics indicate PTSD rates between 12% and 60% among combat veterans. The causes and complications are many. Moreover, a high percentage of veterans continue to experience PTSD decades after the combat experience(s). PTSD has become the most common condition for which veterans seek help from Veterans Affairs.

Contrary to the glamour portrayed on TV and in the movies, the relatively few studies completed to date support very clearly the tragic, debilitating, and often long-term consequences to those who survive combat. In fact, war and related experiences are the most common cause of PTSD among males. It should be remembered also that typically, war represents a long string of traumatic experiences rather than a single horrific incident, including firing upon and being fired upon by the enemy, friendly fire, killing an enemy or an unarmed child thought to be carrying explosives, dangers of guerilla warfare and terrorist actions such as roadside bombs, seeing mates injured or killed and being unable to save them, seeing or handling human remains, and seeing the suffering of civilians, particularly women, children, and the elderly.

In addition, other constant stressors play a part. Among those identified in the 2007 book *Combat Stress Injury* are extremes of heat and cold (temperatures in Southwest Asia can range from below freezing to 50 degrees C), dehydration (combatants can become numb to their internal messages of thirst), hypothermia from winter rains and cold, sleep deprivation, noise and blasts, fumes and smells (including burning trash and flesh), darkness (fear that night time light may draw enemy fire), malnutrition during sustained operations, lack of information, unclear or changing role within a mission, ambiguous or changing rules of engagement, boredom and monotony, isolation from family support, inability to deal with family matters, lack of privacy or personal space, unsupportive media reports, and unfavourable public opinion about the mission or specific actions. The experiences, thoughts, and reactions of military personnel in combat zones can be found on their blogs following a web search such as "Iraq blogs"; sometimes the language is crude and explicit descriptions or pictures are included. Additionally, as described in the 2008 study of Canadian military personnel by Sareen and associates, many soldiers have the same risk factors as civilians including "genetics, childhood adversity, stressful life experiences, social supports, and personality," which might increase the likelihood of PTSD independent of any combat experience.

PTSD is found among current members and veterans of the Canadian Forces as well as in military units from other countries who have been involved in combat. For example, 50 years after WWII, between 17% and 32% who served in the Pacific continued to have PTSD while the rates were between 12% and 22% for those who fought in Europe. About 41% of those serving in the Korean war developed PTSD. Up to 60% of Vietnam veterans developed PTSD and several studies show that from 9% to 15% continued to have PTSD 20 years later in comparison to a general population rate of 2.5%; a further 10% continued to experience partial symptoms of PTSD. Thirty years after combat experience in Vietnam, 10% continued to suffer from PTSD. Researchers studying Israeli combat troops 1, 2, and 3 years after war in the Middle East noted PTSD rates of 16%, 19%, and 9% respectively, even though a great deal of psychological intervention had been provided to the armed forces throughout the war. Five years after conclusion of the Falkland War, 22% of British veterans met the full criteria for PTSD and an additional 38% had some of the symptoms. Studies of U.S. troops showed that 8% of those who served in Somalia developed PTSD. One study indicated a PTSD rate of 15% among troops participating in Desert Storm/Desert Shield, another study pegged the rate at 23%, and a UK study found a rate of 3% among British troops. One study indicated PTSD rates of about 11% among U.S. troops assigned to Afghanistan. In regards the current war in Iraq, various studies have reported PTSD rates between 12% and 20% among U.S. military serving there. A 2008 study by Fontana and Resenheck compared treatment-seeking veterans with PTSD from Iraq and Afghanistan with those who served in Vietnam and found the former group to be younger, to include more females, to be more likely to be single, to have spent less time in jail beforehand, to report less exposure to atrocities, to have fewer diagnoses of substance abuse, but to display more post-deployment violent behaviour.

Caspi and co-authors reported a study in 2008 of 317 community-based Bedouin servicemen who constituted a minority group within the Israeli Defence Forces. Using a very strict definition, 75% reported exposure to traumatic military events and 20% met the diagnostic criteria for PTSD, which usually occurred along with both depression and/or alcohol abuse. Moreover, those with PTSD experienced extensive negative impact on their health as measured by such aspects as diagnosed medical conditions, health-related impairment in daily life, and use of medical services.

At 15% of the total group, PTSD is now the most common psychiatric condition for which U.S. veterans seek VA services. Clearly, as one colonel mentioned at a military seminar I attended, "War is not good for you." In recent years, in the attempt to decrease the embarrassment of PTSD and increase help-seeking among military personnel, a whole host of newer terms have surfaced such as operational stress injury, deployment-related stress injury, combat stress injury, and combat operational stress reaction (COSR).

Studies are emerging to indicate that a variety of health-related conditions accompany PTSD among combat troops. Boscarino (2004) found that military veterans with PTSD were more likely to have abnormal electrocardiographic results including a higher prevalence of myocardial infarction, Q-waves, and atrioventricular conduction defects. In his 2006 study, Boscarino reported increased mortality rates due to cancer and cardiovascular disease. A 6 year follow-up study by Johnson and associates (1994) found a 17% mortality rate among Vietnam veterans with chronic PTSD. The 2008 report by Vasterling and associates indicated that post-deployment PTSD severity among U.S. troops in Iraq was associated with a negative change in various aspects of health independent of such health risk behaviours as smoking and alcohol use.

Also, there is evidence that combat experiences alter basic beliefs. As cited in the 2007 book by Linden and associates, Foster found that Vietnam veterans with PTSD had experienced more changes in basic beliefs than those without the condition. Furthermore, the basic beliefs of those with PTSD continued to change until the last period rated, which was 15 years after their combat experiences.

Moreover, from both my previous clinical experiences and subsequent research studies, there are good reasons to believe that military personnel differ from both those exposed to a single traumatic experience (such as car accidents, civilian assault, and disasters). For example, research indicates that treatment methods such as exposure therapy and EMDR used primarily among civilian populations exposed to single traumatic experiences are much less effective among military personnel. Also, medications (such as *Paxil*) that are effective in treating PTSD symptoms among female civilian victims are often not effective among military people. Generally, I have found both of these two limitations to be true in my work with police officers also. Relatedly, an interesting 2006 review of the literature by Wang indicated that females with PTSD who had lower thyroid levels (free and T3) reported higher general distress and PTSD symptoms while females with higher thyroid levels reported lower levels of clinical symptoms. By contrast, studies of male combat veterans in regards Vietnam, WWII, conflicts in Israel, and conflicts in Croatia indicated that those with high T3 levels reported higher PTSD symptoms.

Greene (2009) offered a perspective on PTSD among military personnel. Countering the view that veterans with PTSD are prone to violent behaviour or exhibit problems in work and relationships, she noted that this was true for only a small proportion. Rather, most veterans with PTSD do well in society. Moreover, they have been strengthened by their experience with combat. Furthermore, most military personnel have the mindset of a warrior wherein combat stress is considered as a desireable personal challenge that will test their skills and discipline. Otherwise they would have remained at home and sought employment in less demanding jobs. Nevertheless, Colonel Green advocates appropriate psychological treatment for those who exhibit significant symptoms of PTSD.

b. Prisoners of War
For troops captured and placed in POW camps, PTSD is a very typical outcome. Six studies show rates of more than 50% and three found rates greater than 70%. Moreover, the symptoms of PTSD persist for decades after the experience. For example, one study of Canadian POWs 40 years after WW II found that 29% of survivors had PTSD. The 2008 research report by Hart and associates of POWs from WWII found that those with PTSD did less well than POWs without PTSD on a variety of tests assessing the functioning of the frontal lobes of the brain as well as tests of psychomotor speed.

c. Peace-keeping/Peace-support Operations
In summary, peace-keeping/peace-support operations are not dissimilar from wartime combat. Moreover, those deployed often have significant restrictions as to what they can do when under fire from one side of the conflict or the other. Thus, those involved in peacekeeping/peace-support operations do not escape from PTSD. In fact, the overall rates of PTSD are reported to range between 2.5% and 20%.

Penn (2000) summarized research to date and discovered PTSD among 2% to 8% of UN peacekeepers within 3 years of returning home. Based on a 2002 survey of about 8,000 members of the Canadian armed forces, between 6% and 9% had PTSD, which for males was mostly the result of participation in international peacekeeping operations. Moreover, there is evidence of a relationship between length of peace-keeping deployment and risk of PTSD; one study of those serving in Bosnia for 1 to 2 months showed PTSD rates of 3% to 6% while those who remained for 9 months had rates between 7% and 12%.

The reasons for PTSD among this population are two-fold. Firstly, contrary to what most civilians believe, those assigned to peace operations experience trauma. For example, results from 2 studies of Canadian peacekeepers in the former Yugoslavia indicated that 66% had experienced incoming artillery fire, 67% had been shot at, 50% had seen a buddy killed, 50% had seen a civilian killed, 28% had handled wounded persons, and between 19% and 28% had handled dead bodies.

Secondly, deployment to UN operations were characterized by stress and conflict. Sometimes they werenot armed. At other times there were difficult constraints on their capacity to respond with the consequences both of disciplinary action for errors and weakening of the UN position. Maintaining the expected stance of neutrality posed additional issues. Conflicting opinions by those in charge, information gaps, rapid changes, and low control were added stressors.

In addition, many have reported incompetent and delayed response by the UN to requests made by on-site commanders during peace-keeping operations. Moreover, those deployed to peace operations who develop PTSD suffer additional consequences. For example, the 2008 research by Richardson and colleagues of Canadian male peace-keepers seeking treatment for subsequent PTSD found an association between PTSD and quality of life; compared to those without the condition, veterans with PTSD had significant impairments in measured aspects of mental and physical quality of life.

d. Police Service

In summary, the few studies available indicate that between 12% and 35% of police officers will qualify for the diagnosis of PTSD at some time. Other studies have indicated that an unknown percentage of officers who leave police work, or who move to "safer" or non-operational assignments in the unit, or who become identified by their colleagues as "slugs" (low performers) are in fact suffering from PTSD and have selected the only coping method they believed was available to them. Others and their physicians focus on common medical complaints associated with PTSD without addressing head-on the possibility of PTSD.

Most of my police clients have been in the field doing operational policing for 15 to 25 years, during which time they have been assaulted and sometimes fired upon, been constantly under real threat of "grievous bodily harm and death," attended death scenes, been first on the scene of gory car accidents, tried but failed to resuscitate victims, assisted at disasters, dealt with mutilated or dismembered bodies, come into direct contact with children who have been battered or sexually assaulted, controlled riots, and dealt with suspects who carry life-threatening diseases such as AIDS and hepatitis C. In his 2007 book, Dr. Rudofossi reported that 25% of uniformed police officers with NYPD were assaulted over a four-year period and physical injury was sustained by one in four officers.

Various research studies as reviewed by McCaslin and colleagues (2006) have surveyed police officers about critical incident stressors and found that exposure to death and disaster, exposure to violence and injury, and encountering victims of crime including child abuse were among the most stressful and also were associated with the greatest likelihood of developing PTSD. Their own research demonstrated that police officers who indicated that exposure to high personal threat (such as threat of death or serious injury to the officer or to someone with whom s/he has a close relationship) or duty-related violence (including encountering death or situations of physical and sexual assault) were most distressing to them also reported high levels of dissociative symptoms at the time of the event and subsequent high levels of hyper-arousal; both of these have been found to be risk factors for PTSD.

After working for 30 years as a police psychologist with the NYPD, Dr. Daniel Rudofossi noted in his 2007 book *Working with Traumatized Police Officer-Patients* that "real police stress is overwhelmingly related to multiple and cumulative experiences of trauma and loss . . . " (p. 10).

Moreover, he notes that many traumatic experiences occur while officers are in their early twenties and before brain maturation is complete, thereby having the highly likely consequence of ". . . increased vulnerability toward development of trauma-response syndromes . . . " (p. 14).

Finally, he reports that

> many ideas and feelings related to one's roles remain guarded and unexpressed because of the fear of legal exposure and prosecution, which is transmitted through institutional and grassroots culture. These antecedents are initiation rites in the experience of public-safety professionals, who are likely to encounter repeated trauma as massive and unexpected loss. (p. 14)

In addition to being important in its own right, information from studies of combat veterans and peace-keepers is important in regards to the police community (officers and 911/dispatch operators). Until research findings prove otherwise, I have taken the position that due to the number and type of trauma to which they are exposed, police people are more similar to those who have experienced military combat or peace-keeping duties than to victims of a single trauma. This decision is important since there is surprisingly little research published on police people but much more on military personnel, as indicated above as well as throughout this book.

Moreover, the presence of PTSD has additional implications. For example, the 2006 study of police officers by Violanti and his research group found that those with the highest as compared with the lowest severity of PTSD were three times more likely to have the metabolic syndrome consisting of combinations of obesity, elevated blood pressure, reduced high density lipoprotein (HDL) cholesterol, elevated triglycerides, and abnormal glucose levels. Other studies have found that such components are related to increased risk for diabetes, future cardiovascular disease, and consequential death.

e. War Experiences of Civilians

In summary, war affects citizens as well as military personnel. Depending on the study, PTSD rates from 2% to 35% have been reported among civilians who have been exposed to war. Moreover, PTSD rates continue to be above 2% up to 50 years later.

For example, studies of Dutch resistance members 50 years after the ending of WW II indicated that 27% of men and 20% of women continued to have PTSD. Six studies of refugees from war found PTSD rates to be in excess of 50%. In a 1960 study, 69% of Dutch civilians who survived a Nazi concentration camp still had PTSD. Of Auschwitz survivors, 65% met all of the criteria for PTSD decades after their release.

In a 2008 investigation, Palmieri and colleagues conducted a stratified random telephone survey of 1200 Jewish and Arab civilians living in Israel during the Israel-Hezbollah war. They reported a 7.2% rate of probable PTSD. The highest rates were associated with being a woman, recent trauma exposure, economic loss, and loss of psycho-social resources. Higher education was associated with lower rates of PTSD.

In 2008 Morina and Ford published research on psychological trauma to civilians in Kosovo. They discovered that 2% met the full diagnosis for the more serious than PTSD ICD-10 classification of Disorders of Extreme Stress Not Otherwise Specified (DESNOS), while a further 24% to 42% experienced some of the symptoms including physical symptoms, changed relationships, and altered systems of meaning. DESNOS symptoms were associated with poorer overall psychological functioning, self-evaluations, satisfaction with life, and social support.

The 2008 investigation of 99 Guatemalans disabled by the civil war conducted by Herra-Rivera and colleagues revealed that 34.3 % had PTSD with a further 16% suffering from some form of depression. The 2008 study by Levine and co-workers of 4,054 Israeli adolescents exposed to terror indicated that 5.5% of them met the diagnostic criteria for PTSD. Feinstein and Botes (2009) had 79 civilian contractors working in a war zone complete an internet-based psychological assessment and found 33% of them to have PTSD scores in the moderate to severe range; 20% exceeded the cutoff for depression, 28% experienced other psychological distress, and 17% over-consumed alcohol.

f. Terrorist Attacks

In summary, studies indicate that between 7% and 40% of victims of a terrorist attack will develop PTSD. Also, 10% or more continue to have PTSD up to 5 years later. Those involved in assisting at the scene of terrorist attacks can also develop PTSD; more than 10% do.

Data from one study collected after 9/11 in the U.S. found that 10% of uninjured victims developed PTSD while 31% of those receiving injuries from the attack had PTSD. Hobfoll and colleagues reported in 2008 that 6.6% of Jews and 18% of Arabs living in Israel during the terrorist attacks since 2000 probably had PTSD. Their report was based on telephone interviews with a random sample of citizens. A 2008 study by Brewin and associates of the 2007 terrorist bombings in London indicated that 30% to 40% of those exposed had probable PTSD based on telephone interviews. The terrorist bombing of the U.S. embassy in Kenya was reported by Njenga and associates (2004) to result in PSTD rates of 35% among Kenyans exposed to the event. As reported in a 2008 paper by DiGrande and her research group, the 2001 attack on the World Trade Center in New York left 39% of those closest to the WTC with PTSD. Two to three years later, administration of the PTSD Checklist found that 12.6% of those living within a mile of the WTC probably still had the condition.

Also, those involved in the rescue, recovery, and clean-up after an attack are vulnerable. For example, Stellman and associates (2008) evaluated 10,132 such folks involved in the World Trade Center (WTC) attacks. Among this group, the following probable rates were discovered: 11.1% for PTSD, 8.8% for depression, and 5% for panic disorder. Probable PTSD rates declined from 13.5% to 9.7% over the five years of the study; note, though, that about 10% still had PTSD five years after the event. Moreover, PTSD was associated with loss of family members and friends, higher rates of behavioural symptoms in children of workers, and disruption of family, work, and social life. A previous study by this group had documented extensive pulmonary dysfunction related to exposure to toxic substances among those working at the scene after the WTC attack.

g. Physical Assault

This term is used when a person is slapped, kicked, stabbed, burned, bitten, pushed, or threatened with harm or death. Studies show that sometime during their lives on this planet, 7% of adult females and 11% of adult males will be assaulted by a stranger. Symptoms of PTSD will occur in more than 28% of the victims soon after, while 7.5% will still have the diagnosis 15 years later. A shocking study by the National Crime Victim Center of American adolescents found that 13% of females and 21% of males had been physically assaulted. Physical assault is a crime and thus those on the receiving end are victims of crime. Moreover, it is reasonable to conclude that a percentage of such victims also sustain traumatic brain injury from the assault. Known as TBI, this is discussed later in this paper.

h. Spousal Assault/Domestic Violence

In summary, while this is a subset of physical assault, it is so prevalent and has such far-reaching and long-lasting consequences that I include it here separately. PTSD rates from spousal assault range from 9% to 29% depending on the study.

Data averaged from careful studies indicate that among those in domestic relationships, 17% (mostly women) report having been assaulted by their partners, and 1 in 200 in a recent study reported severe injuries. Also, any partner who has been physically assaulted is likely to have experienced various forms of emotional abuse as well. Moreover, between 25% and 50% report one or more experiences of spousal rape in addition to physical assault. Thus, domestic violence typically represents multiple trauma over a long period of time and likely results in the same characteristics as the duty-related accumulated trauma experienced by both military and police. Moreover, it is entirely likely that a percentage of victims suffers from traumatic brain injury in addition to the symptoms of PTSD. Only 20% of domestic violence incidents are reported to the police and yet 40% of females reporting to emergency units are there because of injury by their partners. Moreover, even if the victim is able to leave an assaultive relationship, often harassing, threatening, and stalking continue.

Between 9% and 17% of those who have been assaulted by a partner have symptoms of PTSD and this rises to 29% if murder was the intention. The 1999 study by Golding indicated that women exposed to intimate partner violence had PTSD rates from 2.9 to 5.9 times higher than women free of this type of assault. The 2003 meta-analysis of studies to date by Ozer and associates found that interpersonal traumatic events had the highest relationship to the development of

PTSD of any traumatic incidents experienced by women. Of females in prison for committing murder, 40% were jailed after killing an assaultive partner.

i. Rape/Sexual Assault
In summary, non-consensual sexual intercourse is the most common cause of PTSD among females. Studies show that between 13% and 20% of females will experience rape in their lives (most before the age of 18) and the consequences can be tragic indeed.

Evidence from many sources indicates that for those who have been sexually assaulted:
- 65% will have PTSD 35 days later;
- 49% will still have PTSD 94 days later;
- 13% will still have PTSD 15 years later;
- 20% will attempt suicide;
- another 44% will think about suicide seriously;
- 4% to 20% will develop a sexually transmitted disease;
- 38% will avoid sexual activity for 6 months or more;
- those with PTSD are three times more likely than usual to develop bulimia;
- even happy events like childbirth and a new relationship can re-trigger trauma memories;
- most victims will not seek treatment and those who do will not usually do so until 8 years after the rape; and - the chances of another rape experience is 5 times more likely than otherwise.

Unfortunately, it is true also that many aboriginal children were assaulted both physically and sexually during their forced attendance at Indian Residential Schools in Canada. Through the efforts of one survivor whom I met because he experienced additional traumatic events during his military service, I have now seen a number of native people in their fifties and sixties who were victims of this dark chapter in Canadian history. It is abundantly clear that they suffered much from the experience at the time and that it had very negative effects on their subsequent development and functioning. Moreover, even 30 to 50 years after that accumulated trauma, many continue to have symptoms of PTSD.

j. Torture, Political Oppression, Displacement, Ethnic Cleansing, and Refugees
In summary, there is no doubt but that actions of the state can lead to symptoms of PTSD among those who escape with their lives. Studies such as those noted below indicate that PTSD rates of recent refugees can vary widely, from as low as 9% to as high as 88%. For some, rates seem to decrease as the refugees settle into a new country but even then many continue to have symptoms of PTSD as well as co-existing conditions. For others, PTSD rates increase after re-location.

Fazel and colleagues (2005) analysed 20 studies of refugees and found that 9% had PTSD while 5% had major depression. Mollica and associates (1998) reported PTSD rates of 88.2% and depression rates of 56.9% among Vietnam political prisoners living in the U.S. and PTSD rates of 77.3% and depression rates of 34.6% among Vietnam refugees who had not been detained. Of Yugoslavian refugees in Italy, Favaro and colleagues (1999) reported that 50% had PTSD and 35% had major depression. A study by Roodenrijis and associates (1998) of Somali refugees in the Netherlands report that 36% had anxiety disorders and 63% had depressive conditions. As reported by Gernaat's group (2002), among Afghan refugees living in the Netherlands, 35% had PTSD while 57% had depressive disorders. Of Iraqi refugees living in the Netherlands, 37% had PTSD and 22% had depression as reported by Laban and associates (2005). DeJong and colleagues (2001) reported PTSD rates for those who remained in Cambodia at 28.4%, in Algeria at 37.4%, in Ethiopia at 15.8%, and in Gaza at 17.8%. A survey by Mollica and associates (1993) of Cambodian refugees in a camp along the border with Thailand found that 15% had symptoms of PTSD and 55% had symptoms of depression. Tan and Fox (2001) reported that 10% of Senegalese refugees in Gambia had PTSD, 46.3 % had anxiety symptoms, and 58.8% had depression. In a study by Somasundaram and Sivayokan (1994), PTSD was found among 27% of civilians living in a conflict area in Sri Lanka and among 23% of Burmese refugees in Thailand. Two studies reported that about 20% of those reporting torture before escaping Tibet for India had PTSD symptoms that were clinically significant while another study found that rate to be about 1% among those recently re-located to India. A 2008 study by Vjovoda and associates of Bosnian refugees in the U.S. revealed PTSD rates of 76% upon entering the country, 33% at one year, and 24% at 3.5 years. Women had higher rates than men (for example, 44% and 8% respectively at 3.5 years), especially if they had not mastered the English language. Yehuda and co-investigators (2008) completed a 10-year follow-up of 40 of their original group of 63 Holocaust survivors. Those who had PTSD 10 years earlier reported a gradual decrease in symptom severity. However, 10% of those who did not have PTSD previously reported new instances of delayed-onset PTSD with a worsening of symptoms over time.

The research team led by Masmas (2008) interviewed 142 of those seeking asylum in Denmark. Of that group, 45% had been exposed to torture. About 63% of them met the criteria for PTSD as compared with rates between 5% and 10% among those who had not been tortured.

k. Motor Vehicle Accidents (MVA)
In summary, MVAs, from severe crashes to rear-enders, are very common. Survivors can have consequential PTSD, with rates ranging from 8% to 26%. Also, from 7% to 20% will show no signs of PTSD immediately after an MVA but will develop PTSD some time later (delayed onset PTSD). Many with MVA-related PTSD will continue to have symptoms for years after the incident.

Statistics gathered between 1990 and 1995 in the U.S. show more than 2 million severe MVAs resulting in 3 million people injured and between 35,000 and 40,000 fatalities. In the 25 years between 1979 and 2004, the Canadian vital statistics database shows that about 4,000 died each year; extrapolating from U.S. statistics of about 13 injuries per death each year, the estimate then would be injuries to about 52,000 people each year. Thus, in their lifetimes, 14% of females and 25% of males will have at least one MVA, making it the second most frequent trauma for both males and females.

The symptoms of PTSD can persist long after the accident. For example, Ursano's research team found that 17.6% of MVA survivors continued to have PTSD up to 9 months later. A 2008 study in Switzerland by Hepp and colleagues reported that 4% of their sample still had PTSD after 3 years, which represented an increase from 2% 1 year after the accident. Some have reported that 8% still had PTSD five years post-accident. The 2008 report by Kongsted and associates of those who experienced chronic whiplash after an MVA indicated that 13% had PTSD-like symptoms one year later. The PTSD was associated with considerable and persistent pain, neck disability, reduced working ability, and lowered reported general health.

In addition, from 19% to 38% will restrict their driving afterwards or not drive at all.

Also, an undetermined but likely sizeable number will suffer the addition of traumatic brain injury as a result of either banging their heads against something or shearing of brain tissue as a consequence of sudden acceleration/deceleration of the brain. Possible brain injury with significant resulting cognitive impairment is rarely assessed post-accident and the neuro-imaging tools available such as CAT and MRI scans are of little use in this situation, as will be discussed later on.

The 2008 study of 93 Japanese victims of motor vehicle accidents by Fujita and Nishida determined that perceived life risk and persistent medical problems were significantly related to psychological outcome while objective measures of the accident itself, such as barrier equivalent speed and change in velocity during the impact, were not.

l. Plane Crashes
One study found that more than 50% of survivors of a plane crash had PTSD shortly afterwards and between 10% and 15% retained that diagnosis one year later.

m. Other Human-caused Disasters
Events leading to PTSD that are caused by humans are not unusual and include fires, explosions, dam collapses, nuclear accidents, chemical spills, bridge collapses, and work site/occupational accidents. They too can lead to PTSD. For example, shortly after one human-caused fire, 70% of survivors were diagnosed with PTSD, 53% retained the diagnosis 6 months later, and that rate decreased to only 42% one year after the fire. The collapse of one dam left 59% of the survivors with PTSD.

n. Natural Disasters
In summary, earthquakes, avalanches, floods, hurricanes, tornadoes, volcanic eruptions, tsunamis, typhoons, and landslides testify to the forces of nature and can result in PTSD among survivors. A study by the International Red Cross and Red Crescent Societies indicated just how frequent and devastating such natural

disasters can be. Between 1967 and 1991 there were more than 7,700 such disasters in the world and as a result, 7 million people were killed and many more millions affected directly in some way. Depending on the study, resulting PTSD rates between 2.5% and 42% have been reported. Rates remain high even many years after the incident.

A study in Columbia found that between 24% and 42% of the victims of a volcanic eruption developed PTSD, while in Peru the rates were between 32% and 42%.

Researchers of an earthquake in Mexico found that 32% of survivors in the immediate area had PTSD. One group reported PTSD among 5% after an earthquake in Japan while others have reported rates at 10.3% following an earthquake in Taiwan.

Two years after a major flood, 44% of the folks in one study still had PTSD and 14 years later 28% continued to have that diagnosis.

In recent years, bush fires have not been uncommon in Australia. In one such fire, 16% of citizens exposed to such a fire were diagnosed with PTSD. In the U.S., PTSD was determined in 21% of people who survived one tornado and in 59% of those who survived a different tornado.

One group suggested a relationship between PTSD and proximity to Hurricane Mitch among adolescents in Nicaragua; 90% for those living in the most exposed city, 55% among those with mid-level exposure, and 14% in the least exposed city. The aftermath of Hurricane Katrina in 2005 on 810 persons in the southernmost counties of Mississippi in the U.S. was presented in 2008 by Galea and collaborators. Findings included that 22.5% had PTSD with rates dependent on a number of factors such as female gender, financial loss, post-disaster stressors, and low social support. Norris and co-workers (2009) assessed some Vietnamese Americans in New Orleans one year after Katrina and found that 5% had PTSD while a further 21% had partial symptoms.

The 2009 publication by Amstrader and 13 collaborators of a 2006 typhoon in Vietnam indicated that among the 798 studied, rates of 2.6% for PTSD, 5.9% for depression, 9% for panic disorder, and 2.2% for generalized anxiety were evident. Of those who reported any disorder, 70% reported one, 15% had 2, 14% had 3, and 1% had all four of these disorders.

o. Life-Threatening Illness
In summary, there is emerging research literature suggesting that survivors of critical illness, injury, or medical procedures as well as their family members are at significant risk for the development of PTSD. PTSD rates were reported to be from 10% to 35% among cancer survivors, between 16% and 40% among those with heart problems, from 19% to 36% for those admitted to intensive care, and about 9% for those with serious burn injuries. Childbirth and elective abortions can lead to PTSD also. Moreover, parents of a child with a serious illness can develop PTSD.

A 2001 study of breast cancer patients by the Cordova research team concluded that the experience of breast cancer represents a traumatic stressor for some women in that 61% of the group perceived it as a real threat to them and responded with intense fear or helplessness. Other studies (for example Monti and associates 2007) have reported that between 22% and 55% of those diagnosed with cancer report symptoms of traumatic stress and of these, up to 10% meet the diagnostic criteria for PTSD. In their 2008 study of breast cancer survivors, Hara and colleagues reported that the rates of PTSD were much higher after treatment (35%) versus during the early stage of cancer (3%) and proposed that this was so because (a) the stressor never disappears and can cause anxiety regarding cancer recurrence and death and (b) patients are never free from the source of their traumatic fear since it originates within themselves. A 2008 study of breast cancer survivors by Shelby and her group found that over 50% of those with PTSD met the criteria for an anxiety disorder prior to cancer diagnosis, 75% had a prior history of mood disorders, and 33% had a prior substance use disorder. Salsman and co-workers (2008) reported that the frequency of cognitive intrusions among colorectal cancer survivors at baseline predicted other PTSD symptoms 3 months later.

A 2004 study by Lucrubier of hospitalized patients with heart problems indicated that up to 40% of them had PTSD. The 2008 study by Ayers and colleagues studied people who had experienced a heart attack (myocardial infarction: MI) and found that 16% met the full diagnostic criteria for PTSD while a further 18% had some of the symptoms in the moderate to severe range of intensity. After controlling for MI and past history, the authors determined that both perceived consequences of the MI and dysfunctional coping strategies were associated strongly with PTSD symptoms. Also, PTSD symptoms have been reported to continue long after release from hospital. For example, Wikman and associates (2008) assessed 213 patients 12 months after an acute coronary syndrome (ACS) and re-evaluated 179 of them at 36 months; the rate of PTSD at 12 months was 12.2% and at 36 months was 12.8%. At 12 months, PTSD was predicted by both depressed mood during admission and recurrent cardiac symptoms while at 36 months PTSD was predicted on the basis of PTSD symptoms at 12 months and depressed mood during admission. Also, the 2008 research by Ladwig and associates indicated that those with PTSD who were followed after the implanting of a cardioverter-defibrillator experienced more depression and anxiety, had more cardiac symptoms, and had higher mortality rates than those without PTSD.

Davydow and colleagues reported in 2008 a review of 15 research publications regarding survivors hospitalized in general intensive care units. Approximately 19% of the total group of 490 were diagnosed with PTSD and subsequently experienced substantially lower health-related quality of life compared to those without the diagnosis. Being young and female were consistent predictors of PTSD while severity of the critical medical condition was not. Wallen and co-workers (2008) reported the PTSD rate during hospitalization to be 13%, which lowered by one month following discharge. Significant others such as spouses and life partners were investigated by Noble and Schenk (2008) 3.5 months after a loved one survived a subarachanoid hemorrage. About 26% of them met the diagnostic criteria for PTSD, a rate 3 times that found in the general population. In another publication, Noble and colleagues reported that 37% of the patients with subarachanoid hemorraging had PTSD.

A 2008 study by Dyster-Aas and associates of patients in a burn unit in Sweden indicated that 9% met the full diagnostic criteria for PTSD while a further 17% had partial symptoms 1 year after their incidents. Also, 41% had a major depression and 16% had panic disorder. In 2008, Van Loey reported full PTSD in 8% and partial PTSD in 13% of 90 persons 1 to 4 years following discharge after being hospitalized for severe burns. For their 2008 publication, Gaylord and co-investigators assessed 76 U.S. military personnel who had experienced both burn and explosion injuries. Of that sample, 32% had PTSD, 41% had mild traumatic brain injury, and 18% had both.

Parents, especially mothers, of children who have been diagnosed with life-threatening or otherwise serious conditions have been studied. For example, studies demonstrate that if the child received a diagnosis of cancer, then between 15% and 40% of mothers developed PTSD. Nagata and co-investigators (2008) surveyed 145 mothers whose children had undergone surgery for congenital diseases and found that 20% of the total sample likely had PTSD; rates were higher if the child's condition was rated as high in severity as compared to a rating of moderate-to-severe. Horsch and associates (2007) reported that 10% of mothers in their sample met the full criteria for PTSD and a further 15% met partial criteria after their children were diagnosed with Type 1 diabetes. They reported varying levels of PTSD among parents whose children were diagnosed with severe burns, spinal cord injury, or meningococcal disease, or required bone marrow or organ transplants.

Another study showed that between 10% and 20% of spouses/partners of persons with life-threatening illnesses will experience a psychological condition including PTSD.

The prospective 2008 study by the van Emmerik group in Holland found that 19% of women completing elective surgical abortions during their first trimesters had some symptoms of PTSD 2 months later. Symptoms included those in the PTSD criteria of both intrusive re-experiencing and avoidance of measures associated with the abortion. Pre-abortion measures of dissociation predicted both intrusive re-experiencing and avoidance while avoidance was associated also with pre-abortion difficulty of describing feelings.

Ayers and colleagues reported in 2008 that studies from various countries suggest that up to 7% of women develop PTSD related to childbirth. Rates might be higher in developing countries. From 10% to 15% develop depression post-partum. It appears that those with PTSD suffer also from depression. There is little information on the subsequent course of PTSD.

I remember Dr. Basel van de Kolk (1998) reporting studies indicating that approximately 30,000 patients a year received an insufficient amount of anaesthesia during surgery. Thus, they experienced significant pain without being able to prevent it or signal that they were in pain. Although very few studies paid attention to psychological after-effects, it became clear that some patients developed symptoms of PTSD as a consequence.

p. Tragic Death of a Loved One
No matter how carefully and compassionately it is announced, learning that a person you love has died a tragic death is a major shock. Subsequently, and for varying periods of time afterwards, the survivor can experience the symptoms of PTSD. Often this diagnosis is missed and instead terms like "bereavement" are used.

Also, police officers who are required to notify (usually in person) the next of kin after a sudden death are very bothered by this responsibility. The degree of heart-wrenching grief they observe when they give this news takes a toll on them. This is especially true when the victim is a child. In addition, often they view having to give bad news to a loved one as a personal failure since most join police forces to make a difference by saving lives.

q. Severe Psychiatric Illness
In summary, people who experience a psychotic episode involving hallucinations and delusions can develop PTSD. Depending on the study, between 15% and 49% of such patients meet the diagnostic criteria for PTSD. There can be many reasons why this is so.

First, while others will see the hallucinations and delusional thinking as not real, to the person experiencing them they are the only reality there is and the experience goes unchallenged, just as I do not question the reality of the computer in front of me. The realities these patients experience are often terrifying, continuous, and inescapable. As noted by Shaw and associates (1997), commonly such realities can include substantial fears of being killed by others or killing them. They can include perceptions that their bodies have been taken over by aliens who are commanding or making them do things they do not want to do. They can sense that terrible things are happening to their bodies such as disintegrating or exploding. The resulting chaos within them cannot be escaped, leaving them in a state of continual threat and helplessness, unable either to understand what is happening to them or to do anything about it.

Second, there is growing evidence that persons with serious mental illness are at elevated risk for trauma exposure. Women suffer high rates of sexual assault and men are exposed to high levels of physical assault. Relatedly, they are likely to be in other high-risk categories including poverty, housing instability and homelessness, lack of social support, substance use disorders, and engaging in risky behaviour such as prostitution.

Third, between 34% and 53% of those with severe mental illness report a history of childhood sexual or physical abuse and 43% to 81% report some type of victimization over the course of their lives.

Fourth, the person with a severe mental illness can be placed in restraints by police officers and then subjected to involuntary admission in a psychiatric facility. There, such persons may be subjected to involuntary restraints, isolation, and treatments. Also, they may be exposed to other inpatients experiencing their own psychotic conditions.

Among folks with bipolar disorder (formerly known as manic depressive psychosis), the PTSD rates in one investigation was 40% and in another study was 49%.

r. Human Service Work

Those who work with people who are dead, dying, seriously injured, dangerous, or psychological casualties are not immune to resulting PTSD. These can include a range of people such as psychologists, physicians, nurses, social workers, fire-rescue personnel, probation officers, clergy, jail guards, coroners, and ambulance personnel.

Knowing the circumstances and consequences of trauma close-up, being confronted regularly with challenges to a positive world view, facing personal limitations in the ability to alleviate suffering, dealing often with raw emotions, realizing that one could easily become a victim oneself, and dealing with people who threaten you or could hurt you is a very draining set of experiences. Sometimes this experience is referred to as "secondary trauma" or "vicarious trauma." Over time it can lead to symptoms of PTSD although the workers themselves and their colleagues are likely to refer to the symptoms as "burnout" or "compassion fatigue."

The few good studies completed to date indicate that between 16% and 30% of fire fighters and between 10% and 30% of those handling dead bodies develop PTSD at some time in their careers. Also, odds are that a significant percentage of human service staff who report in as being sick are experiencing symptoms of PTSD and the same is likely true for those who leave such occupations.

C-03: Risk Factors for PTSD

Research has indicated that the presence of a number of risk factors increases the chances that a traumatic incident will lead to PTSD. The term "risk factors" is used rather than "predictors" in part because some folks with the factors below do not develop PTSD after traumatic experiences while some of those developing PTSD do not have any of the factors. Also, because two events occur closely together in time does not mean that one causes the other; they may be simply co-related. The majority of the factors noted below have been found in at least one study to have no significant relationship to the outcome of a traumatic event and in some studies the impact of the risk factors have been found to be relatively small. Nevertheless, it is instructive to be aware of factors that may impact on treatment. Generally, I have found that for those exposed to trauma, the greater the number of risk factors present, the greater the chance of developing PTSD. Also, when considering the information below, many clients have a better understanding of some of the reasons that they experienced symptoms of PTSD while work-mates seemed to have escaped this condition.

First, characteristics of the traumatic exposure can be a risk factor, including:

*exposure to multiple or accumulated traumatic experiences compared to a single event. Prior exposure to trauma was found by a number of investigators (such as Dougall and colleagues in 2000) to be highly predictive that PTSD will follow a chronic course upon exposure to any subsequent traumatic event;

*exposure to more severe as compared to less severe trauma. This factor was found in Brewin and associates' (2000) meta-analytic review of studies to be the one most associated with PTSD. Also, many studies such as that of Owens and associates (2009) have found a relationship between amount of combat exposure and the development of PTSD among military personnel;

*exposure to a longer as compared with a shorter-duration traumatic event;

*exposure to purposeful violence by another, as distinct from a natural disaster;

*exposure to the death of another as compared with exposure to the injury of another. Even so, injury is a risk factor; one study of combat troops found PTSD in 16.7% of the sub-group who had been injured as contrasted with 2.5% of the non-injured sub-group;

*exposure to dismemberment or severe disfigurement;

*exposure to sexual abuse as compared with physical abuse; and

*sudden and unexpected exposure to trauma with little preparation time.

*Also, generally both police and military are more bothered by trauma involving children than adults.

Second, variables at the time of the trauma can be a risk factor, including:

*the presence of severe autonomic nervous system hyper-arousal or depression in the acute interval after a traumatic stressor. For example, in their 2008 study of those who experienced a single trauma, Bryant and associates reported that PTSD 3 months later was related to a heart rate of at least 96 beats per minute and a respiration rate of at least 20 immediately after the injury. More about hyper-arousal later on;

*significant dissociation/numbing during the trauma (a sense that things around are strange or unreal, or of being out of touch or in a daze, or perceiving oneself as outside of one's body, or being on automatic pilot);

*initial symptoms of re-experiencing the traumatic event in memories or dreams;

*insufficient prior training for handling the event;

*other stressors/anxieties at the time of the traumatic event including in the workplace;

*job dissatisfaction and/or insecure job future; and

*overwork.

Third, characteristics of the person prior to the trauma can be a risk factor, such as:

>*younger age
>*a history of maltreatment during childhood;
>*a history of low parental or peer support during childhood;
>*a history of prior assault. Smith and co-investigators (2008) assessed 890 women and 4469 men prior to their deployment to Iraq or Afghanistan. About 9% of men and 28% of women reported prior physical or sexual assault. Compared to those without a history of assault, post-deployment rates of PTSD among prior victims were 2.3 times higher for women and 2.0 times higher for men.

*a history of brain injury from such things as blows to the head and respiratory distress;

>*a prior history of psychological difficulties, especially depression and anxiety. For example, the 2009 study by Owens and co-investigators found that prior depression among combat troops was associated with higher rates of PTSD than the rates of those without a history of depression. The same had been reported among civilians by Breslau and colleagues in 2000;
>*a history of psychological difficulties/instability in one's parents;
>*having few friends or social supports;
>*significant introversion;
>*difficulty recognizing or expressing feelings;
>*impaired ability to regulate emotions. Moreover, Cristiansen and Elklit reported in 2008 that among those exposed to either explosions in a firework factory or a high school stabbing incident, men were more likely to respond with PTSD while women were more likely to respond with depression;
>*lower cognitive abilities. A decreased rate of PTSD among Vietnam veterans with higher versus lower pre-deployment cognitive abilities who had experienced low combat exposure was reported, while this relationship was not evident among those with high rates of combat (Thompson and Gottesman, 2008). The 2008 study by Gale and colleagues of 3,258 veterans found that lower pre-exposure cognitive ability was associated with an increased risk for depression, generalized anxiety, alcohol abuse or dependence, and PTSD;
>*reduced capacity to extinguish forehead tension after experimental fear conditioning;
>*exaggerated physiologic reactivity to startling sounds. For example, in 2008 Pole and his group reported a study of 138 police in training who were exposed to 106 db startling sounds under threat of various levels of electric shock. One year after police work with its attendant exposure to trauma, PTSD rates were calculated. It was found that those who reacted most to the startling sounds before becoming police officers had the most severe PTSD symptoms; and

*emotion-focused coping strategy. As reported in 2009 by Taylor and co-investigators, military men taking realistic survival training who used methods such as passive and emotion-focussed coping had more acute stress symptoms than those using active and problem-focussed coping.

Fourth, conditions after the trauma can be risk factors, including:
*insufficient social support including thar provided by supervisors or the unit. Brewin and colleagues (2000) completed an analysis of studies and found that absence of social support after the event was one of the three highest risk factors in the later development of PTSD;
*the presence of additional stressors after the traumatic event. Brewin and associates found this to be among the top 3 greatest risk factors for developing PTSD in their meta-analytic review study;
*chronic pain or significant injury as a result of the incident;
*chronic illness and unemployment (see Harris et al., 2008);
*the presence of sensory hypersensitivity after the trauma;
*insufficient practical assistance after the trauma;
*slow or accusatory actions by legal and administrative agencies; and
*having an unsettled compensation claim (see Harris et al. 2008).

Also, a number of recent studies have demonstrated that those with PTSD who have a greater number of risk factors or who have more severe re-experiencing of symptoms have a more persistent course of the condition than those who do not.

C-04: Prior Childhood Trauma
In summary, the evidence demonstrates that many children and adolescents experience traumatic events. Thus, some adults with PTSD as a result of trauma during adulthood have a prior history of significant childhood/adolescent incidents, often occurring over very long periods of time with attendant emotional abuse/neglect. Consequently, psychological and physiological processes necessary to good functioning are compromised. This appears to make them more vulnerable to PTSD when confronted with traumatic incidents as adults.

In 2005, the U.S. Department of Health and Human Services reported that data collected in 2003 indicated that 12.4 per 1000 children were confirmed victims of child abuse. In a study sponsored by the U.S. Institute of Justice, it was reported that 5,000,000 adolescents in that country between the ages of 12 and 17 had experienced physical assault, 1,800,000 had been the victims of sexual assault, and 8,800,000 had witnessed interpersonal violence. Others have reported that more than more than half of youth between ther ages of 12 and 17 had experienced physical assault in that year, more than 1 in 12 had been victims of sexual assault, and 33% had witnessed interpersonal violence; moreover, they found that there was a 69% rate of re-victimization within one year. Students surveyed students in grades 4 through 12 in New York City prior to 9/11 found that 39% had seen someone killed or seriously injured, 27% to 29% had witnessed the violent or accidental death of a relative or close friend, and 25% reported exposure to two or more traumatic events. In a study of young Japanese women, 12% reported exposure to at least one traumatic event during pre-school years, 21.2% during primary school, 27.5% during high school, and 23.8% during college years. A study of 14- to 24-year-olds in Munich provided data that 21.4% of males and 17.7% of females had been exposed to traumatic events and 17% of the combined sample experienced intense fear, hopelessness, or horror about the event, thus meeting Criteria A-2 of the *DSM-IV* for PTSD. As cited in Courtois and Gold (2009), a total of 1,420 children ages 9, 11, and 13 were followed and by age 16 about 68% of them had experienced significant traumatic events such as violent victimization, attempted kidnapping, or attempted sexual molestation. Clearly, the rates of trauma among children are very high, such that it is reasonable to assume that many adults who report trauma have experienced traumatic incidents before.

Courtois and Gold (2009) remind that many traumatized children with significant symptoms do not meet the diagnostic criteria for PTSD. Also, the researchers cite studies indicating that among adults and adolescents with PTSD related to prior childhood trauma, about 50% show no remission of symptoms when re-assessed 3 to 4 years later.

In addition to the confusion and distress a young person experiences from abuse, research indicates that abuse results in long-term decreases in brain functioning. For example, Andersen and colleagues (2008) used specialized MRI scans from 26 women with histories of repeated episodes of childhood sexual abuse compared to 17 women without such experiences. The victims had smaller volumes of the hippocampal brain structure if the abuse occurred at ages 3-5 or 11-13, reduced corpus callosums if abused between 9-10 years of age, and decreased functioning of the frontal lobes if victimized at ages 14-16. Previous studies had reported similar disturbances in brain functioning but had not tied the specific manifestations to the ages when the abuse occurred. The hippocampus is involved in short-term memory among other things, the corpus callosum allows communication between the right and left hemisphere of the brain, and the frontal lobes are associated with such human aspects as attention, memory, planning, emotional regulation, and impulse control.

The impact of abuse on brain functioning was illustrated also in the 2009 study by Black and colleagues. Comparing those with an abuse history with those without such experiences, moderate to large differences were noted in connectivity between regions of the brain. Also, the Cook research team (2009) found that victims of childhood trauma had different patterns of connectivity (coherence) from those with either adult trauma or no past trauma. Differences were found in frontal, central, temporal, and parietal areas of the brain and suggest that childhood trauma may have a lasting impact on the connections between brain cells. In addition to some of the findings above, Wolf and colleagues (2009) cited a study that indicated that early traumatic stress resulted in alterations in the amygdala (discussed later) as well as pain control functions such as the endogenous opiate system. In comparison with those without a trauma history, previous studies of those with a history of childhood sexual abuse had provided evidence of resulting disturbances in the body's neuro-endocrine system including corticosteroid and thyroid levels, loss of normal EEG synchrony between different areas of the brain, a greater number of glucocorticoid receptors, greater de-regulation of the HPA axis, and more hyper-activity of the autonomic nervous system (all of these are discussed later on).

The 2009 study by Wilcox and associates investigated the risk of suicide following childhood exposure to traumatic events, especially sexual abuse. By following a child sample of 2,311 first-grade children to young adulthood, the authors concluded that those who had developed PTSD after childhood abuse were at higher risk for suicide than those who did not show signs of PTSD. Other studies cited by Wolf and associates suggest that survivors of childhood abuse have a greater chance than others of developing a whole range of psychological conditions including PTSD, other anxiety disorders, dissociative symptoms, personality disorders, aggressive behaviour, and self-mutilation tendencies. Additionally, symptoms of hyper-arousal, deficiencies in such cognitive functioning as attention and memory, and problems modulating intrusive affect have been noted.

Some children and adolescents subjected to multiple occasions of abuse, especially at the hands of care-givers, have additional traumatic experiences during their adult years. If the adult trauma results in PTSD, many experts refer to the condition as *complex PTSD* while others use the term *disorders of extreme distress (DES)*. Many studies including the 2008 one by Briere and collaborators indicate that the severity of the PTSD symptoms among traumatized adults who were abused during childhood or adolescence is much worse compared with those who did not experience abuse in their early years.

Current thinking is that the impact and duration of the childhood trauma caused interference with age-related basic psychological and physiological development. For example, two of the leaders in this area, Dr. Van der Kolk and Dr. Courtois, summarize the impact and noted that such folks experience
> (a) alterations in ability to modulate emotions,
> (b) alterations of identity and sense of self,
> (c) alterations in ongoing consciousness and memory,
> (d) alterations in relations with the perpetrator,
> (e) alterations in relations with others,
> (f) alterations in physical and mental status, and
> (g) alterations in systems of meaning.

My clinical experience is that some of the adults who come to my attention have a prior childhood history of repeated physical and/or sexual abuse. For many of them, the core symptoms of PTSD and life-interference from any of the seven alterations above come to the fore following exposure to one or more traumatic incidents during adulthood. Compared to adults with positive childhoods, those who suffered repeated abuse during childhood require a very long time before symptoms decrease enough to where the victims can function reasonably well.

Also, as cited in the September 2009 issue of the *APA Monitor* (page 35), a 2008 study found that men and women who reported prior assault were twice as likely to report PTSD following military deployment than those who had not been assaulted.

C-05: Factors That Promote Coping
In summary, a number of factors increase the chances that a person will be able to cope with or adjust to traumatic experiences. These include:
> higher versus lower IQ;
> stable and supportive relationships;
> life-long good health practices;
> adaptive coping skills; and perhaps
> the personality traits of hardiness.

As Paul Bartone described in *Managing Traumatic Stress Risk* (Paton, Violanti, Dunning, & Smith, 2004):
> Hardy persons have a high sense of life and work commitment, a greater feeling of control, and are more open to change and challenges in life. They tend to interpret stressful and painful experiences as a normal aspect of existence, a part of life that, overall, is interesting and worthwhile. (p. 131)

Also, enlightened senior administration can act in ways that assist their staff following a traumatic event. As outlined by Dr. Paton and Dr. Hannon (Paton, Violanti, Dunning, & Smith, 2006) in *Managing Traumatic Stress Risk*:
> Senior officers also can facilitate this process by helping personnel appreciate that they performed to the best of their ability, realistically review their experiences relative to situational constraints on performance, understand that little could have been done differently under the circumstances, reduce inappropriate feelings of guilt and responsibility, learn about their reactions, and provide accurate information about what happened and about future issues . . . by assisting staff in the identification of the strengths that helped them deal with the emergency response and building on this experience to plan how future incidents could be dealt with more effectively. (p. 122)

Continuing, this is a far cry from:
> a cultural predisposition within an emergency organization to suppress emotional disclosure, focus on attributing blame, or to minimize the significance of peoples' reactions or feelings (which) can undermine support-provision, heighten stress vulnerability, and extend performance deficits. (p. 122)

Paton, in the 2006 book edited by Calhoun and Tedeschi, noted also that
> . . . bureaucratic organizations increase vulnerability through persistent use of established decision procedures (even when responding to different and urgent crisis demands), internal conflicts regarding responsibility, and a predisposition to protect the organization from criticism or blame. (p. 237)

He concluded that research to date indicates that such organizational variables within emergency professions are better predictors of posttraumatic outcome than the incidents themselves. My clinical experience with police officers has been that often an organizational/managerial stressor was the apparent final straw in their decompensation. It is as if these officers could manage the traumatic stressors until the time came when they no longer believed that the organization was supportive of them.

Moreover, there is reason to believe that good leadership and reliance on comrades during military deployment can decrease the effects of potential stressors (Rona et al., 2009).

Furthermore, based on some of the pre-exposure risk factors noted above, it is reasonable to assume (a) that teaching effective psycho-physiological strategies may decrease the likelihood of PTSD following trauma exposure and (b) that the same strategies may be effective in decreasing symptoms of PTSD in those who develop the condition subsequent to traumatic incidents. Information on psychophysiology is found later in Step F. Relatedly, a current focus of the U.S. military is the development of training experiences to increase resilience to battle conditions. Additional information about prevention is found in Appendix VI.

C-06: *DSM-IV* Diagnostic Criteria B, C, and D for PTSD

According to the *DSM-IV* diagnostic classification system used in North America and many other parts of the world, PTSD is diagnosed if (a) a sufficient number of the following symptoms are present, (b) they were not there before the traumatic event, (c) they have been present for at least one month, and (d) they cause significant distress in the person's life:

Cluster B. The traumatic event is persistently re-experienced in *1 or more* of the following ways:
   1. Recurrent and intrusive distressing recollections of the event;
   2. Recurrent distressing dreams of the event;
   3. Acting or feeling as if the traumatic event were recurring;
   4. Intense psychological distress at exposure to internal or external cues that symbolize or resemble an aspect of the traumatic event; and/or
   5. Physiological reactivity on exposure to internal or external cues that symbolize or resemble an aspect of the traumatic event;

Cluster C. Persistent avoidance of stimuli associated with the trauma and numbing of general responsiveness as indicated by *3 or more* of the following:
   1. Efforts to avoid thoughts, feelings, or conversations associated with the event;
   2. Efforts to avoid activities, places, or people that arouse recollections of the event;
   3. Inability to recall an important aspect of the trauma;
   4. Markedly diminished interest or participation in significant activities;
   5. Feelings of detachment or estrangement from others;
   6. Restricted range of affect (for example, unable to have loving feelings); and/or

7. Sense of foreshortened future (does not expect to have a career, marriage, children, or a normal life span).

Cluster D. Persistent symptoms of increased arousal, as indicated by *2 or more* of the following:
1. Difficulty falling or staying asleep;
2. Irritability or outbursts of anger;
3. Difficulty concentrating;
4. Hypervigilance; and/or
5. Exaggerated startle response.

In addition, associated symptoms are common such as:
*feelings of guilt for what was done or not done;
*feelings of guilt about surviving when others did not;
*other cognitive problems such as with memory and/or difficulty making decisions;
*feeling out of control and unable to do anything about it;
*chronic worry;
*impaired ability to regulate emotions;
*self-destructive and impulsive behaviour (up to 75% with PTSD attempt suicide);
*somatic complaints;
*feelings of ineffectiveness, shame, despair, or hopelessness;
*believing self to be damaged permanently;
*a loss of previously sustained beliefs;
*social withdrawal;
*feeling constantly threatened;
*impaired relationship with others; and
*a change from the individual's previous personality characteristics.

C-07: More Information About Specific Symptoms

a. Symptom B-5: Psycho-physiological Reaction to Reminders of the Trauma

While remembering the traumatic experience(s) during the psycho-physiological assessment that all my clients complete, clear signs of reactions are always evident among those with PTSD. Commonly, heart rate increases, heart rate variation decreases, blood pressure increases, skin conductance increases, peripheral blood flow decreases, aspects of healthy respiration decline, and sometimes blood oxygen levels decrease. These things happen because the sympathetic branch of the autonomic nervous system becomes activated, leading to greater levels of arousal, and/or the parasympathetic branch of the autonomic nervous system de-activates, thereby not putting the brakes on arousal. A growing amount of research literature is reporting similar findings. For example, the 2008 study by Sarlo and associates assessed people who passed out at the sight of blood. Compared to those without such a significant phobia, when exposed to film clips involving blood, those with the condition demonstrated clear increases in heart rate and increases in cardiac output. In contrast to my clients with PTSD, systolic blood pressure among those with the phobia decreased, which accounts for their passing out. Responses to film clips without blood cues did not differ between the two groups.

b. Symptom C-6: Restricted Range of Affect

Alexithymia is not part of the DSM-IV code but is a measurable condition that is defined as difficulties in identifying and labelling emotional feelings and a tendency towards thinking that is directed to external stimuli rather than to internal activity. Thus, at the very least it is consistent with criteria C-6 and may signify an uncoupling of cognitive from emotional processing whereby intense emotional states are not integrated with verbal awareness.

Moreover, studies have found a relationship between alexithymia and PTSD. For example, the 2008 meta-analysis by Frewen and collaborators (2008b) of the 12 available studies to date indicated a large effect size of alexithymia among those with PTSD and particularly among male combat personnel. Their experiment found the same to be true for victims of motor vehicle accidents as well as reporting significant statistical relationships between alexithymia and severity of such PTSD symptoms as re-experiencing, emotional numbing, and hyper-arousal.. Also, in their research report of 2008b, the Frewen group cite evidence that, when exposed to emotional stimuli, the activation or inactivation of brain areas among those with alexithymia has some similarities with the pattern displayed by those with PTSD. Relatedly, in their 2008c study using fMRI, healthy controls had emotional awareness scores that correlated with high activity in the ventral ACC brain area while those with PTSD had scores that were associated with low activation of this region. The role of the ACC is discussed in section K of this manual.

c. Symptom D-2: Irritability or Outbursts of Anger

The symptom of irritability in PTSD (D-2 criteria above) can manifest as demonstrations of frustration or impatience when things are not going as desired. The other half of the symptom, outbursts of anger, can include episodes of significant verbal anger, physical actions such as punching walls and throwing things, and acts of physical violence against animals or people. As discussed in the 2008 review by Miller and co-authors, most typically, such episodes are impulsive rather than premeditated; they (a) are emotionally charged, (b) are uncontrollable, (c) are not planned for beforehand, (d) involve cognitive impulsivity in that they occur very rapidly and without thinking as a "hair-trigger" reaction to a stimulus, (e) sometimes occur without the individual being able to identify the trigger at the time or afterwards, and (f) are disproportionate to the provocation. Afterwards, many people feel shame and guilt, recognize that a loss of control occurred, know that their reaction was disproportionate to the trigger, but cannot explain in any logical fashion why they reacted that way. Others appear to "blank out" during the episode and have no memory of the incident afterwards.

One study of U.S. military veterans with PTSD found that 33% of them were violent towards their partners; this rate was 2 to 3 times higher than veterans without PTSD. Another study found that combat veterans with PTSD reported an average of 20 acts of violence per year compared to less than 1 among veterans without PTSD.

Results of studies in pharmacology, neuropsychology, and neurobiology noted in the 2008 Miller article have demonstrated differences between those with impulsive aggression and others, such as those who contain their anger or those whose aggressive actions are premeditated. Among those with impulsive aggression, these findings include (a) abnormalities in the amygdala (an emotional area of the brain), (b) dysregulation of areas of the brain involved in impulse control (the ACC and OFC of the medial pre-frontal cortex), (c) impaired sensory gating in the thalamus brain region, (d) decreased levels of the brain transmitter serotonin, (e) lower metabolic rates in various areas of the brain, (f) dysregulation in language processing regions of the brain, and (g) a positive response to both Prozac and the anti-convulsant *phenytoin*. All of this means that among those with impulsive aggression, emotions are easily activated in the sub-cortex of the brain while cortical brain regions and cognitive processes that are supposed to put a damper on reactivity are not functioning as well as required.

Also, it is true that the trait of anger can predict later symptoms of PTSD. For example, a 2008 study by Meffert's group assessed trait anger among police recruits during their basic training and then checked for symptoms of PTSD one year later. Those with the high trait anger were more likely to develop symptoms of PTSD than those with low trait anger. The investigators reported also that those with symptoms of PTSD at one year had more episodes of anger one year later than those without PTSD symptoms. Relatedly, a 2008 study by Forbes and associates of Australian veterans of the Vietnam War found that those with the highest levels of anger at intake had the highest severity of PTSD at follow-up.

It is known also that irritability and anger outbursts can be the result of other factors. For example, problems with anger control can be one of the symptoms of traumatic brain injury which sometimes accompanies PTSD. Also, anger outbursts can be associated with seizure activity. Moreover, lack of control over impulses is part of the condition known as attention deficit/hyperactivity disorder. Furthermore, there are circumstances where self-control is difficult to maintain, such as after too much alcohol and when the number of stressors increases beyond the person's capacity to adapt.

Anger is a risk factor for the development of depression as well as a number of cardiovascular conditions such as stroke, myocardial infarction, and high blood pressure, thereby leading to increased mortality. Also, it is related to substance abuse. Furthermore, it is associated with a decrease in both social and occupational functioning. Thus, the consequences to the person with anger are significant.

Accordingly, regardless of origin, clearly clients who display physical outbursts towards others must receive effective treatment sooner rather than later. At the same time, steps need to be taken to maximize the safety of those living with the client. This is the approach I take with clients. Recently, I saw three military veterans and one long-serving police officer who presented with PTSD. With my basic approach described later on, generally the symptoms of PTSD decreased to normal levels. However, the symptom of anger/irritability was still clearly evident. All four have then completed BFT-C following NeuroGuide and low-resolution electromagnetic tomography (LORETA) analysis of the qEEG (all discussed later). For all four, their previous anger/irritability outbursts are now under good control.

An interesting 2008 study published by Dileo and colleagues measured the ability to identify smells among male Australian war veterans with PTSD and matched controls. The PTSD patients exhibited significantly more deficits in smell identification than controls despite no differences in cognitive measures and accounting for levels of IQ, anxiety, depression, and alcohol usage. Also, the deficits in smell identification were significant predictors of aggression and impulsivity. The authors consider these findings as supporting evidence of the role of the orbital-prefrontal area of the brain in the development of PTSD symptoms.

d. Symptom D-5: Exaggerated Startle Response

Most of us have a rapid response to an unexpected sound or a touch. The reaction is generated from an area of the brain known as the pontine reticular formation. The startle response can include a jump or twitch, eye blinks, expanded pupil size, increase in heart rate, sweaty hands, and a sudden turn in the direction of the source. In many studies, the startle response is measured by the muscle activity producing an eye blink. However, the reaction subsides quickly. Also, once the source is identified, our response is markedly less should it occur again; this is known as habituation.

Usually, those with PTSD have reactions that do not subside quickly in comparison to the reactions of average persons, and in fact an exaggerated startle response is one of the symptoms of PTSD. The 2007 review of research publications by Pole concluded that among those with PTSD as compared to those without the condition, the startle response was greater. Also, studies have demonstrated heightened startle reactivity among those with PTSD resulting from military combat and police work as well as among female survivors of sexual assault and female victims of childhood sexual abuse. In addition, a study of rape and physical assault survivors with PTSD by Griffin (2008) indicated that both eye blink and heart rate reactivity increased over time after the trauma along with other PTSD symptoms. In addition, habituation is affected among those with PTSD. For example, a 2008 study by Javanovic and co-investigators compared Croatian combat veterans with a healthy control group when presented with a repeated startle probe. Results indicated that compared to the control group, the veterans' habituation to the probe was impaired. Along with other differences (for example, elevated baseline heart rate and decreased RSA--more about this later on), the authors considered findings to indicate overall hyper-arousal among those with PTSD. I have observed the startle response often in my office among those with PTSD and such clients (and their partners) have mentioned many instances of exaggerated startle in their daily lives.

Studies cited by Payne and Gevirtz (2009) indicate that

Increased acoustic startle has long been recognized as an important symptom of PTSD. Increased eye-blink response to acoustic startle is reported in most but not all studies comparing PTSD and non-PTSD groups . . . . Slowed skin conductance habituation appears to be a consistent correlate of PTSD (19). . . . Shalev and colleagues (2004) have reported greater heart rate response to acoustic startle at 1 to 4 months posttrauma in individuals with PTSD versus those without PTSD. . . . The landmark study of Orr and colleagues supports the secondary acquisition of increased startle through increased heart rate response rate to acoustic startle in Vietnam combat veterans with PTSD, but not their non-combat exposed co-twins or Vietnam combat veterans without PTSD . . . . Evidence also suggests that startle response in patients with PTSD may be related directly to PTSD symptom severity . . . (p. 6)

Also, consistent with the 2008 study by Pole and associates noted above, the authors describe a study of fire fighters who were exposed to an acoustic startle during their training. This was re-assessed within 4 weeks and after the fire fighters had been exposed to trauma. Results indicated that the pre-trauma measurements of eye-blink startle and skin conductance predicted PTSD symptom severity after trauma exposure. Relatedly, Viedma-del Jesus and her team reported at the 2008 Society for Psychophysiological Research meeting that presentation of a startle probe paired with a picture of the feared stimulus to those with a phobia to that stimulus resulted in greater physiological reactivity than when presentation of a startle probe was paired with other types of pictures such as pleasant and unpleasant ones or as compared to presentation of a startle probe to those without a phobia. From this latter finding, it is understandable why those with PTSD respond significantly to a reminder of their traumatic experience (s).

C-08: The Timing of Symptoms
There are differences from one person to the next in the timing of the appearance of symptoms. Sometimes they occur within hours or days of a specific traumatic event, sometimes they are not evident until the client has been exposed to a number of traumatic incidents, sometimes they rise up decades after trauma, and sometimes previously unrecognized symptoms are activated or worsened following a new event or stressor, however minor.

For example, in 2008 Daly and co-workers reported on firemen who had been affected by working at the World Trade Center after the terrorist attacks there. Over and above acute effects at that time, one year later they discovered further increases in mood symptoms and hyper-arousal. Also, a typical finding is that some police officers and many veterans of WWII did not complain of symptoms during their service years but began to experience symptoms sometime after retirement. As stated in a chapter by Campise and associates in the 2006 book *Military Psychology* edited by Kennedy and Zilmer,

In fact the term "reactivated post-traumatic stress disorder" is increasingly used in reference to veterans of WWII and the Korean War who led productive lives until they were in their 60s and 70s, when their ability to function became impaired by a return of their PTSD. (p. 221)

Many of the WWII veterans I have seen suffered in silence upon their return to Canada. They did not discuss experiences or symptoms with anyone but rather threw themselves into their family life, their work, and community activities. From the outside, no one would have suspected they had PTSD. Then, with declining health and mobility, symptoms increased so that they were obvious to others.

For his doctoral thesis, Hiskey reviewed the activation of PTSD in later life. He reported that studies suggest the delayed onset of PTSD symptoms in later life to be between 11% and 34% of military veterans. However, he noted that most cases were preceded by at least some prior symptoms. For some, partial prior symptoms increased so as to meet the full diagnostic criteria. Others with PTSD did not notice, were not significantly bothered, or did not report symptoms to others. Age-related losses including declining physical health, environmental stressors, and the effects of neurodegeneration may all have played a part.

C-09: Associated Psychological Symptoms
In summary, it is rare to see a person with PTSD in a clinical setting who does not have symptoms of additional conditions. Some of these conditions can be reactions to trauma in their own right. Regardless, the presence of these conditions increases both the complexity of treatment and the time required for successful outcomes.

More than 80% of people with PTSD will have one additional psychological condition. For example, in a 2001 survey of 1,126 people in the general population with PTSD, Brown and associates reported that 92% met the diagnostic criteria of another *DSM-IV* psychological condition while the 1996 study of U.S. veterans by Orsillo and associates found that rate to be 82%. Various other studies have indicated that among those with PTSD, 14% to 77% will have a *major depression*, 18% to 28% will have *social phobia*, 16% to 22% will have *agoraphobia*, 15% to 38% will have *generalized anxiety disorder*, 5% to 19% will have *panic disorder*, and about 15% will have *obsessive compulsive disorder*.

As discovered in the 1995 NCS study by Kessler and colleagues, 59% of men and 43.6% of women with PTSD had three or more additional diagnoses. Also, most with PTSD have *sleep disturbances/disorders* and some experience significant and *chronic pain*. In addition, even controlling for depression, studies have indicated that those with PTSD are much more likely to attempt *suicide* than those without PTSD. For example, studying folks with major depression, Oquendo and associates (2003) found that the attempted suicide rate of those without PTSD was 54% and was 75% among those with PTSD.

Also, psycho-physiological conditions can co-occur with PTSD. For example, in their 2008 report on 248 women veterans, Savas and co-workers reported that 38% of the group had irritable bowel syndrome (IBS), which was highly associated with the presence of anxiety, depression, and PTSD. As noted later on, the autonomic nervous system is dysregulated among those with PTSD and one of its three branches, the enteric system, is related to gastro-intestinal (GI) functions. IBS is a disorder of the GI system.

After reviewing the evidence, Clark and colleagues (2009) concluded that "in real patient populations, only those with milder conditions tend to have a single diagnosis" (p. 96).

However, the presence of any co-occurring psychological condition with PTSD increases the time required for treatment. Also, the likelihood of a complete recovery from all symptoms may be compromised. For example, a 2008 report by Zatzick and co-authors examined those who had been hospitalized following a traumatic injury at work. Compared to those without the psychological conditions, those with either PTSD or depression were 3 times less likely to be back to work within 12 months, while that number increased to about 5.5 times if they had both conditions. Also, a 2007 study by Dunn and her co-workers of combat veterans with both PTSD and depression found that cognitive behavioural group therapy that was often successful among folks with a depression made a very minimal difference in her combat subjects. Even this difference disappeared during a follow-up investigation.

It is true also that these additional conditions can be reactions to trauma in their own right without the presence of PTSD; that is, trauma can lead to any of a number of symptom clusters including those diagnosed as depression, panic disorder, sleep disorders, generalized anxiety disorder, and psycho-physiological reactions without symptoms specific to PTSD. In their 2008 study of 228 survivors of motor vehicle accidents, Grant and associates found that PTSD, major depressive disorder, and generalized disorder were distinguishable conditions that were highly correlated and that dysphoria was common to all three conditions. Courtois and Gold (2009) cite a number of studies indicating that depression may in fact be a more frequent response to trauma than PTSD.

The conditions described above are assessed through my structured clinical interview, administration of standardized psychological tests, and other evaluations.

In addition, some client's symptoms can be so marked as to qualify for other *DSM-IV* diagnoses.

First, there is reason to believe that a percentage of those with PTSD will have numbing/dissociative symptoms (Cluster C) such that a *dissociative disorder* could be diagnosed. This is not a condition that I have seen in many clients to date.

Second, undoubtedly some would qualify for the diagnosis of *intermittent explosive disorder*. In this condition the symptom of anger/irritability (Cluster D) is so marked that the diagnosis is warranted. The criteria for this diagnosis include

> (A) Several discrete episodes of failure to resist aggressive impulses that results in serious assaultive acts or destruction of property, and

(B) The degree of aggressiveness expressed during the episodes is grossly out of proportion to any precipitating psychosocial stressors.

Third, the problems with attention/concentration for some with PTSD (Cluster D) are so significant and are associated with other cognitive capacities (memory, information-processing, decision-making) such that a diagnosis of *cognitive disorder not otherwise specified* would be reasonable. I screen for this with the MicroCog.

Fourth, problems with sleep common in PTSD (Cluster D) can be such that a separate diagnosis of *sleep disorder* can be made. For example, for reasons not yet fully known, there is a high relationship between PTSD and sleep disordered breathing such as obstructive sleep apnea (OSA), especially in males over the age of 40.

Depending on results of clinical assessment, treatment for identified co-occurring conditions is provided since it is known that by doing so (a) the quality of life is improved, (b) symptoms of these conditions are reduced, and (c) the chances of a relapse/re-occurrence in this conditions as well as PTSD in the future are lessened.

C-10: Traumatic Brain Injury
In summary, most clients with PTSD experience impaired cognitive functions such as problems with attention/concentration and memory that were not present before the traumatic event(s). This can be the result of being distracted by intrusive thoughts and memories such that the person is not attending to other information, it may be related to the circulation of brain chemicals associated with PTSD that can exert a negative impact on memory systems, or it may be related to sleep loss. However, cognitive problems can be the result also of injury to the brain, which is not uncommon among military and police personnel. The processes involved, ways to investigate the possibility of traumatic brain injury (TBI), and treatment options are discussed below.

Blows to the head such as those that occur in a physical assault, impact injuries to the back of the head (for example from a fall), the sudden acceleration and deceleration and rotation of the brain as in a motor vehicle accident, frequent and/or close exposure to blast/pressure waves such as those related to an explosion or the firing of weapons systems, and any events that involve stoppage of the flow of oxygen can result in diffuse injury to the brain. For example, Mayorga and Reiecher's 2008 study indicated a 30% rate of TBI from all sources among U.S. service personnel, of which 50% were classified as mild, 25% as moderate, and 25% as severe. In 2008, Sayer and co-authors reported on soldiers who had been injured by blast and other injuries and found that "soft tissue, eye, oral and maxillofacial, otologic, symptoms of posttraumatic stress disorder, and auditory impairments were more common in blast injured patients than with war injuries of other etiologies" (p. 166).

Moreover, as Dr. Bob Thatcher wrote in the 2009 volume edited by Budzynski and others,

The duration of reduced brain function following traumatic brain injury can be many years even in the case of mild head injuries in which there is no loss of consciousness" (p. 270).

Also, research to date indicates that the consequences of a head injury may not be noticed until months to years after the trauma. Consequently, Sayer and colleagues indicate that such patients have life-time needs.

Any of the above insults can cause such things as shearing of the axons of the brain cell, death of brain cells, dysfunction of neuronal cell bodies and assemblies, problems with the delivery of blood to the brain, disruption of the essential flow of oxygen to the brain, arterial vasospasm, the release of toxic levels of neurotransmitters, and brain swelling. While such results can stem from a single head trauma, they are even more likely in the person who has had more than one such experience.

The 2008 review of published studies by Nampiaparampil indicated an overall prevalence of chronic headache of 57.8% among those with traumatic brain injury. Headache was more common among those with mild TBI (75.3%) as compared with moderate or severe levels (32.1%).

Still, a firm diagnosis of TBI is difficult to make, especially when the presumed damage is in the mild to moderate range. For a variety of reasons, in the rare occurrence where a CT scan or an MRI is completed after a presumed TBI, generally nothing is detected even among those with clear evidence of cognitive deficits consistent with TBI. Studies to date as reviewed by Belanger and her group (2007) show that CT and MRI scans are not capable of detecting TBI. Also, Thornton and Carmody (2008) cite studies in support of this conclusion. Moreover, Gentry and others have demonstrated that only about 10% of severe TBI cases noted post-mortem were detected by conventional MRI. As reported by Dr. Tanju (personal communication, 2008), almost all patients with TBI in his head injury program had normal MRIs. In his 2009 article, Striefel noted a study by Mauk that indicated that MRI and CAT scans failed to diagnose brain problems in 70% of civilians and 44% of military clients. Dr. Thatcher in a listserv of 2008 (personal communication) explained further:

> The answer to the question why MRIs and CT scans are less sensitive than EEG is that the MRI and CT are the same whether someone is dead or alive because they are structural images. . . . if there is no bleeding then the MRI and CT are usually normal even when a patient is in a coma.

According to Belanger and others, the best direct medical approaches for detecting TBI are such functional neuro-imaging methods as fMRI and PET scans. Thornton and Carmody (2008) cite studies that support this conclusion as well as suggesting that diffusion tensor imaging (DTI), magnetizing transfer imaging (MTS), magnetic resonance spectroscopy (MRS), and single photon emission computer tomography (SPECT) show promise. However, generally functional neuro-imaging and other sophisticated approached are rarely used due in part to their unavailability except in large medical centres.

However, in the 2008 listserv (personal communication), Dr. Thatcher continues

There is a huge scientific and clinical literature demonstrating the sensitivity and specificity of QEEG in the evaluation of the nature and severity of TBI. A search of the National Library of Medicine . . . using the search terms "EEG and Head Injury" shows 2960 citations and 95% of these are QEEG studies. There have been very few non-QEEG studies published since about 1980 . . . . The EEG measures the metabolically dependent electricity of the brain which changes dramatically when the brain is injured . . .

Thus, the approach I take is to consider that there is evidence of TBI if:
a. there is reason to believe that the person experienced a trauma to the head;
b. there is evidence that the person had a concussion as a result of the incident;
c. there is reliable evidence of a deficit in cognitive functioning after the trauma;
d. there is evidence that the deficits were not present before the head trauma;
e. there is evidence that deficits remain after successful treatment of PTSD; and
f. a qEEG is consistent with that of persons known to have TBI.
Some of these factors are discussed below.

a. A Trauma to the Head
The first requirement is that there be some reason to believe that a client has experienced head trauma. The most common sub-type of trauma among my clients is closed-head trauma such as that resulting from the head hitting some surface or from acceleration/deceleration, as during a rear-ender. A sub-type I see rarely is one in which there has been a penetrating injury such as a gunshot to the head. Occasionally, I have seen clients in whom there has not been an impact injury, but there is evidence of bleeding within the brain from an aneurism.

b. Concussion.
In the first minutes to hours after a head injury, clients may have symptoms of a concussion even when there are no focal neurological signs and Glascow Coma Scale scores are as high as 13 to 15. The person may lose consciousness and/or have some of the following symptoms as discussed in the 2005 paper by Nuwer and colleagues. Typical symptoms are presented in a format below so that clients can check any they have experienced.
_____feeling and appearing dazed;
_____vacant stare;
_____delayed verbal or motor responses;
_____inability to focus attention;'
_____disorientation;
_____slurred or incoherent speech;
_____gross observable problems with coordination;
_____emotionality out of proportion to the circumstances;
_____memory deficits;
_____headache;
_____dizziness or vertigo;
_____lack of awareness of surroundings; and
_____nausea or vomiting.

The Nuwer group note also that some symptoms can last for weeks or more which in combination is known as *post-concussive syndrome*. Again, symptoms are listed in an easy-to-check format.

_____persistent low-grade headache;
_____light-headedness;
_____poor attention and concentration;
_____memory dysfunction;
_____becoming fatigued easily;
_____irritability and low frustration tolerance;
_____intolerance of bright lights or difficulty focussing vision;
_____intolerance of loud noises and sometimes ringing in the ears;
_____anxiety and depressed mood; and
_____sleep disturbance.

For most clients, the symptoms improve within a few weeks to a few months. However, headache, irritability, anxiety, dizziness, fatigue, and attention problems are still there 3 months later for some. At that time, 62% of clients reported at least one symptom, and 40% had 3 or more. By 6 months about 33% reported at least one symptom.

By the 1 year mark 10% to 15% still reported significant symptoms and *permanent symptomatic persistent post-concussive syndrome* is the diagnosis used in these situations. Most common symptoms are

_____dizziness,
_____headache,
_____light sensitivity,
_____sound sensitivity,
_____cognitive problems (such as inattention and poor memory), and
_____emotional difficulties (e.g., as anger/irritability, depression, nervousness, and discouragement).

About 70% with this syndrome qualify for a psychiatric diagnosis as well. Also, some of the remaining 85% to 90% are left with symptoms considered to be in the mild range of impairment. The few studies that have examined victims at the 15-year post-injury mark have noted that significant and mild symptoms can still be present. The bottom line is that some are left with mild symptoms that still interfere with day-to-day living and a sizeable percentage struggle with significant symptoms of TBI.

c. Cognitive Difficulties

For the presumption of TBI, there must be reliable evidence that the trauma victim has experienced a rather abrupt change in cognitive functioning. This is best determined by administering standardized psychological tests. I use the MicroCog as a standardized screening test for cognitive functioning and also have the client identify problems by completing up to five different checklists, which include hundreds of very specific cognitive skills. Sometimes referral is made to neuro-psychologists who can administer a battery of standardized tests that can both identify very specific skill deficits and relate these to the functioning of various parts of the brain.

d. Not Present Previously

In order to link any cognitive deficits to the traumatic experience, there must be evidence that the deficits were not present before the most recent head injury. Examination of the results of previous cognitive tests is one way to do this but it is rare for people to have been tested before the event. So, the indirect way used most commonly is either to examine recent records of educational performance or to make reasoned assumptions based on the person's work requirements and performance.

e. Still Present after PTSD Treatment

Because some cognitive deficits such as problems with concentration can be the direct result of PTSD, I have taken the position that treatment of the PTSD must be completed successfully before the possibility of TBI can be considered.

f. qEEG

As described later on, quantitative EEG (qEEG) is a method of assessing brain wave activity in real time. I have the equipment and training to complete a qEEG in my office. Results are then forwarded to an expert in the field, Dr. Bob Thatcher, who examines the EEG visually and then processes it through two software programs. One is NeuroGuide which is the normative sample software which he developed and continues to refine. The other is LORETA which further identifies the location of any dysregulation. In addition, if I indicate that there is evidence of a head injury, Dr. Thatcher subjects the data to the TBI discriminant function, which he and his research team have developed over the years. Essentially, the discriminant program assesses those with a history of trauma to the head on more than 20 separate EEG variables that have been found to distinguish reliably between those with and those without brain injury. The EEG variables were determined from high correlations with known neurophysiological consequences, clinical symptoms, reduced neuropsychological functioning, deviate fMRI measures following trauma to the head, and relationship to scores on the Glasgow Coma Scale. Dr. Thatcher noted on his listserv of 06 May 2009 that

> We explained these findings by arguing that a travelling percussion wave resulting from the impact to the skull disrupted the protein/lipid molecules of the neural membranes thus impeding the ability of the ionic pumps to maintain the proper electrical potential.

As cited by Thornton and Carmody (2008), Dr. Thatcher's TBI discriminant function has shown a sensitivity of 95.4% and a specificity of 97.4%, statistics that are very impressive. The qEEG has been accepted in about 20 recent U.S. civil court proceedings and to my knowledge has never been declined. Moreover, it is identified as an appropriate diagnostic tool for TBI by the Electrodiagnostic and Clinical Neuroscience Society, the U.S. Institutes of Health, and the U.S. Veterans Administration. According to Striefel (2009), the latter two agencies as well as the U.S. Department of Defense considers the qEEG to be "the standard of care" for diagnosing TBI. Moreover, using EEG pattern recognition techniques, a 2008 research study by Baker and associates found that they could predict with 80% accuracy those with mild cognitive impairment who would go on to develop Alzheimer's disease. In addition to determining if my client's qEEG pattern is consistent with TBI, the information I receive back from Dr. Thatcher provides helpful information on the client's brain functioning including EEG wave amplitude, asymmetry, coherence, and phase. This information is used to design treatment protocols (BFT-C) to improve brain functioning.

Within the field of neuroscience, the adaptivity of the human brain is well known. When it becomes deregulated from any of a number of causes, (1) it can self-repair over time, (2) impaired capacities can be taken over by other parts of the brain automatically, and (3) BFT-C can lead to positive changes in brain activity. The net result can be substantial increases in functioning.

Also, a very careful 2008 review of the research literature by Thornton and Carmody concluded that BFT-C has demonstrated effectiveness in the treatment of TBI and in particular with auditory memory while the interventions based on cognitive rehabilitation and the use of compensatory devices have not proven to fulfill their original promise. At the 2008 meeting of the Biofeedback Foundation of Europe, Dr. Surmeli reported a study wherein qEEG-based BFT-C was found to be effective in 22 of 24 clients; among the 22 responders, normalization of qEEG was related to symptom improvement. Moreover, there is reason to believe that BFT-C can lead to significant improvements in cognitive functioning up to even 10 years after TBI (personal communication, Dr. Hammond, ISNR listserv, 2008). As noted by Striefel (2009), the U.S. Department of Defense uses the qEEG to determine whether or not navy pilots can resume flying after suffering a TBI.

At this time, positive result with BFT-C appears to be more likely for those with mild traumatic brain injury than for those with evidence of moderate or severe brain injury. However, at the 2008 conference of the North American Brain Injury Society, Zelek reported a positive outcome after 30 sessions of BFT-C with 10 patients having either moderate brain injury (unconscious for more than 30 minutes) or severe brain injury (unconscious for more than 24 hours). Formal measurement indicated that cognitive abilities were significantly improved after BFT-C as compared to before (scores increased on average from 1.7 to 12.3) and all patients showed improvement in brainwave power and coherence values.

My Sequence

As a result of the above, I have developed a sequence of investigation and treatment for those with PTSD who might have TBI also.

First, I approach the issue from the standpoint of determining what appears to be decreases in cognitive functioning independent of presumed cause.

Second, during my assessment, via client report, clinical observation, information from the client's partner, and results of administrating both the MicroCog and a number of checklists, I determine the presence of any problems in cognitive function such as attention and memory.

Third, since many psychological and cognitive symptoms of TBI are symptoms also of PTSD without head injury, I proceed with the treatment of PTSD. Also, I provide the client with strategies to compensate for any cognitive impairment. Some of these are presented in section "E".

Fourth, following completion of my basic treatment approach, I re-assess PTSD symptoms in part by re-administration of standardized psychological tests and re-assess cognitive processes by a re-administration of the MicroCog and the cognitive checklists.

Fifth, if my basic approach does not result in remission of PTSD, then I recommend a qEEG, which is completed in my office. Likewise, if the PTSD is in remission but the cognitive problems remain, then I recommend completion of a qEEG. The resulting qEEG and information about any history of head injury is forwarded to Dr. Thatcher for analysis. At some future time I would like to obtain the equipment and training so as to add ERP assessment (discussed later) to accompany the qEEG. In respect to ERP, Andreassi (2007) noted that closed head injury was associated with lower amplitude of the N100 and N200 wave as well as delayed latency of the N200 and P300 wave, which might help to explain persistent problems with attention and concentration.

Sixth, based on Dr. Thatcher's analysis and subsequent recommendations, a training plan (BFT-C; discussed later) is started following full explanation to the client and his/her informed consent to proceed. Also, I now have the equipment so I can consider adding HEG (hemoencephalograph) feedback, since there is some evidence that it leads to an increase of blood flow (thus increased oxygen and glucose) to targeted areas of the brain.

Seventh, cognitive functioning is re-assessed and a repeat qEEG is done once a goodly number of BFT-C sessions have been completed. Decisions regarding the details of any further treatment sessions are then based on these results.

My experience to date is that sometimes cognitive functioning improves over time, sometimes some or all symptoms normalize with the successful treatment of PTSD, sometimes functioning improves with BFT-C, and in a small percentage of clients the deficits in cognitive operations are permanent such that life adjustments are required. Also, the 2008 study of attentional processes by Kwok and collaborators indicated that different cognitive functions impacted by TBI can recover at different rates.

To complicate matters further, brain injuries can result from a range of additional factors such as exposure to toxins, having a stroke, infection from a virus or bacteria, respiratory distress, malnourishment, high fevers, degenerative diseases such as Parkinson's, and cardiac arrest. Also, clients with neurological damage can react more than others to medications.

C-11: Pain and Its Management

Summary. Many with PTSD experience co-occurring pain. The pain levels can range from mild to severe, the pain frequency can be from periodic to constant, and the pain history can range from recent to chronic. Depending on location, the most common types of pain among those with PTSD are headache (both tension and migraine), jaw (masseter) pain, back pain, non-cardiac chest pain, irritable bowel syndrome (IBS), Raynaud's syndrome, and fibromyalgia syndrome (FMS). A 2008 study by Defrin and co-investigators compared sensory testing of outpatients with combat or terror-related PTSD, outpatients with a non-PTSD anxiety disorder, and healthy controls. Compared to the other two groups, those with PTSD had higher rates of chronic pain, more intense chronic pain, and more body regions that were painful. Moreover, the more severe the PTSD, the greater was the intensity of pain. Others have reported that chronic pain has the effect of increasing the symptoms of PTSD. Fortunately, in the absence of medical causes and given the collection of appropriate psycho-physiological information, a number of biofeedback training strategies can be helpful in at least decreasing pain levels if not eliminating pain altogether.

As cited in studies such as the 2009 one by Jenewein and associates, over 50% of those with PTSD experience chronic pain. Moreover, the pain is more intense and is associated with higher disability among those with PTSD as compared with those who do not have PTSD. These findings are consistent with my clinical experience. Also, in their own study of those injured in accidents, Jenewein and colleagues report that the intensity of pain among 14% to 19% of those with PTSD increases in the six to twelve months following their accidents.

For the information below, I have relied heavily on the excellent 2004 book by Sherman entitled *Pain: Assessment and Intervention from a Psychophysiological Perspective*. Also, the 2009 presentation on pain at the International Society for Neurofeedback and Research (ISNR) annual conference by Jensen was very instructive.

The sensation of pain is the end result of many processes, in many bodily regions, and often in combination with each other. Thus, there is no known single pain centre. Pain can occur from activation of the hypothalamic-pituitary-adrenal axis, autonomic nervous system hyper-arousal, and electro-physiological activity in any of a number of specific areas of our brain including the insula, the pons, the medulla, the thalamus, and the midbrain periaqueductal gray (PAG) Also, as indicated in the 2008 paper by Afari and co-writers, psychological trauma and pain involve over-lapping areas of the brain. Clearly, some pain can be the result of injury, tissue damage, inflammation, repetitive and unnecessary over-use of specific muscles, constriction of blood vessels, and decreased blood flow to the area of pain. Also, pain can be exacerbated by such factors as stress, the inadvertent tightening of muscles, the lack of appropriate exercise, unhelpful patterns of breathing, poor posture, and the presence of sleep disorders (including insomnia and obstructive sleep apnea). Moreover, anxiety conditions and depression can magnify the experience of pain. In addition, the experience of pain can increase by excessive attention to it as well as excessive focus on bodily sensations. Furthermore, pain can represent side-effects of medication, pain in one body location can be the result of biological issues somewhere else (referred pain), and pain can stem from an undetected and perhaps serious medical condition. Another consideration is trigger points whereby pressure on one area of the body can result in real pain in distant parts of the body in spite of there being no obvious nervous or other physical connection between them. Finally, pain can at least be exacerbated by the attention clients receive from others because of the pain as well as from cognitive habits including that of catastrophizing. Consequently, before making a plan of what to do about pain, all relevant factors such as those above should be considered while recognizing that adequate assessment methods are not always available. The treatment plan can then address the relevant factors.

In regards the intensity of pain, both Sherman (2004) and others ask clients to rate it on a scale that ranges from "0" (no pain) to "10" (you would faint if you had to bear it for one more second). Initially recording pain levels multiple times a day can help to identify stressors and situations that regularly precede the experience of pain. Armed with such information, helpful strategies can be developed. Also, it is known that for those with chronic pain, a sensitizing process in the brain takes place whereby pain intensity becomes high even following causes to which others would not react.

Depending on location and other characteristics, many different terms are used for the experience of pain. Once medical conditions have been ruled out and if there is evidence of contributing psycho-physiological characteristics, biofeedback-assisted training can be initiated. This approach for a number of types of pain has clinical and research support.

Headache: General Information
There are many different types of headache, although only the two below will be considered since there is evidence that psycho-physiological interventions for them can be helpful. However, prior medical assessments are critical since both tension-type and migraine headaches can be caused by any of a number of factors including trigger points, visual problems, sinus conditions, menstrual cycle, tumours, back problems, nerve problems such as trigeminal neuralgia, side effects of medications, aneurism, high blood pressure, dental problems, gastro-intestinal problems, head/neck injury, jaw joint problems, and allergies. Moreover, before employing biofeedback methods, stabilization strategies need to be implemented as required with attention to such relevant variables as sleep, caffeine intake, use of drugs and alcohol, aerobic exercise, and stressor management. Any issues with these can at least exacerbate headache. Particularly in relation to migraine headaches, elimination of contributory foods needs to be considered; this information can be found on the website www.mayoclinic.com.

Tension Headache
Once all of the above have been considered, psycho-physiological interventions

become a possibility since there is ample clinical and research demonstration of its effectiveness with tension headaches. One common cause is unnecessary muscle tension as a result of poor posture. A consistent forward head thrust as well as postures that keep the shoulders elevated unnecessarily can lead to pain in the shoulders, neck, and back of the head. Usually, strategies aimed at correcting these postural issues result in a decrease of both muscle tension and pain. A second common cause is sustained elevated tension in any of multiple muscle systems including forehead, masseter, neck, and trapezius. Commonly, biofeedback is helpful both in learning to recognize when tension levels rise above the norm and in mastering strategies such as specific methods of gentle stretching to decrease relevant muscle tension and pain. Subsequently, clients are encouraged to practice these strategies at home multiple times a day and to implement them additionally whenever they note increases in muscle tension. One study found that those with chronic tension-type headaches had less blood flow to relevant muscles. However, after addressing the problem by following one or both of the first two approaches above, all of my clients with PTSD and co-occurring tension-type headaches were no longer troubled by such pain.

Migraine Headache
Migraine headaches can be steady or pulsating. Typically, they occur on one side of the head. They can be present with any frequency and can last from hours to days. Usually they are accompanied by significant sensitivity to sound (phonophobia) and/or light (photopobia) as well as nausea. They may or not be preceded by visual experiences (aura). Often sufferers will lie in a darkened room for considerable periods of time until the headache passes. Generally, once diagnosed, physicians prescribe medication to either prevent their occurrence and/or to manage symptoms once a migraine starts. The causes of migraine are not entirely known although at least three major theories have been presented. Fortunately, there is evidence that psycho-physiological interventions can have a positive impact on migraines such that for some persons medications are not necessary. This is important particularly among those sensitive to such medications or who cannot take them (for example, pregnant women). Firstly, reports indicate that sometimes migraines can be decreased by correcting poor posture. Secondly, there is evidence that decreasing muscle tension has a beneficial effect, as with tension-type headaches above. Thirdly, for reasons yet unknown, the use of finger temperature biofeedback training has proven to be effective such that various headache associations support its use. In this training, a sensitive thermistor is attached to a finger and its temperature is displayed on a computer screen. Then clients try any of a variety of strategies until they find one that increases finger temperature. Typically, increases are gradual until eventually clients are able to raise finger temperatures to about 95 degrees Fahrenheit. Subsequently, clients practice until they can do this at home consistently and quickly. Finally, they are encouraged to use this strategy at the first sign of a coming migraine. Well-conducted studies report that the decreases in migraine are sustained over a ten-to-fifteen year period. A fourth strategy, biofeedback based on hemoencephalogy (HEG), is being used increasingly following presentation of this successful approach at professional workshops by psychologist Jeff Carmen and scientist Hershel Toomin. In the Carmen system, sensors are placed on the forehead measuring forehead

temperature and in the Toomin system measuring cerebral blood flow. Moment-to-moment information is displayed on a computer monitor and clients try strategies to increase the variable of interest. Generally, clients are able to make desired change without being able to reveal how this was accomplished. Results to date indicate that these two systems have a positive effect on migraine headache.

Jaw/Masseter/TMJ Pain

Clients can have any of a number of conditions involving pain or tension in the jaw area. For example, problems with the temporomandibular joint (TMJ) can cause people to tense their masseter muscles with resulting pain. This and other jaw conditions and dental issues need to be ruled out before appropriate biofeedback training can be considered. The option of biofeedback training arose from many converging lines of evidence. First, just clenching jaws slightly for about 15 minutes a day will result in increases in muscle tension (as measured by sEMG) and pain by the eighth day. Second, those with jaw pain have higher resting sEMG levels than those without. Third, in response to stressors, people with a history of jaw pain experience pain in that area as well as other parts of the face. Fourth, muscle biofeedback training (BFT-M) results in decreases in both sEMG activity and the experience of jaw pain. Accordingly, my approach is to first have medical and dental causes ruled out. Then, should biofeedback training be indicated, via BFT-M, clients (a) learn to recognize sensations associated with different sEMG levels, (b) master the act of lowering and raising sEMG levels, (c) practice strategies for raising and lowering sEMG several times a day outside the office, and (d) in addition, implement these sEMG strategies as soon as they notice an increase in muscle tension.

Back Pain

Chronic back pain is very common and co-occurs often among clients with PTSD who find their way to a psychologist's office. Clients can locate the source of the pain anywhere along the spine. Complicating the clinical picture is that the pain in one area can be referred pain; that is, it can stem from another area. Also, chronic back pain can be the result of many medical factors including arthritis, bone destruction, herniated disk, structural deformities, neurological conditions, infection, cancer, ovarian cysts, muscle strain, muscle spasms, inflammation of muscles or joints, and having one leg longer than the other. Hence, appropriate medical examination is critical prior to starting any psychological treatment. However, (a) in the absence of medical conditions following a careful medical examination and (b) based on relevant observation as well as psycho-physiological recordings, there is considerable evidence in support of the positive effects of biofeedback training. One common psycho-physiological cause of back pain is poor posture such as a pattern of forward head thrust, not holding the back straight when sitting or standing, maintaining the shoulder muscles too high, and leg position while sitting that does not support the back. Correcting postural causes is effective in reducing related back pain. Independent of the original cause, a second common causal factor is the presence of stress in excess of the person's ability to manage it effectively. For example, studies show that stress can increase sEMG levels in the paraspinal muscles among those with low-back pain but not among non-pain controls. In this case the intervention is clear; studies have reported that learning how to manage stress adequately results in decreased pain levels. Occurring for any of number of reasons, a third cause is chronic muscle tension levels greater than the norm. In this regard, both clinical experience and research have supported a biofeedback training approach that includes (a) learning how to tense and then relax the relevant muscle using feedback from sEMG recordings, (b) practising this strategy without reference to the sEMG signal, (c) practising this strategy multiple times a day between sessions, and (d) repeating the strategy any time excess tension levels are noticed.

Non-Cardiac Chest Pain

Chest pain in the area of the heart is frightening to most people. To complicate matters, such pain can be the result of any of a number of medical conditions including heart conditions, pulmonary embolism (chances for a second and deadly episode are high after an initial attack), pericarditis (which can stem from rubbing of the heart against a hardening of the sack around it), gastroesophageal reflux disease (GERB), and hiatal hernia. These must be ruled out by physicians before other treatment is considered. If possible medical conditions are ruled out and the pain persists, the pain is referred to as non-cardiac chest pain. In such cases, a number of psychological treatment strategies with clinical and research support can be tried. If psycho-physiological evaluation indicates anomalies of breathing, then I teach the client what I have labelled "tactical breathing," which includes resonance frequency breathing. Usually, I supplement this with helping the client to identify situations that precede the chest pain.

Irritable Bowl Syndrome (IBS)

The combination of abdominal pain and bloating can indicate the diagnosis of IBS. Additional symptoms can include constipation, diarrhea, cramps, and urgency. However, the experience of abdominal pain can be due also to medical conditions such as appendicitis, infectious diarrhea, perforated appendix, bleeding ulcer, ectopic pregnancy, miscarriage, pancreatitis, bowel blockage, gastroenteritis, kidney stones, stress fractures in the hips or ribs, torn ligaments or tendons, and muscle wall problems. Once (a) these conditions have been ruled out following careful medical evaluation and (b) provided there is evidence of contributory psychological or psycho-physiological causes, clinical and research evidence suggests the value of one of two treatments. For clients with IBS who have symptoms of a co-occurring anxiety disorder, any of a range of anxiety-reduction methods can be utilized. Among them are the stabilization strategies that every client of mine receives, cognitive-behavioural methods of anxiety management, and biofeedback-assisted training for relaxing/self-regulating muscle tension. If these approaches are not optimal in the control of IBS, then I teach tactical breathing, including resonant frequency breathing, since there is a body of clinical and research evidence supporting its effectiveness in IBS.

Raynaud's Condition

Typically, people with Raynaud's condition experience coldness in fingers regardless of surrounding temperature. Objective measurements show finger temperatures well below normal levels and some so low that cyanosis is observed. Additionally, in the presence of cold, those with this condition experience vasospasms accompanied by pain. For some, the vasospasms include other areas of the body such as the toes, tongue, heart, esophagus, and lung. Raynaud's condition can be a serious condition; for example, it is associated with a doubling of the risk of coronary artery disease and stroke. However, symptoms similar to those of Raynaud's condition can be the result of any of a number of medical conditions such as hypothyroidism and obstructive arterial disease. These and other medical conditions need to be ruled out before initiating any other treatment. Regardless of cause, recommendations given commonly include avoiding any precipitating factors, wearing warm clothing, and considering the use of heated gloves and socks. Additionally, physicians can prescribe any of a number of drugs. In my practice (a) once medical causes are ruled out and (b) if the client's psycho-physiological test indicates low finger temperature, then I consider psycho-physiological treatment. The approach with the best clinical and research evidence is biofeedback training (a) to master increasing one's finger temperature up to about 95 degrees Fahrenheit, and (b) practising how to maintain regular finger temperature in the presence of cold.

Fibromyalgia Syndrome (FMS)
Although less now than once, the existence of FMS is controversial among physicians; some consider it a real condition and others do not. The major characteristics of FMS are a diffuse sense of pain/achiness and more pain than normal when a percentage of the body's pressure points are pressed. In addition, problems with sleep, energy, reactivity to chemicals, and depressed mood are common. A large number of biological etiologies have been suggested by at least one study of FMS, so the first recommendation I give to my clients is that they see a rheumatologist who has an interest in this condition. Also, it is true that some studies have documented the co-occurrence of PTSD and FMS while others have found evidence that FMS itself can be a consequence of trauma without co-occurring PTSD. In such situations, there is some information supporting the role of psycho-physiological factors. For example, one study found that, compared to those without the condition, those with FMS showed increased blood flow to brain centres for pain in response to gentle pressure. Another research group reported differences between those with and those without FMS in patterns of both muscle tension and EEG characteristics. Yet others have communicated the success of various psycho-physiological interventions in decreasing the symptoms of FMS. Also, while treating PTSD clients with qEEG-based BFT-C, a number of my female clients reported a corresponding decrease in symptoms of FMS.

My Sequence
As a consequence of the above, over time I have evolved the following approach in the assessment and treatment of pain.

First, I obtain detailed information about the pain: location of the pain, when each area began to hurt, what was happening at the time, description of the pain, what makes it worse and better, how the pain has changed over time, the intensity and duration of the pain, and other important characteristics. Additionally, I assess cognitive and interpersonal factors that might be exacerbating the pain.

Second, I consider the pain in light of the psycho-physiological evaluation mentioned in the assessment section earlier in the book. Taking into account expected psycho-physiological effects of any medication that has been prescribed is part of this assessment. For example, beta blockers can lower heart rate, decongestants can increase sEMG, anti-hypertensives can increase skin temperature, and diuretics can increase blood pressure.

Third, since I am not a physician or medical specialist, my pain assessment is not a complete one. For example, I am not able to assess any medical conditions that might be involved. Thus, the client and his/her physician are alerted to my assessment results so that a proper medical assessment can be completed from which any appropriate medical treatment can be considered. In addition to the factors described above, pain can be caused and/or worsened by a variety of biologic factors including nutritional deficiencies (imbalanced diet and avitaminosis), metabolic and endocrinologic abnormalities, chronic infestations (bacterial, viral, fungal, and immune deficit syndromes), and rheumatologic disorders (osteoarthritis, rheumatoid arthritis, and lupus), which must be investigated also.

Fourth, clients can choose to take pain medications as prescribed by their physicians. A range of such medications is available. Generally, physicians follow a strategy of moving from less to more powerful ones depending on client pain response and the development of any side-effects.

Fifth, following some psycho-education regarding the nature of pain that in itself is helpful, clients are encouraged to implement relevant stabilization strategies as found later in this book. These include sufficient aerobic exercise, appropriate nutritional intake, and effective management of stressors and have been demonstrated to have a positive impact also on pain.

Sixth, biofeedback training is initiated to address findings from the clinical psycho-physiological aspect of assessment.

Seventh, changes in the client's contributory cognitive functioning and the response by others to the client's pain are recommended as required. This can include strategies of distraction, additional ways for coping with pain, letting go of catastrophic thinking, and doing things that decrease feelings of hopelessness and helplessness. Such considerations are addressed at this time since the client now has a system for decreasing pain.

Eighth, both clinical experience and the beginnings of some research studies have noted the reciprocal relationship between pain and sleep; pain leads to sleep problems such as insomnia while non-restful sleep increases the experience of pain. Moreover, randomized clinical trials have demonstrated that the implementation of cognitive behavioural methods can improve pain-related insomnia, which in turn can decrease the experience of pain, although the pain of FMS appears to be an exception. Therefore, these methods are introduced for the improvement of sleep. Then, as required, other methods with clinical and research support for the treatment of insomnia are added such as use of an AVED device. Other studies have found that the sleep medication can be helpful. Although a wide range of medications are possible, some of my clients have noted improvements with trazodone and one study found triazolam to be superior to placebo, and to the hypnotics zopliclone and zolpidem in improving both sleep and pain. However, many prescribed sleep medications now come with warnings about their use. Also, there is evidence that the lack of deep sleep can magnify the sensory experience of pain and this can be addressed through appropriate intervention.

Ninth, a number of lines of evidence have determined that brain electro-physiology is different from typical among those with pain and/or that biofeedback training of the EEG signal can have a positive effect on the experience of pain.

*EEG activity is different among those with and without pain. For example, Jensen and colleagues (2009) cite supporting research that reveals low amplitudes of alpha and high amplitudes of beta among those with pain and high theta amplitudes with low alpha amplitudes among those with chronic pain. Additionally, they review evidence indicating that many brain regions work together in the experience of pain, including the thalamus, the insular cortex, the anterior cingulate cortex, and the prefrontal cortex.

**There is evidence that automatic brain processes known as ERP are involved with differences between those with and without pain in both P100 and P200 components (ERP is discussed later). Also, some findings indicare that decreasing the amplitude of such components reduces the sensitivity to pain.

*As discussed in detail later, there is reason to believe that biofeedback training that targets directly the central nervous system (BFT-C; see Step K) can result in a meaningful reduction of the experience of pain. For example, as discussed later, the NeuroGuide and LORETA evaluation of a qEEG can provide information about dysfunctional patterns of cortical arousal at specific brain sites that can be used for developing an individualized BFT-C protocol for the treatment of PTSD. Surprisingly, a number of my police and military clients have reported meaningful decreases in pain as a result of BFT-C protocols implemented for reducing symptoms of PTSD. For one client, PTSD decreased and he reported relief from significant and chronic back pain by decreasing the amplitude of 20- to 30-Hz brain wave frequencies at sites T3, T4, and C4. Similarly, with other qEEG-based protocols for PTSD, some of my clients have experienced an unexpected decrease in fibromyalgic pain and others have reported a decrease in pain related to rheumatoid arthritis.

* Ibric and Dragomirescu (in Budzynski et al., 2009) provided information from the many clients they have seen who improved with predetermined BFT-C protocols that did not utilize the qEEG. They report such outcomes after BFT-C to increase amplitude in the 12- to 15-Hz frequency range on the brain's motor strip most closely associated with the part of the body experiencing pain. Other studies have reported similar findings.

Some have reported the effectiveness of this approach for fibromyalgia when the site C4 is targeted.

*Jensen and co-authors (2009) cite studies showing that decreasing relative beta while increasing relative alpha was associated with pain relief.

*Other research on the relationship between brain functioning and pain is increasing. For example, Bjork and Sand (2008) reported that migraine is most likely to develop when specific brain events occur (such as increases in the power of delta brain waves in the front part of the brain, increases in power of the alpha waves towards the back of the brain, and differences in the asymmetry of theta brain waves).

*Still, in his presentation to ISNR in 2009, Jensen noted that although BFT-C makes sense from a clinical perspective, rigorous research is required and there are no studies demonstrating clearly that BFT-C is superior to other forms of biofeedback in the treatment of pain.

As a consequence of evidence such as that above, I consider BFT-C for my clients only (a) if more common approaches have proven to not be optimal and (b) if the qEEG suggests anomalies of cortical activation. Even though some studies have reported success with the implementation of BFT-C protocols not based on prior qEEGs, at this time I take the position that it is safer to the client to base any BFT-C strategy on results of analysis of the qEEG.

Hypnosis is a psychological technique with much evidence to indicate its effectiveness in pain management. Also, in his 2009 address to the ISNR Conference, Jensen presented evidence indicating that areas of the brain react differently depending on the wording of the hypnotic suggestions. Since I lack the required training, I do not use hypnosis at this time.

Treatment Step D : Referral For Medical Assessment and/or Intervention

Contents
D-01.   Medical Conditions Relevant to PTSD
D-02.   Medications for PTSD

D-01: Medical Conditions Relevant to PTSD
A number of medical symptoms/conditions can result in some of the characteristics of PTSD without the person actually having PTSD. These include:

*cardiovascular conditions (such as angina, myocardial infarction, arrhythmia, and high blood pressure);
*gastro-intestinal conditions (such as peptic ulcer and Crohn's disease);
*hematological conditions such as anemia,
*hormonal conditions (such as Cushing's, hyper- and hypothyroidism, hypoglycemia, menopause, and porphyria);
*neurological conditions (such as encephalopathies and seizure disorder); and
*respiratory conditions such as asthma.

Thus, I take the position that it is important to rule out such medical conditions before proceeding with psychological treatments. Consequently, a written report of my assessment results is provided to the client's physician that includes medical information from and about the client as well as the range of medical conditions that can produce symptoms similar to those of PTSD. Then the physician is responsible for ruling them in or ruling them out.

D-02: Medications For PTSD
Often physicians rely on various medications to treat the symptoms of PTSD. For example, in the 2008 survey of about 275,000 veterans with PTSD being seen by the U.S. Veterans Affairs, Mohamed and Rosenheck found that 80% were receiving medication; among these, 89% had been prescribed anti-depressants, 61% were taking anti-anxiety or sleep medications, and 34% were on anti-psychotics. The authors report also that "use of anti-anxiety/sleep medications and anti-psychotics in the absence of a clearly indicated diagnosis was substantial . . . use seems to be unrelated to diagnosis and thus is likely to be targeted at specific symptoms (e.g., insomnia, nightmares, and flashbacks)" (p. 961). Additionally, these researchers completed a large-sized 2008 study of VA outpatients over the age of 45 and found similar results: 88% were prescribed anti-depressants, 61% received anti-anxiety/sleep medications, and 33% received anti-psychotics. Among over 10,000 privately insured Americans, the 2008 survey by Harpez-Rotem group discovered that 60% received medication for PTSD; 74% of these received anti-depressants, 74% anti-anxiety/sleep medication, and 21% anti-psychotics. Those with greater use of mental health services and a co-occurring psychiatric diagnosis were more likely to receive prescriptions for medication. As considered in the 2008 review by Bisson, (a) many agents have been found to be superior to placebo, (b) although many receive some benefit from medications, (c) still the effect sizes are relatively small, and moreover (d) few if any achieve remission of PTSD.

Apart from accidental discoveries of the effects of psychiatric medications on PTSD, the major reason for their use relates to their presumed effect on neuro-transmission. Brain cells are separated from each other by a minuscule space called the synaptic cleft or gap. However, one of the things that needs to happen if the brain is to function at all is for these isolated cells to communicate with each other. This is accomplished in part by the appropriate release of a chemical from the end of one brain cell, through the synaptic gap, to the right location on the other relevant brain cell. Afterwards, the neuro-transmitter is returned to the sending cell in a process known as re-uptake. Many neuro-transmitters have been identified including serotonin, epinephrine, dopamine, GABA, neuropeptide Y, the opioid peptide beta-endorphin, and glutamate. Some of the medications noted below seem to have their effect by blocking the re-uptake so that the neuro-transmitter remains in the synapse long enough to activate the receiving brain cell.

Based on my reading of the current professional literature on medication management, the following is a summary of both the rationale for medications as well as research results on their effectiveness.

Selective Serotonin Reuptake Inhibitors (SSRI)
On the basis of research reviews, groups such as International Consensus Group on Depression and Anxiety consider the class of medications known as SSRI to be the first-line of medical treatment for PTSD. Also, according to the 2005 review of the research by the British Association of Psychopharmacology, most of the evidence to date supports the use of SSRIs. The 2004 review of controlled clinical trials by the U.S. Veterans Administration and U. S. Department of Defense concluded that only SSRIs have "significant benefit" in the treatment of PTSD (their 175 page document can be found at www.oqp.med.va.gov/cpg/PTSD/PTSD_Base.htm).

However, as reported by Dr. Turner and his colleagues at the 2005 International Society for Traumatic Stress Studies (ISTSS) conference, a comprehensive review of drug studies by the UK group charged with determining best practice recommendations did not find convincing evidence that average effects of the medications were significantly different from placebo although it was clear that some clients did improve. In their 2006 review of the research, Ipser and colleagues found significant symptom reduction in about half of the 35 short-term studies they located; among those who responded in 13 of the studies, about 59% had positive results to medication while 38% improved with placebo. Relatedly, Gao and associates (2006) concluded that response rates with SSRIs rarely exceed 60% and that only 20% to 30% achieve remission. Other studies indicate that about 30% do not achieve any lasting remission.

Zoloft with civilian populations, Paxil for females with PTSD, and Prozac with combat veterans living in the community are three of the SSRIs whose effectiveness has been reported in at least one placebo-controlled study during the acute phase of PTSD. Thus, on the basis of multi-site trials, federal regulatory agencies in both Canada and the U.S. have approved Zoloft and Paxil in the treatment of PTSD; as I understand it, the manufacturers did not submit Prozac for the trials required to obtain such approval. In the trials of Zoloft and Paxil, 20% showed little or no improvement, 50% showed notable improvement but only partial remission, and 30% achieved complete remission. The 2008 study by Simon and co-investigators did not find that Paxil had any additional benefit over prolonged exposure therapy among civilians with PTSD.

Often the decision of which medication to begin with is made on the basis of which side-effects are most desirable; Paxil seems to help best if symptoms include insomnia, while lethargy seems to respond better to Zoloft. On the other hand, side-effects can be negative. For example, the 2007 review of Paxil by Pae and Patkar and the 2008 review by Marks and his group have noted frequent undesirable side effects such as constipation, sexual dysfunction, and weight gain.

Using uncontrolled research methods, favourable results in general populations with PTSD have been reported in studies of two other SSRIs--Celexa and Luvox. Also, a 2006 uncontrolled study by Hamner and associates of the SSRI Cipralex as used by combat veterans reported that PTSD scores decreased from 79.4 to 61.2, which is a significant decrease but still not to the point of remission.

Tri-cyclic Antidepressants (TCAs)
There is some evidence from good studies that amitriptyline and *imipramine* from the class of anti-depressant medications known as TCAs can lead to symptom reduction in treating PTSD among the general population. In these studies, *desipramine*, another TCA, was not found to be effective.

Monoamine Oxidase Inhibitors (MAOIs)
A good study of phenelzine, a medication within the class of MAOI inhibitors, found it to be very successful in reducing the re-experiencing and arousal symptoms of PTSD among Vietnam War combat veterans. Results were mixed in uncontrolled studies. An uncontrolled study of the MAOI medication *brofaromine* yielded positive results for PTSD symptoms. Another MAOI type of medication, maclobemide, reportedly improved all three PTSD clusters in an uncontrolled study.

Serotonin-2 Antagonist/Reuptake Inhibitors (SARIs)
A good research trial and some uncontrolled studies of Serzone, from the class of medications known as SARI, showed it to have positive results in the treatment of PTSD. However, due to concerns about death from liver impairment, Serzone is no longer available in Canada or the U.S.

Selective Serotonin Norepinephrine Reuptake Inhibitors (SNRIs)
A 2006 24-week, double-blind, placebo-controlled study by Davidson and associates of the SNRI *Effexor* (mean dosage 221 mg/day) among adult outpatients with PTSD showed an improvement of about 52 points in symptoms while those taking a placebo improved by 44 points. By the end of the study, 51% of the medication group and 38% of the placebo group were considered to be in remission. In a previous 12-week study, those receiving Effexor improved

by 42%, those on *Zoloft* improved 39%, and those taking a placebo improved by 34%; remission rates were 30%, 24%, and 20% respectively. The excellently controlled 2008 multi-site study of the extended-release form of Effexor by Rothbaum and colleagues reported significant reductions over placebo on the total as well as all cluster scores of PTSD. Almost 700 clients were included in the study with a range of traumatic experiences including accident, sexual abuse, and combat.

Noradrenergic Specific Serotonergic Antidepressants (NaSSAs).
One good study and one uncontrolled one demonstrated that Remeron reduced PTSD symptom severity. This medication is from the class known as NaSSA.

Alpha-1 Antagonist
As presented by Dr. Raskind at the 2005 ISTSS conference, a number of small studies with combat veterans have demonstrated that the alpha-1 antagonist prazosin, originally developed as a medication for high blood pressure, taken three times a day can reduce nightmares and other PTSD symptoms. Mechanisms seem to be related to decreasing CRF and normalizing REM sleep. The 2008 research paper by Thompson and colleagues reported reductions in frequency and intensity of nightmares, night wakening from nightmares, and sleep difficulty among 22 veterans with PTSD after they were given an average of 9.6 mg of prazosin a day. The concurrent use of Viagra or other medications that affect blood pressure is contra-indicated.

Beta-Adrenergic Antagonist
In small studies with civilian populations, the beta-adrenergic antagonist propranalol showed some beneficial effects.

Anti-Convulsant Medications
There is research suggesting that some anti-convulsants can reduce some of the symptoms of PTSD. One good study with *lamotrigine* reported a decrease in PTSD among 50% of the experimental versus 25% of the control group but the conclusions were challenged based on a re-analysis of the data presented. In three small open clinical trials with veterans and adolescents, carbamazepine resulted in reductions in traumatic memories, flashbacks, insomnia, irritability, impulsivity, and violent behaviour when other medications were not completely effective. Four open-label trials indicated that Valproate had a generally mild positive effect on the PTSD symptoms and in particular those of hyper-arousal and anger/irritability, as did one single-blinded study (for a review, see Adamou and associates, 2007). In another open-label trial, about 71% had their PTSD re-experiencing symptoms eliminated with topiramate while a further 21% reported a partial reduction. Three case reports have indicated positive results on some PTSD symptoms with Gabapentin and tiagabine. Vigabatrin improved the startle response, anxiety, and insomnia among 5 patients with PTSD. Davis and associates reported in 2008 that Divalproex as a single medication was ineffective in treating hyper-arousal symptoms of PTSD among combat veterans. In summary, anti-convulsants, usually in combination with other medications, may lead to some symptom-reduction but not to full remission of PTSD.

However, the relative gains and losses of using anti-convulsants needs to be weighed carefully since this class of medications can include side-effects such as impaired concentration, memory problems, sedation, and confusion. The UK survey noted above did not recommend their use as a best practice. The 2004 review by the Veterans Administration and Department of Defense in the U.S. characterized this group of medications as having "unknown benefit" in the treatment of PTSD.

Neuroleptic Medications
There is some evidence, mostly from studies of combat veterans, that adding a neuroleptic when an SSRI is having a partial positive impact may be helpful. Neuroleptic medications were initially intended for the treatment of psychosis. As reported at the 2005 ISTSS conference by Dr. David and again in the 2006 review paper by Gao and his group, well-conducted studies to date demonstrated that a low dose of respiridone decreased core PTSD symptoms as well as reducing irritability and aggressive behaviour. Similar findings were reported in a single well-controlled study of olanzapine (Zyprexa) while another good study failed to find any benefits. Pae and colleagues completed a 2008 review of seven RCT studies of respiradone and olanzapine involving a total of about 200 patients and found that they had a beneficial effect over placebo conditions, mainly associated with decreases in intrusive re-experiencing. Quetiapine (Seroquel), used in conjunction with other medications or alone after failure of SSRI medications, was found to be similarly effective in an open trial, a retrospective client review, and in several case reports. In recent studies, both Zyprexa and Seroquel were shown to improve sleep among those with PTSD. Overall, the positive effects were independent of the presence of psychotic symptoms.

Also, Gao and colleagues (2006) remind that the side-effects can be considerable with these medications and should be monitored carefully. Such side effects can include movement disorders, elevation in hormone levels, and complications within the metabolic system. Other potential side-effects are increased rates of diabetes, obesity, and dyslipidemia.

In summary, there is evidence of the positive effect of antipsychotic medication, usually as an augmenter of other medications. However, this prescription strategy does not result in full remission. As of their 2004 review of studies, the Veterans Administration and Department of Defense determined that this class of medications provided neither benefit nor harm in the treatment of PTSD.

Cyclocorine
Dr. Russell reported at the 2005 ISTSS conference on the results of one animal experiment and one human study with people who had a phobia for heights that cyclocorine was effective in extinguishing fear. The medication potentiates the brain chemical NMDA, which is believed to affect the amygdala (discussed later).

Duration of Medications
Most of the medications above were studied in the acute phase of PTSD, generally in the first 12 to 14 weeks after diagnosis. After that period, all studies

on the consequences of discontinuing the medication among those for whom it was effective have demonstrated a very significant relapse rate. Accordingly, a number of research projects have been completed regarding the value of continuing an effective medication for up to one year after remission of symptoms. As reviewed in a 2006 article by Dr. Davis and her colleagues, all three published studies of Zoloft, the one study of Paxil, and the one study of Prozac indicate that

> long-term treatment of PTSD with SSRIs effectively maintains the previous treatment response and improvement in quality of life, converts more patients to responders status and accounts for one-third of overall treatment gains...long-term treatment with . . . respiridone and clozapine . . . nefazodone . . . AEDS (and) valproate also appears to result in significant improvements in PTSD symptoms (but) required further controlled study. (p. 468)

The review by Veterans Administration and Department of Defense in the U.S. established guidelines recommending continuing SSRIs for one year after remission was obtained.

Medication with Combat Veterans
Evidence from many studies indicates that most SSRIs have a less robust effect on males and combat veterans than on females or civilians. Average remission rates for PTSD among combat veterans are in the 20% to 30% range. One reason may be that PTSD among military is related to multiple or accumulated traumatic experiences and another may be the intensity of the stressors. Thus, medication is likely to be less effective than among non-military populations.

Side-Effects
It is important for clients to be aware that the first expected result of any of the medications noted above is the development of side-effects. Only if the person can persist in spite of unwanted side-effects (which often can be managed) will it be possible to determine if the medication has any positive impact on PTSD symptoms. On the other hand, some of the side-effects are desirable; for those with problems falling asleep, a medication with a sedating side-effect can be a godsend.

Benzodiazepines
This class of drugs, also known as minor tranquilizers or anxiolytics, does not appear to have specific efficacy for PTSD although when added to an SSRI as required they can improve sleep and decrease both generalized anxiety and panic attacks. These medications are quick-acting (usually 20 to 30 minutes) and work by modifying the GABA neuro-transmitter system to decrease anxiety. All the clinical guidelines I have seen recommend against prescribing benzodiazepines for PTSD and the 2004 review by Veterans Administration and Department of Defense in the U.S.A depicts them not being harmful but as not beneficial either.

Summary

A fair summary of the research information on medications for PTSD is that (a) there are still too few good research studies from which to draw firm conclusions, (b) the placebo response can account for up to 38% of benefits, (c) SSRI medications (including the placebo effect) may provide some degree of help in up to 60% of civilians with PTSD, (d) only a minority, perhaps 30%, of clients, will achieve remission if medication is the only treatment available, (e) psychological interventions are more effective generally than medications, (f) dropout rates of 32% can be expected if medication is the only treatment available while 14% will stop treatment if only psychological treatment is available, and (g) response rates with medication are lowest for military clients.

My clinical experience is that while medication was helpful for some clients, most clients decided against medication or, when they took medication, it did not have much of a positive effect. Also, most of my clients achieved remission without the addition of medications. Moreover, my experience has been that many who started with medications were able to decrease them significantly or stop them all together once psychological interventions were well in place. Clients I have seen who were on medications were taking either an SSRI or an SNRI: none were taking medications intended initially for either seizures or psychosis, none were taking benzodiazepines, and none were on combinations of medications. However, it is also true that I have no idea as to the number of police or military personnel not seeing me who were doing fine; it is likely that some of these were on medications that were effective for them.

Also, just as with some of my methods of psychological treatment which do not yet enjoy overwhelming research support, it is entirely likely that those colleagues in psychiatry who focus primarily on PTSD have developed the clinical experience and expertise to yield positive results greater than those available to date in the pharmacological studies.

Treatment Step E: Implementing Relevant Level 1 Stabilization Strategies
There is evidence that what I refer to as *stabilization strategies* (others refer to it as behaviour therapy, or health psychology, or life style psychology) can make a difference, since those with PTSD often have life habits contrary to research-based recommendations.

Thus, stabilization strategies relevant to each client are introduced one or two at a time in each treatment session while elements of *Step F* below are underway. The rationale for this is that (a) people can make only so many changes at a time without getting overwhelmed and (b) elements of *Step F* are important to promoting as rapid regulation as possible of psycho-physiological systems known to be critical in the treatment of PTSD. The order in which strategies are presented is determined by (a) what appears to be a more pressing issue and (b) what the client is able to manage. In respect to manageability, generally relevant strategies from among the first 14 are selected before those later on in the list. Also, not every client requires attention to all of the 28 below.

The range of stabilization strategies are listed below and then each is discussed in turn.

Contents

E-01: Introduction

Following traumatic incidents, people can develop symptoms of post-traumatic stress disorder (PTSD). Sometimes, symptoms of additional psychological conditions develop as well including depression, panic attacks that occur during the day and/or night, and significant disturbances of sleep. Also, for some folks, PTSD is accompanied by any of a variety of psycho-physiological reactions such as tension and migraine headache, high blood pressure, non-cardiac chest pain symptoms of fibromyalgia syndrome, and symptoms of irritable bowel syndrome. Moreover, others experience chronic pain from the injury that has given rise to PTSD, while yet others find that cognitive processes like attention/concentration, short-term memory, information processing, and decision-making are affected. Furthermore, some with diabetes, arthritis, heart conditions and other pre-existing medical problems find that their illnesses worsen after a trauma. Usually, the folks I see with PTSD have a combination of the associated conditions noted above, especially following exposure to multiple or accumulated trauma.

The recommendations in Step E are consistent with results of research in various professional fields including clinical psychology, health psychology, and affective neuroscience. Folks have found them to be helpful in lessening the severity of symptoms, in gaining some control over reactions instead of continuing to feel powerless in the face of symptoms, in having a plan of effective action rather than following a path that is counter-productive, and in providing loved ones with ways they can be helpful.

Sometimes clients have put a number of these strategies into play on their own before ever seeing me for additional methods of treatment. At other times, some of our session time is devoted to relevant stabilization strategies. Regardless of how they originated, I encourage clients to utilize these strategies on a permanent basis, since they are key ingredients in a healthy lifestyle as well as one aspect of a treatment plan.

Typically, the following is what I say to clients about implementation:

> During weekly sessions over the first few months, I will give you one or at most two recommendations at a time that are relevant to your situation. Providing them all at once will only lead to failure, thereby increasing your stress. This is not productive and in fact is counterproductive. Read the recommendations over many times throughout treatment since your memory is likely to be somewhat impaired for a while. Make specific plans of action and revise them as needed. Write the action plans and any changes in the space provided. Remember, though, that these recommendations are only one part of the plan and that all parts are important.

Information about the relevant stabilization strategies are provided to clients as follows.

E-02: Regular Appointments

Research has demonstrated that regular appointments are critical in recovering from conditions such as PTSD. There likely are a number of reasons for this.

First, appointments provide you with a therapeutic environment so you are able to talk as needed *and in confidence* about matters that are important to you. Often, just getting them off your chest is helpful. As well, we consider possible solutions to the issues you bring forward.

Second, appointments are the opportunity for you to learn all you need to know about your condition. The diagnosis and treatment of psychological reactions to traumatic stress is what I do and have done since 1985. It is the only condition I treat at the present time and is the focus of all of the conferences I attend and the reading and writing I do. I spend about 500 hours a year in these activities of professional continuing education so that I am up to date with research and practice in this area. Also, I present my current approach via lectures and travel drives to other professionals. As I am on the cutting edge of the treatment of PTSD, a number of professional colleagues consider that I am an expert in this area even though I do not live in Vancouver, New York, or London. Lynne and I choose very deliberately to live here and we have no intention of leaving.

Third, appointments are directed, step-by-step, to the effective decrease of symptoms.

You are well-advised to *avoid any psychologist, counsellor, psychiatrist, or other helper who has no therapeutic approaches other than to tell you that you must talk about the trauma again and again until symptoms subside.* This was the approach most of us helpers were taught during our graduate training. However, my clinical experience since 1985 in working with people traumatized by multiple events (as with police and military personnel) has taught me that the talk-and-more-talk approach is not only counter-therapeutic but, worse yet, usually results in an increase in the number and severity of symptoms as well as their continuation for very long periods. Many of the colleagues I meet with during international conferences report the same experiences. Moreover, as you will learn later on, we now have the knowledge of body and brain functions to understand why approaches other than repeatedly re-telling the trauma story are more helpful.

To facilitate your recovery, purchase a one-inch thick three-ringer binder and a package of dividers. Keep it in a place at home where you can get to it easily but away from any prying eyes. The binder will be helpful in at least three ways. *One*, it is a place to keep the information and assignments I will give you during our sessions. *Two*, it is a single place where you can record matters of importance that may develop between sessions—questions that arise, observations you make, symptom changes, successes or difficulties with assigned tasks, and so on. *Three*, it is an aid to memory and other cognitive processes that can be impaired early on. With the binder, you will be able to find the information you require quickly and easily and to review aspects you may have forgotten. Bring the binder to each meeting.

E-03: Support to and from Family Members and Colleagues

a: Psychological Health of Family Members
The psychological health of family members and relationships within the family often are affected adversely when one person has PTSD.

Below is a poignant quote from the wife of a traumatized Israeli veteran reported in the 2007 book *Combat Stress Injury* by Figley and Nash:

> I'm afraid for myself, of what I've become since the war. Our entire life is lived under pressure. I feel I'm in a war day in and day out to keep my sanity. I married a man who was happy, sociable, and diligent. . . . Today, he's impatient, tired, depressed, anxious, and vulnerable; he can't hold a job, yells at the kids, and is indifferent towards me. I feel rejected and socially isolated. I'm angry. Sometimes I feel that I don't want to live. (p. 137)

In addition, spouses commonly report a lower sense of well-being, a greater sense that they are on the verge of a nervous break-down, somatic problems, sleep difficulties, depression, over-sensitivity, demoralization, and loneliness. Such symptoms are even more marked when their soldier spouses have significant symptoms of PTSD but deny their presence or their effects.

The reasons for such reactions are many. As cited by Laffaye and her team in 2008, relationship difficulties among male veterans with PTSD have been well documented and this would clearly include greater spousal conflict than before deployment. Also, compared to pre-combat levels, other reasons include less family cohesion, greater parental burden with the low support in running the household and managing the children, seeing the pain and suffering of their partners, watching the partners struggle with PTSD symptoms, worrying what will happen after the primary victims make threats of suicide, social isolation since the veteran does not wish to meet other people, and seeing the look of pain on the faces of their children when they try unsuccessfully to interact with the primary victims. In addition, less physical intimacy given the partner's emotional numbing and avoidance is likely to be a factor. For example, in their 2008 report on Croatian war veterans, Anticevic and Britvic noted that, compared with controls, (a) those with PTSD had less sexual activity, hypoactive sexual desire, and more erectile difficulties and (b) some of these characteristics may have been related to side effects of the antidepressant medication they had been prescribed; thus another reason for careful consideration of medication for those with PTSD.

Moreover, spouses are in a number of binds. First, they often have a terrific sense of commitment such that they are unable to leave the marriage. Especially in military communities, there is a strong and shared conviction that one does not abandon an injured soldier in the field. Second, often they do not know who to consult for help for themselves or for their children or are ashamed to do so.

My experience to date with police and military clients is that things at home improve once treatment begins and usually problems at home return to pre-PTSD levels once the client's symptoms are under control. Also, often families report that, once PTSD is in remission, things at home are better than before the onset of PTSD. So time to complete the treatment is one of the things required.

However, I am willing and able to help partners through this difficult experience. Supportive counselling, stress management strategies, and other help is available. It is just a matter of a phone call to arrange an appointment with me (250.374.3215). For police or military veterans whose application for a disability status has been approved, VA Canada allows their partners free access to me for up to 20% of their allocation for psychological services. Also, with the client's consent, partners can attend any of our sessions without cost; matters of importance to the client and partner can be addressed at that time. This can be important since the Laffaye research of 2008 indicated that for veterans with PTSD, spouses were seen as both interpersonal resources as well as sources of interpersonal stress.

b: How Partners Can Be Helpful

Partners of individuals with PTSD cannot go wrong if they assume nothing and treat their partners with respect and compassion. Moreover, a number of small studies have reported positive results in both PTSD symptoms and relational satisfaction when the spouse of a single-episode trauma victim is informed how best to help (for example the Billette and colleague 2008 study with female victims of sexual assault).

Below are some tips on how to help someone who has been involved in a traumatic incident. Much of this is adapted from *On Combat* by Lt. Col. Grossman, formerly of the U.S. Marine Corps.

___i. Ask the victim if you can read the papers on stabilization that I provide. That way, your contact with the victim will be consistent with my recommendations.

___ii. Don't ask for an account of the incident but let the person know that you are willing to listen to whatever s/he talks about. People often get tired of repeating the story and they find idle curiosity distasteful. Also, for many, retelling the story makes the symptoms worse.

___iii. Ask questions that show support and acceptance, such as "How are you doing?' and "Is there anything I can do?" If offers of help are declined, accept that. Ask again at a later time, without insisting.

___iv. Unless it clearly is harmful, accept the person's reaction to the event and avoid suggesting how s/he should be feeling or acting. Remember, people have a wide range of normal reactions to trauma.

___v. Apply non-judgmental listening. Monitor your facial and postural expressions and simply nod your head at what s/he tells you.

___vi. Do not encourage the use of alcohol or caffeine. If you go out together, restrict yourself to a soft drink, juice, or decaf. Better yet, go for a walk together.

___vii. Do not call the victim "killer," "terminator," or other such terms (even as a joke) or make light-hearted comments about the incident.

___viii. Although you may find yourself second-guessing the victim's actions during the traumatic event, keep these thoughts to yourself. You were not in the situation and the victim likely has forgotten some relevant aspects of it.

___ix. Be supportive if s/he needs to take time off work. Police officers and others are bothered about taking time off work since they are loyal to their colleagues and know that their absence causes a redistribution of their workload onto other members of the team.

___x. Encourage the victim to maintain regular appointments with me. Offer to go with the victim if s/he is resistant.

___xi. Do not refer to the victim as having psychological issues or "mental" problems or use any such terms. Such stigmatizing will encourage the victim to deny problems and not get the help s/he needs.

___xii. Gently bring to the victim's attention any negative behavioural or emotional changes you notice, especially if they persist for longer than 2 to 4 weeks; for example, "Bill, I've noticed that you are yelling at the kids probably more than you realize. How can we deal with this in a helpful way?" Odds are that s/he may not be aware of changes, in part because trauma symptoms are interfering with thought processes. Usually, others notice changes before the victim does.

c: Additional Suggestions for Colleagues
___i. If you are a colleague, initiate contact in the form of a telephone call to let the person know that you are concerned and available for support or practical help. If the person does not want to talk to you at that time, respect that decision. Try again a week later.

___ii. If the person lives alone, offer to stay with the victim for the first few days after the event or invite the person to stay with you.

___iii. Let the victim decide how much contact s/he wants to have with you. Don't take any refusals personally. S/he is handling the trauma as best as possible and the victim's desires will change from day to day.

___iv. Do feel free to offer a brief sharing of any similar experience you have had to help the victim to feel that s/he is not alone and that you have some understanding of what s/he has been through. However, this is not the time for you to work through any unfinished trauma of your own, which can easily be re-triggered by the experience of your colleagues. If you are re-triggered, see a psychologist with interest and competency in PTSD.

___v. Normally, the victim will want to return to work as soon as possible. When s/he returns, do not pretend that the event never happened, do not avoid the person, and do not treat him/her as fragile or otherwise change your behaviour in any way. Welcome him/her back (for example, "Good to see you back, Bud" and then continue to treat him/her as you always have. The victim will be very sensitive to any changes in actions towards him/her.

E-04: What to Say to Others

Decide beforehand how you will reply to relatives, friends, co-workers, neighbours, etc., who ask about your experience. Share only the details you feel comfortable about sharing, and only with people who need to know and whom you can trust. For others, the simple statement, "I don't want to talk about it" will suffice.

Relatedly, have a couple of scripts prepared in your mind should people ask why you are not at work. You can choose to tell the truth ("I am being treated for a medical condition and I don't want to talk about it"), joke about it ("I want to get a preview of life for after I win the lottery"), or lie ("I have been asked to burn off the excess annual vacation time I have accumulated").

Outside of close relatives and friends you trust, generally it is not a good idea to share that you have a psychological condition. In my office it is important that we talk openly and honestly. Outside of my office, many people, including those who should know better, have a false and warped sense of what a psychological condition is. Also, knowledge of your psychological status can have negative effects on your career, even though this should not be the case.

Remember, you are under no obligation to share what you do not want to share and others do not have a right to know. People who care about you will understand and you do not need the others in your life.

If the following people ask, I will say:

_____

_____

_____

E-05: Remember: "A Primary Objective Is to Regain and Maintain My Health"

Remind yourself frequently that in order to get better, you need to focus seriously and consistently on regaining/maintaining your physical and psychological health. This objective *must* be one of the top three priorities in your life at this time. The other two might be family and work. Carry an index card with you and read it several times a day so that this objective remains in consciousness; the card should read "*My primary goal is to regain and maintain my health.*"

Also, you need to dedicate the time and energy to doing what is required. Only you can take charge of your life in this way. Others can support you, but *you* have to do it. As the Dalai Lama has said, "The point is that in your life, unless you make specific time for something that you feel committed to, you will always have other obligations and you will always be too busy" (1996, p. 26).

In the early stages of treatment, you will need 1 to 2 hours each day to put in place those recommendations we will discuss.

The journey to improved health is like any other trip. It requires setting goals, organizing the journey into steps or stages, thoughtful planning, purchasing the supplies needed, taking things one step at a time, monitoring progress, making changes as needed, and so on.

Gather_support by letting the caring people closest to you know what you are trying to accomplish. One way is to ask them to read this and the other papers you will receive. Ask for their understanding and their support. Then, tell them exactly how they can be part of your growth towards health.

Remember to *pace yourself*. There are many aspects of re-stabilization that you will need to put in place. You will not be able to do this all at once. Trying to do so will only result in failure and frustration. By proper pacing, you will accomplish goals, feel good about progress, decrease the bad feelings that come with unrealistic expectations, and regain the sense that you do have control over important aspects of your life.

Cognitive over-ride is a good strategy to use if you lack the motivation to put the stabilization steps into place and follow them frequently. You need to do something about that because not implementing the steps delays progress on your journey. Basically, you may need to say to yourself:
>What I feel about doing the steps is irrelevant. I need to do this.
>I will do this. This is part of my path towards regaining my health.
>I will feel better by doing it.
Then, as the Nike ad once advised, *just do it.*

Sometimes the rate of progress is slow. When I am bothered by the pace, I go to the following quote from the Tao te Ching (64):
>A tree that reaches past your embrace grows from one small seed.
>A journey of a thousand miles starts with a single step.

Be on the look-out for wisdom wherever you find it. For example, consider the following advice:
>live in rooms full of light; avoid heavy food; be moderate in the drinking of wine; take massages, baths, exercise, and gymnastics; fight insomnia with gently rocking and the sound of running water; change surroundings and take long journeys; strictly avoid frightening ideas; indulge in cheerful conversations and amusements; and listen to soothing music.

These suggestions for health were written by A. Cornelius Celsus, a Roman physician, in the first century.

This is what I will do to both remember and implement the primary objective:

_____

_____

_____

E-06: Make Healthy Choices
This section is based on the previous stabilization strategy.

Do those things that promote physical and psychological health and avoid those that make you feel worse or otherwise interfere with your recovery. In other words, eliminate those factors that lead to suffering and cultivate those that lead to calmness, contentment, and peacefulness.

As Sharon Salzberg (1997) once wrote:
When you know for yourself that "these things are unhealthy, they incline to harm and suffering," then you should abandon them. When you know for yourself that "these things are healthy, they incline toward welfare and happiness," then having come upon them, you should stay with them. (p. 65)

For example, if watching the news bothers you or increases your symptoms, then stop watching the news. The universe will not collapse if you avoid TV news. Besides, you can control what you are exposed to by going to internet news (try www.cbc.ca).

Or, if talking to a friend perks you up, then spend more time with that friend and less time with people who leave you feeling agitated or depressed.

If you find that listening to some kinds of music leaves you feeling crummy, switch to relaxing music. I avoid most popular and country music.

So before making a choice, ask yourself the question "Is this healthy for me?" It is okay to take some time to decide; you do not need to rush into make decisions. When you have thought it over, if the answer is "no" (it is not a healthy choice), then don't do it. If the answer is "yes," then *just do it*.

Healthy Choices (I will do more of these):

_____

_____

Unhealthy Choices (I will do fewer of these):

_____

_____

_____

E-07: Identify and Reduce Other Stressors

Experienced clinicians have long noticed that the presence of additional stressors among those with PTSD increases both symptom intensity and treatment difficulty (in fact, that is true for most conditions including chronic pain). I came across a study by Solomon and associates (2008) supporting this clinical experience. Israeli war veterans were assessed at two diffeent times for PTSD symptoms as well as for postwar negative life events. Results indicated that stressful events throughout the life cycle had a significant impact on PTSD symptoms. The authors referred to one study showing that life events after trauma can facilitate the onset of PTSD and to another study suggesting that postwar life events had a negative effect on recovery from PTSD. One study did not find such connections.

I suppose that stress can impact on PTSD for a number of different reasons. *First*, chronic stress affects brain processes. For example, the ability of the amygdala to learn and express fear associations can be increased. At the same time, the ability of the prefrontal cortex (PFC) to control fear can be reduced. The resulting vicious cycle of increased fear and anxiety leads to more stress, which leads to further dysregulation in many systems including the autonomic nervous system. Parts of this scenario were supported in a 1993 study by Schaubroeck and Ganster, who reported higher resting levels of blood pressure, lower cardiovascular responsivity to acute challenges, and delayed cardiovascular recovery following acute challenges among those employed in chronically demanding occupations. Other parts were confirmed in a recent study (Liston et al., 2009) of healthy adults. Experimental subjects were exposed to one month of chronic stressors. Results via fMRI and testing indicated consequential impairment of both the PFC and attentional processes among the experimental group but not among control subjects. The good news is that following one month of reduced stress, all effects were reversed; the PFC and attentional processes returned to normal such that the experimental group no longer differed from the control group.

Second, PTSD among the clients I see often taxes cognitive and emotional resources very significantly. Additional stressors decrease coping capacity even further. The analogy that comes to mind is the bucket of water. When it is filled to the brim, the addition of any more water results in spilling over.

Third, coping capacity is reduced by the realization that one cannot cope with even the normal stressors and strains of daily living. Police and military personnel pride themselves in their ability to cope with significant stressors and still carry on with their duties. When they cannot, then negative self-statements creep in accompanied by increased feelings of hopelessness and helplessness.

Hence, one early treatment objective is to reduce the number of concurrent stressors as much as possible. Delegating responsibilities to others, backing off from previous commitments, and putting major decisions on hold are among the strategies that are discussed in therapy sessions. Most often, once clients develop control over their PTSD symptoms, portions of therapy sessions can be directed to elements of effective stress management such as identifying stressors, determining what is and is not within the client's control, prioritizing them, attacking each of the stressors within their control in turn using effective problem-solving techniques (with attention to gains and losses associated with potential resolutions), and learning how best to cope with those stressors that are beyond client control.

However, early on in treatment, encouragement to reduce concurrent stressors is a hard sell, especially to police and military folk. For a variety of reasons, many are very much focussed on what they should be able to do rather than on what they are capable of doing at that time. Understandable core self-definitions of capability and the need to remain in control are contributing factors that I have found must be addressed before many can let go.

Also from both clinical experience and research, it is clear that stressful life events occurring prior to the traumatic events have an impact on PTSD. This was re-affirmed in the Liston et al. study. Stressful childhood events and in particular violence within the family were major risk factors for later PTSD. In fact, the authors discovered that such experiences played a more significant role in the development of future PTSD than events during combat. After approaching stressful childhood experiences in various ways and at various times during treatment, generally with police and military personnel I have found the most successful strategy to be acknowledging the experiences during the assessment, noting the role they could have played in the development of PTSD, re-affirming that the client can bring this subject up at any time in our sessions, and then moving on to the rest of the assessment and resulting basic treatment approach. In this way, we are able to bring PTSD into remission. Sometimes, but rarely, clients will return to childhood issues later in the treatment process, but not usually until after symptoms are under control. Even then, what clients seem to want is a simple review of the impact of those events and seem satisfied with the realization that they cannot change the past but have to use the techniques learned to let these events go and move past them.

So, the presence of other stressors in your life (both minor and major) can have the result of increasing your symptoms and delaying your recovery. Also, there are many other consequences of excessive stressors, including the death of brain cells, a decrease in immune functions with a resulting increase in infections, unproductive levels of irritability or impatience, sleep problems, a decrease in energy levels, and so on.

First, take time to identify other stressors in your life. Do it now, and also notice other stressors as they develop.

Second, write them down in the space provided.

Third, take steps to reduce their number as quickly as you can.

    a. Since many stressors do not need immediate attention, postponing decisions on them is one sensible way to reduce their number. For example, you could write them on a calendar indicating some future date that you will deal with them, thereby eliminating the need to think about them until then.

    b. Another way is to delegate the tasks to others: family members, friends, people you hire.

    c. Dealing, one at a time, with tasks/demands that must be addressed sooner rather than later is a third part of reducing their number.

Before each stressor/task/demand below, place a P (postpone), a De (delegate) or a Do (do).

Fourth, pay attention to common stressors/irritants. If you are not feeling 100%, they can be upsetting. So take corrective or preventative action about them sooner rather than later. Based on client experiences, some of the solutions to irritants are noted below. If these irritants impact on you, take the action needed. At the end of the list is space for you to record how you can best manage similar stressors. Reducing these common irritants will reduce stress in the same way that the back of the camel is less indented with each straw removed.

____a. Do nothing which, after it is done, leads you to tell a lie. The energy needed to remember lies is very draining.

____b. Make duplicates of all keys. Bury a house key in a secret place (not under the mat—everyone knows that).

____c. Have timer switches connected to various indoor and outdoor lights that work whether you are home or not. This will help keep the non-impulse burglar away. Also, leave a vehicle in your driveway while you are away. Also while you are away, arrange to have someone pick up the mail and newspaper daily as well as check your place.

____d. Plan ahead. Don't let the car gas tank get more than a quarter empty before refuelling, keep a well-stocked "emergency shelf" of home staples, buy stamps in self-stick booklets and replace when you are down to 5 stamps, etc.

____e. Don't put up with something that doesn't work right. If your alarm clock, shoe laces, windshield wipers, etc., are a constant irritation when you go to use them, get them fixed or get new ones.

___f. Always set up contingency plans. For example when out, agree to meet at place X in Y minutes and have place Z as a back-up. If memory is a problem, write this information down (as well as where the car was parked in the lot).

___g. Relax your timing standards on unimportant stuff. The world will not end if the grass does not get cut this weekend or if the sheets are changed Tuesday rather than Monday. As the title of one little book I recommend says, *Don't Sweat the Small Stuff.*

___h. Turn *needs* into *preferences.* Our basic needs are quite limited: food, water, sleep, shelter, exercise, illness-control. Everything else is a preference. *Don't* get over-attached to any particular preferences but *DO* remember both (1) to honour your body by taking care of it and (2) to maintain basic responsibilities to other members of the family.

___i. Create order out of chaos. Organize home and work space so you can find what you are looking for quickly. When you put things away where they belong, the stress of not being able to find what you want is lessened.

___j. Learn to live one day at a time.

___k. Keep proper change in your vehicle for such things as parking metres and shopping carts.

___l. Become more flexible. Not everything has to be done with incredible precision and perfection. Some things are okay even if not done perfectly and some issues are okay to compromise on.

___m. Do one thing at a time.

___n. Examine your life and consider other specific things you might do to reduce stress.

_____

_____

_____

_____

_____

_____

Fifth, when concentration, thinking, and emotions are improved enough to undertake solution-finding, you can deal with any unresolved matters. With the treatment plan I have outlined, you will improve so that you can resume a normal life (which always includes dealing with stressors, tasks, demands, and so on—that is just how life is).

E-08: Eliminate Illegal Drugs

Studies show that 34% of men and 27% of women with PTSD have a drug use problem. The 1995 National Comorbidity study by Kessler and associates indicated that women with PTSD were nearly 4.5 times more likely to have a diagnosis of drug abuse or dependence compared to women without PTSD. Sometimes, use of these products began well before any trauma, sometimes they were begun as a way to blot out the thoughts and overwhelming anxiety that are part of PTSD, and sometimes they were used to help the person fall asleep. However, such innocent and understandable reasons for starting drugs can lead ultimately to an addiction with all of the added problems that brings.

Drugs can have any of a number of negative consequences. *First*, because drugs interact with and can impair the brain's mood regulating systems, they can drastically alter the way we feel. Although the drug may have temporary pleasant effects (which is why some folks self-medicate with them), the brain's chemistry can be altered forever, leaving the person with a form of brain damage; for example, the brain may respond by ceasing to produce the chemicals we require. *Second*, brain imaging studies indicate that repeated drug use causes disruptions in that part of the brain regulating cognitive activities such as impulse control, planning, and memory. *Third*, some drugs such as cocaine can aggravate PTSD symptoms, for example making users more reactive to traumatic reminders. *Fourth*, attentional processes are affected negatively, a phenomenon that is related to alterations in the brain wave pattern known as P300. *Fifth*, those on drugs have a more persistent course of PTSD than those who are drug-free.

Contrary to popular belief, *marijuana is not a harmless drug*. Today's marijuana is 7 to 10 times stronger than what was used in the 1960s. Marijuana impairs brain and reproductive function, is harmful to the immune system, and is more damaging to the lungs when smoked than tobacco smoke. Moreover, the sellers of drugs (who want you to be hooked) are known to add even more addictive substances to the weed without telling you (such as crystal meth).

So, wean yourself off any illegal drugs as quickly as you can.

I will become drug free by:

_____

_____

_____

E-09: Decrease Alcohol Intake

Among those with PTSD, studies have revealed that about 52% of men and 28% of women use excessive amounts of alcohol. Moreover, a 2008 study by Jacobson and others of 55,000 U.S. military deployed to war zones found that combat exposure was associated with new-onset binge-drinking in regular forces with the addition of both heavy drinking and alcohol problems in reserve units. Also, it has been found that excess alcohol intake preceding the traumatic event increases the likelihood of exposure to traumatic incidents among civilians.

Alcohol use is reinforcing since consumption has measurable pleasurable effects including euphoria, disinhibition, anxiety-reduction, and sedation. The neuro-biological basis of these effects may be that alcohol increases the amount of the brain neurotransmitter GABA and perhaps also dopamine. GABA increases feelings of calmness while dopamine increases the sense of pleasure. Relatedly, neuro-imaging methods such as fMRI and PET show changes in both oxygen and glucose utilization leading to increased activation in such areas of the brain as the orbital frontal cortex, the pre-frontal cortex, and the anterior cingulate cortex. Activation of these areas serves to dampen signals from the brain's emotional centres. Also, alcohol intake results in an increase of alpha brain waves. On the other hand, during acute and chronic withdrawal from alcohol, there is a decrease in the metabolic activity of these brain areas, which may help to explain why it is so difficult to break the addiction.

However, alcohol remains the most dangerous legal socially sanctioned drug available, especially among chronic users. *One*, research consistently demonstrates its major negative effects on just about every body system and mechanism including reduced blood flow to the brain, increased blood pressure, increased heart rate, and earlier death with moderate to high consumption. *Two,* alcohol decreases restful sleep by increasing night wakings, altering sleep architecture such as suppressing REM, reducing the amount of slow-wave sleep when body repairs are done and from which the sense of restful sleep emerges, and decreasing the amount of oxygen to the brain. *Three,* alcohol can continue to have negative effects on sleep for months to years after a chronic drinker has stopped; this may be due to the long-term effects on brain transmitters such as GABA and serotonin or inhibition of the release of melatonin by the pineal gland. *Four*, it reduces the effectiveness of some medications including those used in the treatment of trauma-caused conditions. *Five*, among chronic users, it increases treatment time for those with PTSD, probably due in part to compromising the frontal areas of the brain. This area is heavily involved in basic processes of concentration, memory, decision-making, and impulse control, which may already be affected by PTSD. *Six,* alcohol abuse can have very negative effects on family relations, employment, and interactions with the law. *Seven*, overindulging in alcohol undermines motor functioning in older adults more than in younger ones even when blood alcohol levels are the same.

Since a couple of drinks works so well temporarily to make people feel better and because alcohol is so easily available, people are tempted into consuming more and when under stress and often use it to medicate themselves. However, ultimately alcohol is a central nervous system depressant and thus increases symptoms in those whose reaction to trauma includes depression and sleep problems. Among those whose reaction is acute or post-traumatic stress disorder, it simply delays dealing with the symptoms. So, alcohol definitely is not recommended for those suffering from trauma.

Successful plans to stop drinking often include components such as the following:

    ___a. get into a healthy lifestyle (as in other parts of this series);

    ___b. identify stressors that are followed by bouts of drinking;

    ___c. use the ways you will learn in therapy rather than alcohol to both manage these stressors and regulate emotions;

    ___d. identify the excuses used to justify drinking and do not let yourself be fooled;

    ___e. let family and friends know how they can help;

    ___f. develop interests and friendships that do not support drinking;

    ___g. keep no alcohol in the house;

    ___h. avoid people and situations formerly associated with drinking; and

    ___i. consider seriously joining AA. I can help with that.

In addition to the methods I checked above, I will decrease alcohol intake by:

_____

_____

_____

E-10: Decrease Tobacco Use

About 45% of those with PTSD smoke, which is about twice the rate among the general population. Also, smokers with PTSD are more likely than smokers without PTSD to smoke during exposure to stressful situations. Moreover, they report that smoking reduces tension and PTSD symptoms. Furthermore, there is evidence that nicotine reduces negative mood states (common in PTSD) and can exert an anti-depressant and/or an anti-anxiety effect. In addition, it acts as a stimulant and can improve cognitive functions including attention; usually clients are tired from lack of sleep and have problems with cognitive functions. Relatedly, there is research evidence that nicotine tweaks the brain's reward systems through changes in the levels of the dopamine neurotransmitter as well as impacting on other neuro-transmitters such as serotonin, norepinephrine, GABA, and glutamate. For all of these reasons, typically daily nicotine usage is high among those with untreated PTSD symptoms and quitting is made more difficult by a (short-term) increase in negative mood when trying to stop the addiction.

On the other hand, health problems associated with tobacco use are well documented and include emphysema, asthma, bronchitis, increases in insulin secretion (which accentuates the swings in blood sugar levels), a 10- to 20-point increase in blood pressure with each inhale, greater demands on the heart, an increase in potentially fatal heart conditions (disease, heart attack, and sudden cardiac death), an increased chance of rupturing blood vessels, decreased blood flow to the brain, changes in brain wave rhythm from decrease in theta and alpha waves to an increase in beta activity, a 50% increase in the rate of spinal disc degeneration, delay in wound healing, increased risk of complications during surgery, negative effect on sleep (problems with onset and/or maintenance, increased snoring, increased arousal levels, dangerous decrease in oxygen saturation, and increased daytime sleepiness), an increase in various forms of life-threatening cancer, and a decrease in the effectiveness of medications. Moreover, stress levels are felt more acutely between cigarettes so that another cigarette is needed sooner, leading to high rates of nicotine use. For those with panic disorder, tobacco can cause another attack and, relatedly, smoking among people with PTSD may be associated with the development of panic attacks (Feldner and colleagues, 2009). Therefore, reducing nicotine intake is a healthy choice. However, to be successful, it is critical that you have available other effective methods to reduce negative mind states.

Many have found that the graduated strength decrease of nicotine patches and the use of the prescribed medication Zyban are helpful in controlling the habit. Engaging in activities where smoking is not involved, modifying habits associated previously with smoking, developing health life-style habits, increasing the intake of fruits, and eliciting the support of loved ones are helpful also. In addition, replacing the use of nicotine with ample opportunities for high-pleasure activities has been shown to activate the brain's reward centres. Being faithful in implementing the relaxation strategies you will learn is helpful also. Still, early on in controlling the addiction of chronic smokers, there may be a temporary deterioration in sleep as well as increases in irritability, anxiety, and tension, which lead some to resume the habit.

I will decrease tobacco use by:

_____

_____

_____

E-11: Decrease Caffeine Intake
Caffeine is the most widely used psycho-active substance in the world to change energy and is found in home-brewed coffee (80 mg per cup), tea (40-60 mg per cup), cocoa (20 mg per cup), chocolate (45 mg per regular bar of dark chocolate; 10 mg for milk chocolate), cola (45 mg per bottle), some medications (for example, Anacin contains 32 mg, Excedrin contains 65 mg, and Midol contains 50 mg), and "energy drinks" (50 to 145 per 8 oz serving). Regular coffee at places such as Starbucks contains twice the caffeine in a cup of coffee made at home. Caffeine blocks adenosine A1 and A2a receptors in the brain, thereby producing a psychomotor stimulant effect. Used appropriately, moderate amounts of caffeine are useful for keeping people awake when they need to stay alert, such as during a night watch.

Caffeine has major and rapid effects on a number of systems including the central nervous system (such as decreased blood flow to the brain), the cardiovascular system (such as increased heart rate and blood pressure), and the immune system (decreased functioning). An early effect of caffeine is a fairly dramatic increase in energy and a corresponding increase in tension. Unfortunately, in the long run, energy is actually decreased to levels below those before the drug was consumed and the temporary energy boost is replaced by both fatigue and irritability. In chronic users, this leads to requiring another shot of caffeine with subsequent greater and greater physiological arousal, fatigue, and tension. Excessive use of caffeine is thought to account for as much as 14% of coronary heart disease and 20% of stroke-related deaths. Caffeine can lead to sleep problems even 10 to 12 hours after ingestion. Generally, caffeine leads to an increase in anxiety symptoms for those with acute and post-traumatic stress disorder, greater lethargy in those with depression, and a single cup for those with panic disorder can cause a full-blown panic attack. A recent finding is that caffeine increases sugar readings dramatically among those with diabetes.

Hence, although a little caffeine a day is probably not a problem for many people, removing caffeine from your diet if you have PTSD is a healthy choice. With heavy users, it may be necessary to withdraw gradually in order to minimize withdrawal symptoms such as throbbing headache and the jitters. If becoming alert or staying awake is the reason you drink caffeine, a brisk 5-minute walk works as well.

If you need caffeine at all, you might consider replacing coffee with weak tea. It contains L-theanine, which protects neurons, increases neurotransmitters (such as GABA, dopamine, and serotonin), may decrease blood pressure, and increases alpha waves in the brain, which are associated with calm focus. Additionally, tea improves mood and thinking without increasing anxiety. Also, green tea contains anti-oxidants.

I will decrease caffeine intake by:

_____

_____

_____

E-12: Put Regular Routines into Place
If you are off-duty, on day shift, or retired, get into regular daily routines, since doing so helps us feel in control and able to function effectively, both of which are important after trauma.

For example, a morning routine should include rising at the same time each day regardless of the type of sleep you had the night before, washing, making yourself look good, getting dressed, and joining the family for a healthy breakfast while keeping the conversation light, such as remembering positive things from the past and funny/interesting things that happened yesterday. Positive meal-time with family will also reduce the levels of worry they have about you.

Then plan a general outline for your day, which must include the home practice of the recommendations I make if you plan to get better; early on you will need to devote a good 90 to 120 minutes per day. Set some chores to be done, keeping their number within your capabilities such that you can complete them and experience success.

You might want to use the space below to develop a plan for the week and then revise it for the next week etc.

The following is my plan for a regular daily routine:

_____

_____

_____

E-13: Eat Healthy Foods

Unhealthy eating can put undue stress on your body and slow any healing process. Thus, eating reasonable amounts of food within a balanced diet is generally best in order to maximize chances of getting sufficient amounts of the key nutrients that your body and brain require. The standard recommendation is to eat a variety of foods from each of the major food groups: complex carbohydrates (grains, rice, and potato), fruits and strongly coloured vegetables, items containing calcium such as dairy products, and meat/fish/protein alternatives as well as nuts. Grains, meats, and dairy products are good sources of tryptophan, which synthesizes into serotonin, a critical neuro-transmitter in the management of mood. A registered dietitian can help you to achieve a balanced diet and *this is something I recommend strongly for all clients.* That professional will likely begin by having you record your intake for one week.

Make sure that some omega-3 is included in your diet. Two 3 oz (85 grams) servings a week of fatty fish (salmon, herring, and tuna) are recommended.. Some omega-3 is found in trout and less yet in walnuts, dark leafy greens, and flax seed/flax cereal. Taking a supplement such as the one from USANA is an alternative. Studies show that omega-3 is helpful in lowering blood pressure, reducing blood clotting, improving immune functions, reducing the effects of rheumatoid arthritis, and reducing the risk of heart disease. Also, there is evidence that omega-3 decreases the symptoms of PTSD and in conjunction with standard medications improves mood among those with depression. However, the 2008 review by Stahl and co-authors did not find a depression effect in 3 of the 7 studies. As reported by Pawels and Volterrani in 2008, omega-3 plays a major role in brain development and functioning; components such as docasahexaenoic and eicosapentaenoic acids provide essential fluidity to the cell membrane, thereby facilitating neuro-transmission and ion channel flow.

The healthiest foods are either fresh or frozen (without additives); generally avoid what comes in a can, a box, or a jar, since these things are often chock full of additives, most of which have never been tested. Nor do we know the consequences of combining various additives. Junk food and foods high in sugar will ultimately leave you feeling sluggish. Eating foods you are sensitive to will cause discomfort and drain energy.

Taking the time to plan a weekly menu and then listing what you need for the menus is a good strategy. Then purchase the nutritious foods to fit that menu. Resist the temptations of the marketplace to purchase what is not on your grocery list; aisles, colour, lighting, product positioning, packaging, etc., are all set up to make you want to put extra products into your cart. Like many, Lynne and I lead busy lives and do not have the energy at the end of the day to start preparation of a healthy meal. So, we got into the habit of cooking the week's dinners on Sunday mornings together (I am a great sous chef), refrigerating or freezing them, and then slipping them into the microwave at dinner time.

Also, although still debated vigorously, there is some reason to believe that the foods we eat do not contain sufficient levels of the nutrients we need; for example, foods often are harvested before they are fully saturated with nutrients. Moreover, there is some reason to believe that it is not possible to obtain sufficient levels of all the nutrients we need from the food we eat, especially if we suffer from medical/psychological conditions or are getting up there in age.

If you are unable to obtain the essential nutrients from the food you eat, consider a supplement containing the recommended daily dosages of essential vitamins and minerals. One recent study in the U.S. indicated that only 22% were getting at least 67% of the recommended daily amounts of nutrients while a UK study indicated that 50% were deficient in the B vitamins, which are critical for cell functioning, energy levels, and mood control. As near as I can tell, we require the following amounts of the B vitamins daily: B-1- 1.2 mg.; B-2 - 1.1 mg for women and 1.3 mg for men; B-3 - 14 mg for women and 16 mg for men; B-6 - 1.5 mg for women and 1.7 for men; B-9 - 400 mcg; and B-12 - 2.1 mcg. Extra levels of the B vitamins are mostly excreted in the urine but high dosages of B-3 or B-6 may cause side effects.

Personally, I have been persuaded of the need for supplements. After considering the various companies that produce supplements, I chose USANA (http://www.usana.com/dotCom/index.jsp) for a variety of reasons including the good range and dosages of nutrients they provide and the fact that they are delivered to my door monthly. A certified nutritionist I know who is associated with USANA can help assess your needs and get you started - Wanda DeCosta @ (250) 878-8235. I have no financial interest in this organization.

Some information now on the use of "natural" food supplements. The bottom line is that caution is indicated for a variety of reasons. *First*, their production is not subject to the same rigorous standards that are used for medications. So, from one batch to the next, you do not know how much, if any, of the alleged key ingredient is present. *Second*, you do not know what other ingredients have been added, nor their effect, nor the effect of their combination with the "active" ingredient. *Third*, many health foods can have serious negative effects when combined with prescribed medication, as found in the August 2009 health letter from the Mayo Clinic. For example, St John's wort will affect anti-depressants, anti-blood-clotting medications, some asthma medications, and some immune-suppressing medications or steroids. SAM-e can cause serious side effects if you are also taking SSRIs. Garlic, ginseng, ginger, and feverfew can increase risk of bleeding in people taking aspirin or blood-clotting drugs. Ginko can increase risk of bleeding if taken with blood-clotting medications, can counteract the effects of medications designed to lower blood pressure, and can increase the frequency of seizures among those taking anti-seizure medication. Kava can cause serious liver problems even with short-term use and even more so if you are already taking cholesterol-lowering drugs.

My Health Eating Plan Includes the Following:

_____

_____

_____

E-14: Increase/Maintain Aerobic Exercise
There is research indicating that among those with post-traumatic stress disorder, participation in physical activities declines significantly compared to levels before the traumatic experience(s). Lack of time and lack of motivation were the most common reasons given for the change. For some, physical activities did not play much of a role in their health plans before exposure to trauma. The elderly may be more reactive to stressors if the ability to move about is compromised among those who previously depended on physical exercise as a method of stress management.

However, there are good reasons to believe that maintaining or gradually bumping up your amount of aerobic exercise is helpful in decreasing the symptoms of any traumat-induced psychological condition. The 2008 review by Daley concluded that exercise is more effective than no treatment and often is as effective as traditional treatment in decreasing symptoms of depression. Others have reported that exercise is effective in maintaining a positive mood for up to 12 hours afterward. Also, there is support for the notion that exercise is especially important as we get older; both good fitness levels and significant education seem to protect against cognitive impairments in aging. While the criterion for fitness is exercising four days a week, for psychological purposes the objective is moderate or high intensity aerobic activity for at least 45 minutes a day, 6 days a week.

In addition, there is evidence that a program of regular aerobic exercise has many benefits, including increased energy and calmness, increased stamina, increased blood and oxygen flow to the brain (which increases the effectiveness of neurotransmitters such as serotonin), protection of memory structures in the brain, stimulation of the brain's ability to generate new cells and new connections between cells (which may reduce age-related memory decline), reduction of damage to brain cells from toxic substances, improvement in the regulation of caloric intake and weight, increased positive body image, decreased blood pressure, decreased risk of heart disease and stroke, decreased risk of colon and reproductive cancer, reduced risk of breast cancer by 26%, increased slow-wave restful sleep, a modest positive effect on sleep-disordered breathing, increased sexual pleasure, decreased chance of osteoporosis, decreased symptoms of irritable bowel syndrome, increased immune functions, decreased pain from fibromyalgia and other chronic pain conditions, decreased risk of diabetes, a heart that pumps with increased efficiency, increased ability to remain active as you age, improved balance and coordination, maintenance of clear arteries, increased tolerance to both cold and warm temperatures, reduction in muscle tension, increased effectiveness of many medications, lower levels of depression and a greater sense of well-being, improved muscle strength and tone, and longer life span.

An activity is aerobic if it produces a sustained increase in heart rate. Thus, it is a healthy choice to include aerobic exercise in your daily plan through such activities as dancing, power walking, cross-country skiing, rowing, in-line skating, taibo, biking, climbing/hiking, running, aerobic movement classes, and swimming.

Step-wise, the recommendations are to:

_____a. see your physician as required;
_____b. select one or two aerobic activities to do;
_____c. decide on the conditions; for example, do by myself, join a gym, buy some equipment;
_____d. decide on the length of time. Start small and then increase gradually to 45 minutes a day;
_____e. put the exercise schedule into your weekly planner;
_____f. ask friends and family to support you. Tell them how;
_____g. just do it whether you want to or not; and
_____h. change aerobic activities now and again. Doing so may help to keep you motivated.

I will increase aerobic activity by:

_____

_____

_____

E-15: Expose Yourself to Enough Sunlight
Spending thirty minutes early each morning in direct sunshine is part of a healthy lifestyle for those suffering from depression and can also boost spirits for those afflicted with an anxiety condition such as PTSD.

Firstly, sunlight stimulates the skin to synthesize vitamin D. Since few people get enough vitamin D from their diet, insufficient exposure to sunlight can lead to insufficient levels of vitamin D. There is some evidence to suggest that lower levels of vitamin D can result in weakened bones, arthritis, and some forms of cancer. Moreover, unlike supplements for vitamin D, you cannot overdose on the vitamin D that comes from sunlight. Also, during the summer months, the body stores vitamin D from sunlight, which compensates in part for the decreased sunlight in the winter months.

Secondly, most of us sense an increase in levels of energy and feel happier on sunny days than on cloudy, dreary ones.

Thirdly, there is much evidence that exposure to morning sunlight (with eyes open but not looking directly at the sun) has a therapeutic effect on folks with the form of depression known as SAD (seasonal affective disorder). Those with this condition who live in the northern hemisphere note that the symptoms of depression appear around November and disappear in late March. It now seems that this is so because during this period of time:

    a. the amount of sunlight is lessened;
    b. we are inside most of the time in buildings and houses where the light levels are far below that of natural sunlight (rather like the dwellings of the cave people of old); and
    c. we do not expose ourselves to outside sunlight even on the sunny days because of the cold temperatures.

Moreover, my own clinical experience is that many folks with the other forms of depression suffer an increase in symptoms during November to March, probably for the same reason. Sunlight exposure helps them, too.

Why sunlight has such a positive impact on us is not completely clear; theories to date with some support include:

    a. that the sun stimulates melatonin levels;
    b. that it sets circadian rhythm (including the sleep-wake cycle via the suprachiasmatic nucleus, a bundle of neurons situated in the hypothalamus of the brain);
    c. that it increases the levels or functioning of some neurotransmitters; and
    d. that the photons make the rods in our eyes better able to gather what light there is.

Regardless of the specific mechanism, it is clear that (a) those suffering from SAD or other forms of depression, (b) those folks struggling with an anxiety condition, and (c) those facing stressors that are overwhelming all experience a decrease in symptoms after increasing the amount of sunlight entering their eyeballs during the winter months. This effect is maximized when it is part of an overall treatment plan. The trick then is to develop as many strategies as you can for increasing the amount of light entering the eyeballs.

One way is to win the lottery and become a sunbird, escaping to places like southern California, Arizona, Florida, or Hawaii for the winter months. I suspect that getting the sunlight, rather than simply escaping the snow and cold, is the biological reason that the snowbirds travel.

Secondly, if you have worked long enough, you could retire or relocate to a sunnier climate.

A third way, perhaps more possible for most of us, is to take winter vacations in southern, sunny areas instead of taking time off during the summer months. With careful planning, which includes searching the web for condos, the vacation does not have to cost an arm and a leg, although it is still a considerable expense.

A fourth way is to take advantage of whatever natural early morning sunlight there is. For example,
*bundle up and go for a 30-minute walk on those morning when the sun is out;
*sit for 30 minutes near the window when the early morning sun streams through;
*go for a drive when the sun is out; and
*get above the clouds and spend time just sitting or skiing down the slopes.

A fifth way is to increase indoor light levels. Among the things you can do are the following:
a. trimming hedges or branches near windows that block natural light from entering;
b. constructing skylights or sun rooms;
c. painting the interior of your home in light/white colours;
d. replacing dark wood panelling; and
e. choosing white or off-white carpeting or highly-reflective wood flooring.

A sixth way is to use a SAD light since there is lots of good evidence that doing so is effective for those with SAD and I suspect will cheer up almost anyone. These lights are known also as light boxes, bright light therapy boxes, and phototherapy boxes (you will need to insure that the device is designed specifically for treating SAD). Moreover, there is now scientific evidence as well as clinical impression that this approach, which goes by the name of "light therapy," helps most folks regardless of their type of depression. A recent review of the research literature on folks with non-seasonal types of depression found that light therapy resulted in symptom improvement in the 12% to 35% range. The positive effects are noted as early as 2 to 4 days after starting and the degree of improvement continues over the next several weeks. Also, studies of women with severe pre-menstrual symptoms indicate significant reductions in depression, irritability, and physical symptoms with light therapy.

The original studies called for 2 to 3 hours of daily exposure to a light that was about 2500 lux (a measure of light intensity). More recent studies suggest that sitting about 1 to 3 feet away from a light at 10,000 lux for 30 to 90 minutes is as effective (an interesting bit of trivia is that 10,000 lux is the amount of light coming off the sky at sunrise as compared with 100,000 lux at noon). Some have found that increasing the exposure time is helpful as the winter deepens and that the effect is maintained when decreasing exposure time when the days begin to get longer and brighter.

The device you choose should filter out harmful ultraviolet light to avoid eye and skin damage. So it is best to look for a unit that produces as few UV rays as possible at high light intensity or a unit that shields carefully any UV light it produces. The light should be in one's peripheral vision so it is not necessary to look directly at the source. Thus, one could read or eat at the same time as being exposed to the light. Although other formats are available including attachments to hats, the best evidence of effectiveness has been with box-like units.

The device should be positioned so that the light comes from above your line of sight, not directly at it. In using the device, keep your eyes open and without sunglasses. Doing this at the same time in the early morning (about the same time you normally wake during the summer months) seems to work well for most people, although some report better results by doing it in the evening. Some studies suggest that exposure both in the morning and in the evening may add to the effectiveness of the procedure while other studies have not found this to be the case. Starting the procedure by mid-October and continuing it to the end of March seems to work best.

If the recommendations above are followed, light therapy is generally well-tolerated and when side effects do occur, they tend to be mild. As noted in the recent book *Winter Blues*, the most reported (although still rare) side effects are headache, eye strain, irritability or anxiety, over activity, insomnia, fatigue, dryness of the eyes, dryness of the nasal passages, and sunburn-type reaction of the skin. Headaches and eye strain can usually be managed by decreasing the duration of exposure, irritability by decreasing the duration of exposure or increasing the distance from the light, insomnia by switching to morning exposure, dryness by artificial tears or use of a humidifier, and reddening of the skin by use of sunblock creams. However, the development of any significant side effects should be a message to either alter the procedure or to stop it altogether. Folks with vision difficulties other than needing corrective lenses or those with a history of skin cancer should consult their specialists before starting light therapy. There is no evidence that light therapy is unsafe during pregnancy but it probably should not be used while nursing since the bright light may not be safe for baby's eyes and might alter her/his daily rhythm. Since it works on a different system than vision, light therapy is helpful also for those who are blind.

Since I have many of the features of SAD (low motivation and energy, low mood, irritability, carbohydrate cravings), I have found that following my own advice in regards healthy living as well as using a combination of the methods in this paper helps me through the November to March period.
1. I use a SAD light at 10,000 lux for about 1 hour early each morning while working at my desk.

2. During other times of the day when it is sunny out, I go for a brief walk outside or sit for a while in my sun room. This is a closed in sundeck with lots of windows facing north, east, and south.
3. My office has a large window facing south and without obstructions. I let the sun in most of the time I am at my desk.
4. Over the years, Lynne and I have lightened up the house following the suggestions noted earlier in this paper.
5. I escape south for 10 days or so in late November and again in February and cut back on vacations during summer months when there is lots of sun in my part of the world. My experience is that my mood and energy improve significantly while I am away. With the addition of other components such as a SAD light when I am back home, the positive effect of winter vacations lasts pretty well until the next trip. Doing this represents a healthy choice for me.

Often I have a supply of light therapy devices that I sell at the actual cost to me. Often this is less than you can find at a store, since for every four units I purchase from the manufacturer in Alberta, I receive a fifth free of charge.

My plan for increasing exposure to sunlight is:

_____

_____

_____

E-16: Relax Your Large Muscles

Some people retain tension in their muscles and joints. They may experience unusual stiffness, movement difficulty, headache, other pain, or sleep problems. Or they may adopt postural positions that might be considered either defensive or in preparation for attacking; if maintained for long enough, muscle tension and fatigue can result with the additional possibility of lessened energy levels. Also, tension seems to increase other symptoms of PTSD, likely due to the activation of the sympathetic nervous system and the reticular formation in the brain.

Research evidence indicates that there are many successful ways to reduce muscle tension. Some of the options are described below.

One is to practice the movement aspects of hatha yoga, which one can learn by taking an instructional class or purchasing a video/DVD. A study reported in the 2007 book *Principles and Practice of Stress Management* edited by Lehrer and colleagues compared aerobic swimming with low-intensity yoga and found greater reductions in tension, fatigue, and anger among those who practiced yoga. In the same volume was a summary of many studies, which concluded:

Studies have shown that both short-term and long-term practice of yoga techniques are associated with a number of physiological and psychological changes including reductions of basal cortisol and catecholamine secretion, a decrease in sympathetic activity with a corresponding increase in para-sympathetic activity, reductions in metabolic rate and oxygen consumption, salutary effects on cognitive activity and cerebral neurophysiology, and improved neuromuscular and respiratory function. (p. 450)

Hatha yoga involves the use of many body postures called asanas, which are held for various lengths of time. The asanas create a dynamic tension between muscular energy, lengthening, and relaxing. This leads also to gains in flexibility. Studies have demonstrated that yoga is effective in reducing blood pressure among those with hypertension, in increasing lung capacity, in both arresting the progression of and reversing coronary artery disease, in improving bone density, in decreasing headache including the migraine type, in reducing stress and a number of stress symptoms (such as anxiety, depression, and insomnia), in lowering both the frequency and severity of PTSD symptoms, in elevating one's sense of control over life, and in increasing strength and flexibility. In addition to gentle movement, both yoga and tai chi involve breath retraining, which could account for some of the positive results. Later on significant attention will be devoted to what I have termed *tactical breathing*.

Two is to do the same as above with tai chi, which combines rhythmic movements, breathing techniques, and focussed attention. Tai chi also addresses core stability, flexibility, and balance. Positive results similar to those with yoga noted above have been demonstrated.

Three is to obtain a physician's referral for massage therapy. Sessions with a registered massage therapist are known to decrease muscle tension, increase blood flow, and decrease emotional difficulties.

Four is to soak in hot water (with or without bubbles) as often as needed. A hot tub at home or elsewhere, a soaker tub at home, and planning vacations to hit as many hot springs as you can are among the ways to do this. So that it does not create sleep problems, a common recommendation is to exit the hot tub at least 2 hours before bedtime.

Five is to learn stretching exercises such as those used n the method known as progressive muscle relaxation (PMR). These can be done for many skeletal muscles or for a select few. Over the years, I have learned and experimented with ways to decrease tension in forehead, face, jaw, neck, shoulders, back, abdomen, arms and legs, hands, fingers, ankles, and toes. Performing the stretches at least once a day for life as well as when noticing any unnecessary increase in muscle tension serves also to decrease sensory messages to the amygdala, which will prevent it from over-reacting as it does among those with PTSD. More about the amygdala will be discussed later in this manual. Research indicates that PMR decreases levels of circulating norepinephrine levels, myocardial contractilty, electrodermal activity, heart rate, and heart reactivity. A review of well-conducted studies as reported in the 2007 book edited by Lehrer and associates summarized some of the positive effects of PMR as

. . . improvement in the quality of life for cancer patients; reduction of symptoms of posttraumatic stress disorder; increased sleep quality; reduced stress and anxiety in persons with generalized anxiety disorder, women with urinary dysfunction, and people without dysfunctions; reduced blood pressure in persons with hypertension; and reduction in headache symptoms. (p. 639)

Six is to begin a graduated program of resistance training (weight lifting) with the advice of someone knowledgeable in the area. The objective is not to build gigantic muscles (although you can do that too), but to work out the kinks and keep the muscles functional. In addition, the evidence is that resistance training leads to a decrease in anxiety, depression, and hostility.

Seven is to become familiar with a method known as autogenic training (AT), which the originator, Schultz, described as a self-hypnotic procedure. In one variant of AT, the trainee concentrates on bodily sensations in a calm and passive way. Then, going through one muscle system at a time, you can implement the following as found in the 2007 book *Principles and Practice of Stress Management* by Lehrer and co-editors):
the left for arm is heavy (repeated slowly 6 times);
I am very quiet;
the left arm is pleasantly warm (repeated slowly 6 times);
I am very quiet; and
the heart is beating calmly and regularly (repeated slowly 6 times).

Then you concentrate on another body part. This is repeated until all body parts have been involved in this process. Depending on results, over weeks or months, the process can be done on several body parts at a time; for example, both arms. Then other aspects can be added. Studies have documented significant results from AT including reduced heart rate, reduced cardiac output, improved utilization of carbon dioxide, stability of the ECG, reduced breathing rates, normalization of stomach and intestinal functioning, improved blood flow, and positive changes in the EEG.

Five or six times a week, my pattern is to do PMR-type stretching exercises throughout my 30-minute jog and follow this with about 15 minutes of resistance training. This leaves me feeling great.

Eight is the movement or exercise element within the tradition known as health qigong. The 2009 research review by Ng and Tsang reported a range of positive effects including
increasing the numbers of white blood cells and lymphocytes, stroke volume, peak early transmitral filling velocity, peak late transmitral filling velocity, forced vital capacity, and forced expiratory volume, and, conversely, lowering of total cholesterol, systolic blood pressure, diastolic blood pressure, and depressive mood scores. (p. 261)

A sensible way to proceed involves the following:

a. Learn about one or more of the options above that appeal to you. Pick one you can do anywhere;

b. Obtain the support of family; some clients have made this a family project;

c. Start small and increase gradually;

d. Schedule in your planner regular times you will engage in the muscle relaxation option you chose;

e. Just do it;

f. Also, do the relevant exercises as soon as you become aware of muscle tension;

g. Be aware of how the option affects you. My atrial fibrillation is triggered if I use a hot tub;

h. If you tire of any option, then try another one.

I will relax my large muscles and joints by:

_____

_____

_____

E-17: Obtain Restful Sleep

Most of us have no idea what happens during the roughly one-third of our lives that we spend in sleep. This is true in spite of the considerable amount of research-based information that is now available on such topics as why we need sleep, what happens to us during sleep, what we need to do in order to obtain restful and refreshing sleep, what the common sleep disorders are (there are 78 of them), how sleep disorders can be treated effectively, and what the relationship is between sleep and psychological health. Moreover, although up to 60% of patients seeing their physicians have sleep problems, only half of them raise the subject.

Sleep problems are common among those who have been exposed to trauma and occur at least twice as often as among those who have not experienced trauma. As discussed by Kendall-Tackett (2009), problems in sleeping occur in as many as 68% of those who are sexual abuse survivors. A study she quoted determined that 80% of sexual assault victims had either sleep-disordered breathing or sleep-movement disorders.

Take the necessary steps to get adequate amounts of the restful sleep you need. From 7 to 9 hours of sleep per 24 hours is the average healthy range.

However, for the first 7-10 days after a single trauma, some may require more than this and acquiescing to the demand for more sleep is recommended. Thereafter, limiting sleep to the healthy range should be the objective.

Following trauma, others may experience a significant loss of restful sleep. For example, they may have significant difficulties falling or staying asleep, they may be in a state of anxiety/hyper-arousal, they may find themselves being very watchful and over-sensitive to any sound at night, they may have intrusive thoughts about the trauma, they may have or wake from distressing dreams, or they may experience panic attacks while asleep.

Unfortunately, multiple nights of insufficient restful sleep for any reason lead to negative consequences including decreased daytime alertness/concentration, increased daytime fatigue, slowed reaction time, decreased stamina, increased stress/tension/anxiety, increased moodiness, more episodes of anger/irritability/impatience, health problems (high blood pressure, coronary heart disease, higher chance of stroke, and increased rate of diabetes), greater susceptibility to infections via decreased immunological resistance, inadequate cell repair and wound healing, more sensitivity to pain (including fibromyalgia), and impairment in cognitive activity (such as memory and decision-making). The consequence of some of these is increased daytime accident rates. As well, they contribute to a shorter life span compared to those who enjoy sufficient restful sleep.

The sleep hygiene suggestions below are where we start. Place a check mark beside each as you implement it.

____a. do regular daytime aerobic exercise ending no less that 2 hours before bedtime;

____b. do gentle stretching before bedtime such as yoga or tai-chi;

____c. avoid alcohol and caffeine altogether or at least in the 6-10 hours before bedtime;

____d. limit use of the bedroom to sleep and sexual intimacy (not problem solving or TV time);

____e. have a relaxing pre-bedtime routine;

____f. go to bed at the same time each night;

____g. buy a comfortable bed if yours is not;

____h. keep the bedroom cool, comfortable, quiet, dark, and humidified;

____i. get up until you feel sleepy again if you don't fall asleep in 15 minutes;

____j. get up at the same time each morning regardless of the amount of sleep you had the night before;

____k. do not nap for more than 30-45 minutes during the day.

I recommend you begin with the suggestions above. Once all of the above are well in place, if sleep is still problematic, then later on we can implement a number of additional strategies.

E-18: Laugh More

A regular dosage of laughter truly is good medicine, especially for any psychological condition. Research has demonstrated that laughter increases blood circulation, decreases the likelihood of a second heart attack, lowers blood sugar in those with type-2 diabetes, works the abdominal muscles, decreases blood pressure, decreases arterial inflammation, increases levels of good cholesterol (HDL), enhances interpersonal relations, reduces feelings of anger and irritability, relieves fear, enhances our immune system, lowers the flow of dangerous stress hormones, burns off calories (40 calories per 15 minutes), lowers the sensations of pain by relaxing muscles, and leaves us feeling good.

Therefore, I recommend that you find many daily opportunities for laughter.
  a. seek out friends who have a funny side to them;
  b. watch comedies on TV rather than violence or the news;
  c. enjoy joke sites on the internet (try www.dogpile.com and key in "jokes" for access to hundreds of hyperlinks from the tame to the gross). I encourage people to go to a joke site for a laugh when things are not going well in the same way that one would take an aspirin for a minor pain;
  d. select books or magazines with humour in them;
  e. if you read a newspaper, go to the comic section first;
  f. encourage your friends to send you jokes via e-mail and then forward them to others;
  g. select humourous offerings at the video store. Some of the old videos that I watch again when I need a laugh or a smile are below
      Fawlty Towers, A Fish Called Wanda, Grumpy Old Men, The Survivors, Hopscotch, Mrs. Doubtfire, Once Upon a Crime, Tootsie, Victor Victoria, Dirty Rotten Scoundrels, What's Up, Doc?, The Inlaws, Real Genius, Gotcha, Groundhog Day, Without a Clue, Nine to Five, any of the *Pink Panther* series, *Foul Play, Seems Like Old Times, The Man With One Red Shoe, Blame It on the Bell Boy, Miss Congeniality, Meet the Parents,* and reruns of *Archie Bunker, Mary Tyler Moore,* and *Seinfeld.*

I will increase laughter in my life by:

_____

_____

_____

E-19: Smile More
This one will make you scratch your head a little in disbelief. But it is true and it does work: *several times each day, smile even if you do not feel like it.*
a. After washing your hands, look into the mirror and smile.
b. Smile at the people coming towards you on the street and in your mind wish them happiness.
c. Do the same when you are driving.
d. Go through your photo album and smile back at the happy scenes.
e. Look at baby pictures and wedding photos in the newspaper and smile.
f. Listen to music that puts a bounce in your step and a smile on your face.

Why would you do this? Well, smiling stimulates the sensory and motor nerves related to our face. These are connected to a system that regulates such things as our heart, lungs, and gastro-intestinal tract. By the simple act of smiling (even a fake one), we promote calm states in our nervous system. So, we can use facial muscles to calm down. This might help use to understand why putting on a happy face can actually help us to feel better.

I am going to increase smiling by:

_____

_____

_____

E-20: Increase Positive Daily Activities

First, make a list of the things you have enjoyed doing in the past that you could do again.

_____

_____

_____

_____

_____

_____

Second, select a couple that you want to do again (or some new ones) and plan how you could do them.

_____

_____

Third, schedule some time in your day for the activities you choose.

_____

_____

_____

Fourth, remember why you are doing this. If nothing else, doing enjoyable activities is distracting. Also, this can give you a sense of control/accomplishment; few of us enjoy activities we do not do well enough to satisfy us. Therefore, we can experience positive feedback when we do what we enjoy. Moreover, there is evidence that one reason why this works is that increasing time spent engaged in activities you enjoy leads to the release of the neuro-transmitter dopamine in an area of the brain known as the nucleus accumbens; the release of dopamine is associated with feeling good. Besides, engaging in pleasant activities is the spice of life.

I will make time to enjoy life. It is a healthy choice. I will facilitate my healing.

E-21: Spend Quality Time with Your Partner

Spend quality time with your partner doing things you both have enjoyed in the past or try new possibilities. Time spent with your partner or a close friend has the capacity to reduce fear and both biological and psychological over-reactivity in the face of stressors. Depending on who it is, invite her/him on a date, play Scrabble, sign up for dancing lessons, go for a drive, put the tent up in your back yard and sleep there overnight, look over photo albums together, go to a funny movie, make lunch together, go out for fries in the middle of the afternoon, put a love message on a sticky note and put in the drawer of underwear, paint a room together, park the car on a new street and then walk the street together, go the park and sit on swings, get up at 3 a.m. and go to a place where you can see the stars clearly, be standing in the nude when s/he comes home from work, build a snowperson . . . the possibilities are limited only by your imagination.

First, write options in the space below:

_____

_____

_____

Second, with your partner, select a few to do each week:

_____

_____

_____

Third, put them into your respective weekly planner:

_____

_____

_____

Then just do it.

Something else to ponder. As needed, try to clarify your feelings and assumptions about your partner, and then check out to see if they are accurate or not. Many problems can be avoided by not jumping to conclusions or assuming that you know what the other is thinking or feeling. Remember that generally men and women react differently. Women tend to be caretakers and to put others first. Men tend to have more difficulty acknowledging and expressing feelings of sadness and helplessness and believe in "toughing it out." We all like our partners to say and do things that show they value and care for us–make an effort to do this for your partner. (Taken from the VA Canada brochure on PTSD).

E-22: Devote Quality Time to Your Children
Spend quality time with your children enjoying their activities with them.

Depending on their age(s), read stories to them, wrestle, build with their blocks, play board games, throw the ball, take them to their activities and stay to watch, go camping, sleep out in the front room with a fire in the fireplace, jump with them into piles of leaves, surf the net together, select together a video to watch, play a video game with them, play paint ball together, dance with them, play "I spy." These and other activities build good relationships and help to reassure them that you are okay. Also, they are the things they will remember when they grow up and also the kinds of things they will do with their children.

First, list with them activities you both would enjoy:

_____

_____

_____

Second, choose a few together from the list above:

_____

_____

_____

Third, write the chosen activities into your weekly planner:

_____

_____

_____

Remember:
1. Talk to your children; share interesting or funny things that have happened during the day.
2. Try to be supportive and patient, Obviously, this is not always easy, but getting angry and losing control only makes things worse for all (their reactions and your guilt). Set an example by expressing your concerns gently, controlling your anger, and showing them the skills that lead to solving problems.
3. Most likely your children will be your children for the rest of your life. And as one pastor who worked with the dying said, "I have never heard a dying person say that they ought to have spent more time at the office."

The things you spend time doing with your children are what your children will remember when they are no longer children and when you pass on.

E-23: Do More Touching
Most of us do better when we have a sense of connection with significant others. Appropriate touching is a physical indicator of connection with others. And such physical contact is good for us humans. For example, many studies have demonstrated that babies who are not touched or held enough do not do as well as those who are held and touched. Other investigations have concluded that touch is a biological need.

So, touch your partner frequently. Snuggle while watching TV. Hold hands while walking. Give each other foot massages. Sneak up behind and give a hug. Make love with your partner; it is okay even if the traumatic event involved serious injury or death to another.

Touch your kids too. Stroke their backs. Wrestle with them. Hug them each night before they go to bed. Put your arm around them while you watch TV. Tickle them.

Time with pets is helpful too. Studies have shown that people with pets had lower resting levels of blood pressure and heart rate compared to those without pets. Also, in comparison, those with pets had less of an increase in such variables when engaged in stressful tasks.

Consider getting a massage regularly; research has demonstrated consistently that doing so decreases the buildup of lactic acid, dilates blood vessels, stimulates blood circulation, soothes aching muscles, and reduces stress and tension.

Also, there is some interesting research about why touch works. Two areas in the hypothalamus of our brain can secrete the neuropeptide oxytocin. This substance plays a central role in reproduction, birth, and lactation. As importantly, when released, it reduces stress hormones in our bodies and decreases the reactivity of our autonomic nervous system (such as reducing heart rate and blood pressure), thereby buffering the anxiety we experience in relation to stressful life experiences. This helpful neuropeptide is released by social stimuli, especially touching.

I will Increase physical contact with loved ones by:

_____

_____

_____

E-24: Increase Social Contact
It is normal to withdraw from others in the early stages of PTSD. However, doing so for long periods will not help in decreasing symptoms. So, direct some of your energy to really connect with those relatives and friends who are *positive, supportive, and have their own lives in order*. Because of your motivation, energy levels, attention, and emotional reactivity, you may want to start with one or two regular contacts and then increase them.

Social contact with and support from positive people have been demonstrated to reduce symptoms of both post-traumatic stress disorder and depression, decrease hyper-arousal of the autonomic nervous system (for example, decrease heart rate and blood pressure as well as minimize the normal blood pressure increase that is associated with aging), decrease thoughts of suicide, increase health in part by improved immune functioning (for example, by promoting higher natural killer cell activity), increase ability to cope with stress, decrease other physical symptoms, and increase the life span (for example, by decreasing mortality associated with cardiovascular dysfunction). Other studies comparing lonely versus connected people found chronic loneliness to be associated with an increase in the stress hormone cortisol, an increase in the activation of the hypothalamic-pituitary-adrenal (HPA) system, and a decrease in immune functions. A 2009 study by Cluver's team found that AIDS-orphaned children in South Africa with high social support had fewer symptoms of PTSD after both low and high levels of trauma exposure than those without such support. In their 2008 study comparing war-traumatized women in the former Yugoslavia with those less affected, Klaric and co-investigators reported that low level of perceived social support was a significant predictor of PTSD.

Additionally, being with trusted others can provide information that is helpful, assistance that is tangible, emotional support, distraction, constructive feedback/advice, and the sense that you are not in this alone. For colleagues who do not know what to do or say, encourage them to obtain a paper I wrote on *supporting colleagues after trauma;* it is written for them.

As best you can, avoid unsupportive, negative, crabby, whiney, critical, and unhappy folks, since they will drag you down. Also, studies show that contact with hostile or non-supportive others in times of stress can increase stress responses.

If you are not sure how to determine who is supportive and who is not, complete the social support scale at the end of this section. I modified it somewhat from a 1992 publication by Ray.

Also, because of your symptoms, people sometimes withdraw from those with PTSD. So it is probably best to wait until symptoms of anger and agitation are under control before extending your social contacts.

I will increase my contact with the following by:

_____

_____

_____

Social Support Scale

Beside each statement below, place a 1 for never, a 2 for almost never, a 3 for sometimes, a 4 for quite often, a 5 for very often, or a 6 for always. By completing this for a number of people you know, you will soon see who you need to avoid and who you need to spend time with.

_____01. Can you lean on and turn to them when things are difficult?
_____02. Can you get a good feeling about yourself from them?
_____03. Do they put pressure on you to do things that you cannot do at this time?
_____04. Do they take over your chores when you feel ill?
_____05. Do they express concern about how you are?
_____06. Do they misunderstand the way you feel and think about things?
_____07. Can you trust them, talk frankly, and share your feelings with them?
_____08. Can you get practical help from them?
_____09. Do they argue with you about things far too often?
_____10. Do you feel they are there when you need them?
_____11. Do they press you to say that you feel better when you do not?
_____12. Do they listen when you want to confide about things that are important to you?
_____13. Do they express irritation with you in ways that are not helpful?
_____14. Do they accept you as you are, including your failings as well as your Strong points?
_____15. Do they help out when things need to be done?

_____16. Do they show you affection?
_____17. Do they make helpful suggestions about what you could do without pressuring
        you?
_____18. Are they critical of the way you respond to illness?
_____19. On important matters, do they do things that conflict with your values?
_____20. Do they give you useful advice when you ask for it?
_____21. Do they express frustration with you in a manner that is not at all
        helpful?
_____22. Do they treat you with respect?
_____23. Do they often disagree with you about what is best for you?

If you want a box score, subtract the sum of your ratings for 3, 6, 9, 11, 13, 18, and 21 from the sum of the rest. The higher the remaining score, the more supportive the person is. Choose people with higher final scores and decrease time spent with the others.

E-25: Accept or Solicit Help from Others
Let trusted relatives and friends know how they can help. They probably want to be a support to you but they likely do not know what to do and are afraid they will do or say something that will make things worse for you. So you will need to tell them directly. There is nothing wrong about asking for help. Besides, you likely would help friends in distress, so why not accept their help at this time in your life?

If I need help with the tasks below, I will ask:

_____

_____

_____

E-26: Make Home a Place of Calmness and Contentment

When your energy begins to return, consider how to make your home a place that gives you as great a sense of calmness and contentment as possible. Often we get caught up in making our homes look and feel as others think they should rather than doing what is good for us. So pay more attention to what works for you. Begin with small steps.

For example
a. paint walls with soothing colours that also brighten the house by reflecting outside light;
b. hang pictures that give you a good feeling;

c. increase the availability of natural light within;

d. gradually change to more comfortable furniture;

e. play only that music which is calming or uplifting (such music also decreases blood pressure and heart rate);

f. add plants or cut flowers;

g. light aroma-producing candles and other items that contribute to calmness; and

h. create a special corner where you can sit quietly without interruption to practice meditation.

I will increase my sense of calmness and contentment in my home by:

_____

_____

_____

E-27: Consider Spiritual Practices

Maintain spiritual practices or, if you have stopped because of lack of interest or because other things crowded them out, consider returning to your spiritual roots. Studies have shown that spiritual practices at the very least are a useful buffer against the effects of stress. Also, one interesting study found that praying for others led to improvements in well-being in the person offering the prayers.

Alternatively, spend time on a regular basis doing those things that bring calmness, contentment, and meaning to your life. These can include gardening, gazing at the stars, walking through the woods, sitting meditatively in a quiet place, looking at snow-capped mountains, or listening to the waves of the ocean or the sounds of a running brook.

Also, studies have shown that listening to the right kind of music can be soothing. Many people find that slow, melodic music with fewer rather than more notes per minute works best. For me, that means listening to composers such as Mozart and his contemporaries. The opposite type of music has been shown to lead to internal discomfort associated with such things as increased heart rate and blood pressure.

Whatever your choice, it is important to put it into your weekly schedule. If you do not, it will get crowded out.

Consider repeating my version of the Serenity Prayer:

I will learn to change what is in my control.
I will learn to cope with what is not in my control.
I will develop the wisdom to know the difference.

Repeating this modified version on at least a daily basis will help it to get into your head and become part of how you live.

I will bring calmness, contentment, and peacefulness into my life by:

_____

_____

_____

E-28: Notice and Remember Blessings

When things are going badly, it is easy to see and remember the negative parts of life while ignoring the many positives that surround you. The end result is a view that is out of balance as well as an increase in feelings of depression. Thankfully, there is much you can do to reverse this one-sidedness.

First, at the end of the day (or better yet, as they happen), write in your binder three good things (big or small) that happened during that day. Encourage other members of your family to do the same and then share these with each other. Review by yourself once a week what you have written during that week. Also, review your writings when you are feeling down. Most of us are surrounded by blessings that we take for granted and rarely consider. Personally, I am constantly amazed by being able to hear (music, ocean sounds, the pitter-patter of falling rain, children laughing), being able to see (mountains, the moon above, streams flowing from glaciers), enjoying restful and refreshing sleep, appreciating a system that turns food into energy that makes my body go without any effort on my part, eliminating what my body does not require again without my conscious effort, being able to think and remember, sunshine on my body, the smell of flowers, being able to walk, medicines that heal illnesses, the fact that the earth is able to produce food for me, friendship, children, grandchildren, the experience of love, pleasant dreams, clothing, shelter, the vastness of the universe, telescopes, microscopes, houses, cars, the opportunity to learn from others, the growth of wisdom (which was a long time coming for me), and the blessing of simply being alive. Add to this list some of your own blessings.

Second, go for a 15-minute walk a couple times a week and notice all the things around you that make life a pleasant experience.

Third, once a week remember and record two life situations in which you were grateful for the help, love, and companionship of others. Consider writing a brief note to one of them expressing thanks for them being there; you can send the note or not; up to you.

Fourth, write a one-page positive description of yourself, noting your strengths and accomplishments as if it were to be presented by a friend introducing you to a group. If you have trouble doing this task, ask others (such as your partner) for help. But do not disagree with their assessment of your positive qualities. Read this introduction once a week.

Treatment Step F: Components of Applied Psycho-Physiology
Simply put, psycho-physiology involves the relationship between environmental or cognitive events and our physiological responses. All my clients complete a psycho-physiological assessment which typically indicates dysregulation of the the autonomic nervous system (ANS). Variables of interest include blood pressure, heart rate, heart rate variation, blood oxygen levels, circulation, sweat gland activity, and aspects of respiration. In addition, levels of muscle tension are monitored. In the format I employ, applied or clinical psycho-physiology has the objective of changing undesirable baseline ANS activity (hyper-arousal) as well as reactivity to reminders of the trauma (hyper-reactivity). Similarly, an objective can be to decrease muscle tension. To accomplish such goals, I clients utilize a number of strategies as noted in the sub-sections below.

Contents

F-01: Optimal Regulation of the Autonomic Nervous System (ANS)
During a perceived threat to one's safety, a number of biological *systems are activated automatically* with the overall purpose of responding appropriately. These systems function without conscious awareness or control and evolved over tens of thousands of years, allowing our ancestors to survive in a physically dangerous environment. Those with biological processes allowing them to perceive and then respond appropriately to life threats survived, while those who did not react in time died. Those who lived passed on their genes while those who succumbed did not.

One of the biological systems involved is the HPA (hypothalamic-pituitary-adrenal) axis. Among the components of this system that is activated during stress or threat is the hypothalamus, which releases the neurohormones CRH and AVP. The hypothalamus is a key brain area and is known to be more reactive among those with a history of trauma. These neurohormones from the hypothalamus are transported to the pituitary gland, which releases chemicals such as corticotropin and endorphin. These chemicals travel to the adrenal glands atop the kidneys, from which cortisol is released, which activates the autonomic nervous system (ANS). Simultaneously, the adrenal medulla secretes epinephrine and norepinephrine, which reinforces activation of the SNS portion of the ANS, especially to the cardiovascular system. Another brain area, the hippocampus, has the capacity to inhibit many aspects of the HPA. Additional processes are involved and are discussed later on.

One easily measured result of the threat reactions is *hyper-arousal and/or hyper-reactivity of the autonomic nervous system (ANS)*, which is associated also with the elevation and loss of normal modulation of the brain chemical norephinephrine. The ANS governs such internal functions as heart rate, blood pressure, circulation, breathing, heart rate variation, and digestion. Alterations in such functions in the absence of a real emergency is one of the most consistent findings in studies of both PTSD and depression and this dysregulation of the ANS is typical of other anxiety disorders as well. In the normal course of events, once a threat is over, studies indicate that the ANS returns gradually to baseline levels. However, among many with PTSD, the research evidence and clinical experience is that the ANS does not return to baseline levels after the threat has passed; rather, it remains elevated (acting instead as if the emergency were continuing) and/or is hyper-reactive. Moreover, the baseline levels and/or hyper-reactivity often increase further with each successive trauma, and often increase further yet with each exposure to reminders of previous trauma. This occurs because ultimately either the sympathetic branch of the ANS becomes more activated, leading to increased arousal, and/or the calming parasympathetic branch of the ANS becomes less activated, leading to less inhibition of arousal, and/or the enteric branch governing the organs of digestion becomes dysregulated.

Dr. Pole (2006) reviewed over 100 studies (mostly with combat veterans) and found significant differences between those with and those without PTSD (a) in resting heart rate, skin conductance, and both systolic and diastolic blood pressure, (b) in startle studies of heart rate, and habituation slopes of skin conductance, and (c) in response to personal trauma cues in heart rate, skin conductance, and diastolic blood pressure. Such results are consistent with the view that psycho-physiology is affected by PTSD.

In their review chapter in the 2007 book *Handbook of PTSD: Science and Practice*, Southwick and his group conclude that (a) the most consistent effect has been demonstrated for accelerated heart rate in individuals with PTSD during presentation of trauma-related sounds, videos, and imagery, (b) pooling subjects from all published studies, approximately two-thirds of subjects with PTSD have demonstrated exaggerated psycho-physiological reactivity to trauma-related cues, (c) the percent appears to be even higher in subjects with severe PTSD, and (d) most studies have found that subjects with PTSD do not experience exaggerated physiological reactivity in response to generic, non-trauma-related stimuli.

In their chapter in the 2007 *Handbook of PTSD: Science and Practice*, Dr. Terence Keane (a well-known and respected expert in PTSD related to combat veterans) and his colleagues wrote that

> studies that have compared psycho-physiological reactivity of individuals with and without PTSD to trauma-related stimuli have consistently shown that groups of individuals with the disorder exhibit greater mean levels of reactivity than do trauma-exposed controls. These effects have been observed in populations ranging from combat veterans of the Vietnam

War, Korean Conflict and World War II, to survivors of childhood sexual abuse. (p. 292)

In citing various studies, Payne and Gevirtz (2009) wrote:

Increased trauma-specific physiologic reactivity has been associated with poor outcomes since at least WWI. For example, WWI soldiers who demonstrated increased physiological reactivity to combat reminders were unable to return to the combat zone…World War II soldiers with increased psychophysiologic reactivity had worse mental health outcomes…More recent studies found that physiological reactivity…to audiovisual cues of combat experiences discriminated between combat veterans with PTSD, veterans without PTSD, and veterans with other psychiatric disorders…Even 20 years after combat exposure, physiologic reactivity measures…correctly classified two thirds of veterans with and without PTSD…(p. 20)

In my clinical work with police or military folk with PTSD since 1990, pretty much all of them show signs of dysregulation of the ANS with elevated heart rate, decreased heart rate variation, decreased finger temperature, increased skin conductance, and respiration anomalies usually during both baseline and when recalling trauma. Even if the traumatic incident(s) occurred decades ago, the nervous system remembers trauma and reacts to reminders of it. This is consistent with the findings reported above.

Regardless of the mechanism and in addition to the effect on symptoms of PTSD, multiple studies have demonstrated that with prolonged ANS over-arousal and/or hyper-reactivity comes an increase in the likelihood of serious health consequences including high blood pressure, heart disease, irritable bowel syndrome and other gastro-intestinal conditions, diabetes, skin problems, illness associated with suppressed immune functions, slower wound healing and slower recovery from surgery, increased vulnerability to viral infections as the immune system becomes impaired, diabetes, asthma, osteoporosis, headache, and other musculo-skeletal pain. Also, since the ANS is a multi-component system, sometimes the activity of just one aspect can be associated with health issues. For example, anomalies with TB-2 discussed below are associated with irritable bowel syndrome (IBS) and there is research to demonstrate that righting TB-2 can lead to a decrease in IBS symptoms. Interestingly, studies have demonstrated that 8% to 9% of those seeing a physician for what is believed to be a strictly medical problem are suffering from PTSD.

Consequently, using the specialized electronic/computer hardware I have purchased and employing the skills gained from the hundreds of hours of specialized post-graduate training I have taken, one year of supervised practice (for example, with Dr. Tony Hughes), regular reading of newly published information, and attendance at relevant conferences, I include in my assessment *direct measurement of the ANS*. Baseline levels of hyper-arousal, indications of hyper-reactivity, and assessment of one's capacity to recover from stressors are assessed. To do so, sensors are placed on bodily surfaces (such as fingers and palm), the signal is amplified, artifacts are removed electronically, and the resulting data are recorded and saved through connection to a software program operating on my computer. The process in neither embarrassing nor painful.

Subsequently, depending on results of measurement, *biofeedback-assisted training in relaxation/self-regulation of the ANS (BFT-A)* is provided, since well-conducted studies as well as my clinical experience have indicated clearly that this method of treatment is effective in teaching clients (a) how to alter ANS functioning in a positive direction, (b) how to maintain a state of relaxation in the face of traumatic cues, and (c) how to regain that state of inner calmness should arousal levels increase. Thus, BFT-A is related to stress inoculation training (SIT), which has been a long-time component of CBT (cognitive behavioural therapy). Both BFT-A and SIT focus on managing and extinguishing PTSD symptoms rather than on processing the traumatic experiences. However, as noted below, there is ample evidence that my biofeedback methods provide additional positive effects as compared with SIT. Very importantly, implementing my BFT-A approach consistently has proven that for many clients, the link between exposure to reminders of the trauma and the previous resulting ANS hyper-arousal/hyper-reactivity is weakened or broken such that the trauma reminders become like any other memory. In addition, there is some research indicating that when ANS hyper-arousal/hyper-reactivity is reduced, other symptoms of PTSD also decrease. Moreover, as indicated in the section below under Tactical Breathing-2, many other symptoms and conditions are improved that have not normally been thought to be associated with PTSD and thus additional advantages result.

F-02. Tactical Breathing Overview

Four of the major specific biofeedback training methods I now use with all clients are Tactical Breathing-1, Tactical Breathing-2, Tactical Breathing-3, and Tactical Breathing-4. Training sessions are held every other week thereby giving clients plenty of time to master the methods. Generally, sessions are about 1.5 hours long and include also er discussions and recommendations regarding stabilization strategies as well as issues brought forward by the client.

Although most of us take it for granted, in fact breathing involves a complex interplay among a number of brain areas including the forebrain, the brainstem, and the spinal cord. Additionally, the pattern and rhythm of breathing is influenced by a large number of brain chemicals such as serotonin, epinephrine, dopamine, the peptide known as substance P, and neuropeptide Y.

As presented in the 2007 book *Principles and Practice of Stress Management* co-edited by Lehrer and associates, breathing acts also as a central pump in moving various fluids as needed including those associated with venous blood, lymph, and cerebrospinal functions. In addition, the components of respiration are related to such facets as posture, walking, and lifting.

Related to these and other factors, breathing can become dysregulated. The most common form of dysregulated breathing is known also as hyperventilation or hypocapnic hyperventilation. Many symptoms have been associated with this including those listed in the 2007 book edited by Lehrer and co-editors: dizziness, faintness, muscle stiffness, cold hands or feet, shivering, muscle cramps, fatigue, tightness in the chest, hot flashes, headache, muscle weakness, stiffness around the mouth, warm feeling in the head, sweating, blurred vision, non-cardiac chest pain, chronic pain, chronic fatigue, attention problems, rapid heartbeat, unrest/tension, anxiety/panic, depression, and feelings of unreality.

Fortunately, with the electronic tools available to me, it is possible to alter breathing as a fundamental method to achieve proper regulation of the autonomic nervous system, thereby reducing the symptoms of PTSD, other anxiety disorders, and depression as well as other symptoms noted above. Moreover, in a 1999 book, Dr. J. H. Austin, a professor of neurology, noted that the tactical breathing mentioned below has a direct therapeutic effect also since it slows firing of the nerve cells in the amygdala, thereby having a basic calming effect. The role of the amygdala is explored later on.

F-03: Tactical Breathing-1 (TB1)
TB-1 begins by sitting with back straight, connected to biofeedback equipment, with the Thought Technology ProComp Plus and other software running. Stabilization of good posture has a positive effect. Thereafter, clients learn particular elements of breathing, since it is known that this can be done with conscious awareness in order to change mental and physical states.

Clients are then encouraged to begin breathing in through the nose and out through the mouth (unless there is a nasal obstruction). Nasal intake breathing likely increases nitrous oxide (NO) levels, which are involved in a large number of physiological responses including broncho-dilation, oxygen transport, neuro-transmission, insulin response, memory, mood, and learning (Courtney, 2008). As well, breathing in through the nose serves to filter the air and warm it.

Next, clients observe their own actual breathing characteristics on the monitor, such as type of breathing, rate of respiration, amplitude of each breath, form and timing of each breath, interval between each breath (for example, breath holding), and blood oxygen levels. The end result is learning which breathing characteristics lead to observable and measurable decreases in sympathetic nervous system arousal and/or increases in parasympathetic nervous system activity with the accompanying internal state of calmness. Parts of this process are known as diaphragmatic breathing, deep breathing, or belly breathing.

For example, clients practice lowering respiration rates to about 6 breaths per minute while increasing tidal volume. Also, we aim for an inhalation time of about 40% of the total cycle length, an exhalation time of about 60%, and slight breath pauses at the end of each phase. One of the measurable results is production to about about 100% in the amount of oxygen bound to hemoglobin molecules in the blood (SpO2).

Tactical Breathing-1 is a pre-requisite to the next phase and can only be learned satisfactorily with the use of the kind of biofeedback equipment I have. Usually, this is accomplished in a single treatment session.

Between this first TB-1 session and TB-2, clients are encouraged to practice the new method of breathing at home 3 times a day for 5 to 8 minutes a time and additionally as stress/anxiety/arousal increases. Sitting straight on the floor or chair with one hand over the belly button and the other in the lap is one way to practice. Another is to sit straight with hands clasped over the head, since this promotes diaphragmatic breathing. Yet another way is to lie flat in bed with two books of equal weight–one on the chest and one over the belly button–and try to make the belly button book move more and the chest book move less.

F-04: Tactical Breathing-2 (TB-2)

This is also known as HRV (heart rate variation breathing), RSA (respiratory sinus arrhythmia breathing), CVR (cardiovascular reflex regulation), and now most often as RF (resonant frequency breathing). I have attended a number of presentations and professional workshops with world leaders (for example Dr. Rosenberg, Dr. Gevirtz, and Dr. Lehrer) to obtain mastery in this area and I continue to keep abreast of developments in this field. As well, early on when using this approach, I consulted on every client with a pioneer in the field.

In TB-2, sensors are attached so as to measure respiration, aspects of heart functioning (via a three-channel electrocardiogram), other sub-systems within the autonomic nervous system (ANS), and relationships between these variables. Ultimately, information from the sensors is processed by a sophisticated software program (CardioPro 2 by Thought Technology) and the results are displayed on a computer monitor.

The basis of TB-2 is this. When measured in milliseconds, heart rate is found not to be constant from beat to beat. Rather, normally it increases and decreases in a fairly regular pattern, depending on the relative activity of the sympathetic and parasympathetic branches of the ANS. The sinoatrial node on the rear wall of the right atrium of the heart (also known as the pacemaker) produces a heart rate of about 120 beats per minute at normal body temperature. The sympathetic branch of the ANS can increase heart rate via the sinoatrial node through the release of norepinephrine. The parasympathetic branch of the ANS via the vagus nerve (the 10th cranial nerve) tends to hold the rate down to about 70 to 80 beats a minute by influencing both the sinoatrial node and the atroventricular node through the release of acetycholine. In healthy people, the increase-decrease pattern activated by the interplay between sympathetic and parasympathetic branches occurs about 6 times a minute (referred to as 6 cycles per minute or 0.1 Hz) with a normal range of about 5.5 to 7.5 cycles a minute. The actual heart rate cycles per minute is referred to as the "resonant frequency."

Moreover, when healthy people are breathing at about 12 breaths a minute, the average difference between the highest and lowest heart rate (referred to as heart rate variation, or HRV) is between 10 and 15 beats. Young, healthy hearts may show as much as a 30 beat-per-minute peak-to-peak difference. Trained athletes have higher HRV than sedentary persons. Those on medications such as beta blockers will have lower HRV and this is true also for most after the fourth or fifth decade of life. However, exercise training increases HRV in healthy older adults as well as among those with coronary artery disease. Also, breathing in synchrony with the normal heart rate cycle usually increases HRV.

So why is HRV so important? As discussed in the 2008 paper by Lagos and colleagues (including pioneer Dr. Paul Lehrer), essentially HRV is considered as a measure of the continuous interplay between sympathetic and parasympathetic influences on heart rate. The activation of the sympathetic branch of the autonomic nervous system (ANS) increases heart rate, while the activation of the parasympathetic branch, primarily mediated by the vagus nerve, slows it. Variation in heart rate can be caused by a variety of factors, including breathing, emotions, and various physical and behavioral changes. The heart rate changes as well in response to internal body rhythms...In general high HRV represents a flexible ANS that is responsive to both internal and external stimuli and is associated with fast reactions and adaptability. (p. 110)

In other words, generally high HRV is an indication of good ANS regulation; that is, there is a good balance between sympathetic nervous system influences and the activity of the parasympathetic branch of the ANS. Good ANS regulation is seen also as a general measure of adaptability.

Furthermore, in healthy people there is a relationship between respiration and normal heart rate variation such that increasing heart rate and inhaling are associated while decreasing heart rate and exhaling are related. Thus, again according to Lagos and co-workers (2008):

> In a phenomena known as "respiratory sinus arrhythimia"(RSA), the vagus nerve shows a rhythmic ebb and flow associated with the rate of respiration. Breathing at about six breaths per minute activates these resonant properties and induces high amplitude oscillations in heart rate at 0.1 Hz . . . (six cycles a minute). (p. 112)

Relatedly, in their 2008 paper, Jonsson and Hansson-Sandsten indicate that "RSA is suggested to be associated with behavioral flexibility, attention and emotion regulation, and with rapid regulation of cardiac output to promote engagement or disengagement with the environment" (p. 124).

Also, there is a similar relationship between heart rate variation and the rise and fall of blood pressure via the baroreceptors that regulate it. Thus, increases in both heart rate and blood pressure are considered to be the result of increases in the activation of the sympathetic branch of the ANS and/or decreases in the activation of the parasympathetic system. Conversely, decreases in both heart rate and blood pressure stem from decreases in the activation of the sympathetic nervous system and/or increases in the activation of the parasympathetic branch.

When the patterns of respiration, heart rate variation, and blood pressure are in synchronization, the ANS can be said to be in a state of proper regulation.

Sometimes, this is referred to as being in phase. When RSA data indicate that this is not so, the ANS is said to be dysregulated. Many authorities consider that lowered RSA among clients with PTSD is an indicator of dysregulation of the ANS.

The software program I use provides much relevant statistical information about many ANS variables related to HRV and RSA from which conclusions about the functioning of the ANS can be made. One key statistic is the peak frequency of the power spectrum of the heart rate variability analysis; generally, an ideal spectral value is dominant at about 0.1 Hz, which suggests a good balance between sympathetic and parasympathetic influences on the heart. Other important statistics are the extent of heart rate variation, the standard deviation of intervals between successive heart beats (SDNN), and the time domain of heart rate variation (such as the RMSSD, the root mean square of the standard deviation between successive R waves of the cardiac cycle). Also, generally, the high frequency (HF) component of the heart rate variability power spectrum is considered to reflect predominantly parasympathetic vagal activity while the low frequency (LF) component likely represents a mixture of sympathetic and parasympathetic influences.

During baseline periods, those with PTSD experience anything but being relaxed and the various CardioPro-2 statistics above confirm the experience. For example, ANS dysregulation is noted by such statistics as low levels of HRV, a peak frequency markedly different from the desirable one of about 0.1 Hz, and low HF amounts. Moreover, by watching the computer monitor, the client and I can observe that the client's ANS is sent into further dysregulation when the client is exposed to reminders of traumatic experiences. This is associated with the experience of high levels of anxiety and even panic. Furthermore, once hyper-aroused by reminders of the trauma, often those with PTSD are unable to return biological variables to their previous baseline levels, which were higher than normal in the first place. Other studies have demonstrated that the above pattern is linked to deficits in emotional and behavioural self-regulation.

The positive in all of this is that knowing what is happening within the ANS allows us to devise psycho-physiological strategies for improved regulation between the interplay of the sympathetic and parasympathetic branches; in other words, for improving HRV/RSA. Then, the strategies are implemented for a reasonable period of time during our session and we are able to see clearly the impact they have on the ANS via resulting changes in numerics as well as by client evaluation of relaxation levels.

Generally the strategies involve determining the heart rate cycle (resonant frequency) and changing the breathing pattern subsequently, such as by learning to inhale while heart rate is increasing and to exhale while heart rate is decreasing. As detected by the computer program, resulting changes in HVV/RSA readings indicate an optimal relationship between sympathetic and parasympathetic activation of the ANS; in other words, improved regulation of the ANS.

Then, the client strategies found to be effective are practised as "homework" and incorporated into daily life. The easiest way is to down-load *EZ Air* from the Biofeedback Foundation of Europe web site (www.bfe.org). EZ Air was developed by Thought Technology, whose hardware and software I use in psycho-physiological training. As a starter, I encourage clients to practice three times a day for 5 to 8 minutes at a time *and* also as soon as tension is detected. Many of the stabilization strategies recommended earlier in treatment are helpful in improving regulation of the ANS such as decreasing stimulants, increasing aerobic exercise, and taking steps to obtain restful sleep. The 2009 Sherlin and co-worker study found that home practice with the stress eraser (Helcior Inc., New York), a portable device designed to increase RSA, was effective for about 70% of their sample.

Most published studies have demonstrated that those with PTSD usually have a dysregulated ANS characterized by an over-activated sympathetic nervous system and/or deactivated parasympathetic nervous system even when simply sitting quietly. For example, a 2007 study of Bulgarian peace-keepers in Kosovo by Nikolova found "reductions in parasympathetic function and baroreceptor modulation of heart rhythm" (p. 46). Also, a study of Vietnam combat veterans presented by Aitkens at the 2008 Society for Pychophysiologicl Researsh (SPR) conference revealed that among military veterans with PTSD, those with the lowest vagal tone (a measure of parasympathetic activity) had the highest symptoms of trauma re-experiencing and the greatest amount of distress over trauma cues. The 2004 study by Sack and colleagues reported decreased RSA to trauma reminders in 29 outpatients with PTSD as compared with baseline and neutral script imagery. The 2008 study of Croatian military veterans by Javanovic and colleagues found decreased RSA among those with PTSD as compared to a control group. The Ardity-Babchuk group (2009) studied those who had and had not been exposed to trauma when presented with both traumatic and pleasant memories; those exposed to trauma had lower parasympathetic activation and lower recovery following trauma recall. A few studies such as the 2008 one by Woodward and co-investigators have not found RSA magnitude to differ between combat veterans with and without PTSD.

Other studies point to the relevance of RSA. For example, at the 2008 SPR conference, Melzig and her team of researchers reported a decrease in startle responses following the threat of an shock among students with high HRV; students with low HRV did not decrease startle responses, suggesting that low HRV is a marker of prolonged anxious responding. At the same conference, Pu and associates demonstrated that non-clinical subjects with higher baseline RSA recovered more rapidly from the stress of a paced arithmetic test than those with lower baseline RSA levels. As noted in the 2007 book *Principles and Practice of Stress Management* edited by Lehrer and associates, babies born prematurely have low HRV, which leads to vulnerability to stressors later in life. Their book reports also (a) that adults with panic disorder have higher heart rates and lower HRV than non-anxious controls, (b) that the same hold true for those with various levels of depression, and (c) that low HRV is characteristic of those with other forms of anxiety or worry. The 2008 study by Park and co-investigators found high levels of obstructive sleep apnea (OSA) to be associated with low levels of

RSA; OSA is discussed in the section on sleep and many with PTSD suffer also from OSA. The 2008 investigation by Licht and colleagues of over 1800 patients found that their depression was associated with significantly lowered heart rate variability compared to the 524 controls. Some of the lowered heart rate variabiliyy appeared to be due to anti-depressant medication. In 2008, the Spetalen research team found that as discomfort increased, patients with irritable bowel syndrome (without constipation or psychiatric conditions) had increased heart rate and skin conductance as compared to healthy controls. This condition can co-occur with PTSD as well as being a response to trauma in its own right. In his extensive 2008 review of studies to date, Amstadter observed that many studies have found lowered HRV/RSA levels among those with panic disorder and those suffering from generalized anxiety disorder. These psychological conditions co-occur frequently with PTSD, can result from trauma without additional symptoms of PTSD, and can occur in the absence of any known trauma history.

However, the evidence is that by regulating the ANS with RSA, the sub-cortical limbic system of the central nervous system involved in emotional reactivity is better modulated. With reference to the brain's anterior cingulate cortex (ACC; more about that later on), it was interesting to hear Dr. Sasks at the 2005 ISTSS conference report research findings that positive changes in the ACC were associated with an increase in RSA; studies have shown decreased ACC activity among those with PTSD. This was confirmed in the 2008 study by Woodward and colleagues wherein RSA power was found to be correlated with the magnitude of blood oxygen levels in areas of the ACC. A recent small sample research project by Sherlin presented at the 2008 ISNR conference found that improved RSA led to an increase in the ratio of alpha to beta waves at brain sites O1 and O2; this finding suggests that RSA feedback may decrease cortical arousal. A further study by Sherlin and Wyckoff (American Association for Psychophysiology and Biofeedback Conference, May 2008) revealed that the alpha frequency produced by RSA breathing was localized in the brain area known as BA 24 associated with the cingulate gyrus, which is involved in emotional regulation and in the affective dimensions of pain. The Longhurst chapter in the 2008 volume by Squire and colleagues noted that cardiovascular reflex regulation involves brain regions such as the pons, medulla, and hypothalamus. Relatedly, Nash (2009) concluded that HRV is a powerful measure of brainstem activity and that increasing HRV will make positive changes in the functioning of those areas.

Also, there is now direct evidence that HRV biofeedback can affect PTSD. At the May 2008 AAPB conference, Zucker reported psychological and physiological benefits among those with PTSD who completed RSA training. At the 2008 Biofeedback Foundation of Europe conference, Zucker reported his study of 38 adults with PTSD who received either RSA biofeedback or progressive muscle relaxation. Both groups demonstrated a significant reduction in symptoms; for the RSA group, increases in measures of ANS regulation (e.g., SDNN) were correlated with a decrease in PTSD symptoms.

Since many clients with PTSD suffer also from depression, it is instructive to remember that a number of studies have demonstrated that those with depression have increased heart rate and decreased HRV. Also, cardiac patients with severe depression were found to have less HRV relative to those with less severe depression. Moreover, case studies have shown that HRV biofeedback may improve symptoms of depression. All the preceding was discussed in the 2008 research paper by Siepmann and associates, which included also a good study comparing those with and those without depression. The authors confirmed increases in HRV after biofeedback among those with depression but not among control subjects. Moreover, depression scores decreased significantly after HRV biofeedback, from a starting score of 22 (moderate level of depression) on the Beck Depression Inventory to an average score of 6 (no depression) after intervention. Similarly, anxiety scores decreased among those with depression. The authors note also that HRV among those with depression does not improve with the taking of the commonly prescribed SSRI medications.

RSA is important also in the fields of behavioural cardiology and respiration. For example, in 2008, Peltola and colleagues reported a study in which 1631 patients with a previous myocardial infarction had RSA evaluated and were then followed for about 40 months. Low RSA was a strong predictor of later sudden cardiac death. Also, a 2008 ten-year follow- up study by Lykke and collaborators of 391 individuals with type-1 diabetes found death among these patients to be associated with low HRV. Those with both abnormal HRV values plus prolonged QTC intervals had an even poorer prognosis compared to those with values within the normal range; of the 34 with both abnormalities, 15 died, 14 of them from cardiac causes. Additionally, in a 2008 study of those newly diagnosed with high blood pressure, Pavithran and colleagues found that high blood pressure was associated with low HRV; those with high blood pressure are more at risk for medical conditions and mortality than those whose blood pressure is within the normal range. The value of laboratory-based assessments has been demonstrated in studies such as those reported by Sherlin and associates in 2009: higher heart rate reactions to laboratory stressors are associated with increased likelihood of developing cardiovascular disease. Other studies have linked low HRV to such conditions as ventricular arrhythmia, ischemic heart disease, rejection after heart transplant, and sudden infant death.

Other studies where clients learned to increase RSA have reported significant reductions in panic attacks, chronic pain including fibromyalgia, blood pressure, irritable bowel syndrome, inflammation, asthma, anxiety, and functional heart problems such as non-cardiac chest pain and arrhythmia as well as improvements in aspects of respiration (efficiency, gas exchange, and oxygen saturation), and performance at work and in sports activities. For example, in the 2008 study by Pavithran and co-workers comparing what they termed HRV deep breathing between those with normal blood pressure and those newly diagnosed with high blood pressure, RSA levels were significantly lower in the second group. A study by Lehrer and associates reported at the May 2008 AAPB conference found that RSA breathing decreased sensitivity to light after injection of a substance that produces flu-like symptoms. At the sameconference,

Lehrer's group reported that the 10-week training combination of muscle relaxation, autogenics, and RSA led to improvement in illness severity and degree, depression measures, and hyperventilation among those with multiple physical symptoms for which no medical cause could be found, while those waiting for treatment did not show such improvements. Yet other studies have found that RSA training increases essential gas exchange (oxygen in and carbon dioxide out), improves overall oxygen saturation levels in the blood, and facilitates desirable distribution of blood in the brain. Also, according to Lagos and associates (2008), RSA "results in a system-wide energy efficiency and metabolic energy savings that has been demonstrated to enhance athletic performance" (p. 109). Furthermore, the 2006 study by Suvorov indicated that RSA training resulted in fewer mistakes and an increased rate of information-processing among operators. RSA biofeedback resulted in increased tolerance for physical exercise among those whose heart failure was due to high left ventricular ejection fraction, as reported by Swanson and colleagues in 2009.

Sometimes, 10 or more weeks of daily home practice are required before benefits are apparent. On the other hand, in a well controlled 2009 study, Sherlin and colleagues used a single episode of RSA training among those reporting stress and demonstrated significantly reduced heart rates and levels of state anxiety as a consequence.

F-05: Tactical Breathing-3 (TB-3)

I added TB-3 to my basic approach in July 2008 after purchasing the necessary hardware (capnometer) and software and completing some training with an expert in the area, Dr. Peter Litchfield (see www. better physiology.com). Essentially, carbon dioxide levels are measured ($ETCO_2$, the level of carbon dioxide released at the end of expiration) during breathing since there is evidence that exhale levels of carbon dioxide outside a particular range (for example, less than 37 mm of mercury) can have a negative effect.

First, less than desirable $ETCO_2$ levels can result in hyper-arousal of the ANS via either activation of the sympathetic nervous system branch and/or less activation of the parasympathetic branch. Also, as these levels increase, the overall circulating amount of CO_2 decreases (called hypocapnia) leading to vasoconstriction of the cerebral blood vessels, an increase in blood pH levels, less tissue oxygenation, and ultimately the production of any of a large number of symptoms.

Second, among the symptoms related to $ETCO_2$ levels outside a specific range are symptoms of anxiety including palpitations, tachycardia, chest pain, shortness of breath, dizziness, gastric pain, tension, fatigue, weakness, and sleep disturbance.

Third, levels of $ETCO_2$ outside of a specific range can lead to counter-therapeutic activation of some key muscle systems such as the trapezius as well as muscle pain and tremors. The importance of this to the treatment of PTSD and related conditions will become apparent when muscle relaxation is discussed.

Fourth, studies have demonstrated that $ETCO_2$ levels outside of a specific range are associated with a decrease in the amount of oxygen in the brain and a decrease in the brain's capacity to utilize glucose as measured by brain imaging devices such as PET scans. Our brains require high levels of oxygen and proper utilization of glucose for optimal functioning.

Fifth, there is evidence that when clients learn to improve $ETCO_2$ levels, then various brain conditions improve; for example, when desirable exhale levels are initiated in response to the aura that sometimes precedes a seizure, some clients experience a decrease in the frequency and intensity of seizures (personal communication, Dr. John Nash, listserv communication, 2008).

Sixth, although I cannot locate the study, Meuret in 2008 demonstrated that clients who learned how to achieve appropriate levels of $ETCO_2$ had a decrease in panic attacks with improvement still evident at follow-up 12 months later. Relatedly, the 2008 report by Van den Bergh and co-workers at the SPR conference found that tidal volume doubled during sighing and was accompanied by a sense of relaxation. Physiological experiences of panic are common among those with PTSD, the diagnosed condition of panic disorder co-occurs often among those with PTSD, and my observation is that many with anxiety disorders sigh frequently.

Seventh, there is clear evidence that low CO_2 plays a role in broncho-constriction and, that in people with dysfunctional breathing, fluctuating levels of CO_2 are more likely to be present than consistent levels.

My first client for TB-3 was the victim of a motor vehicle accident. He continued to experience light-headedness after mastering TB-1 and TB-2 even though he felt relaxed and the data on the computer monitor were consistent with this. However, $ETCO_2$ levels were about 25, whereas the desirable range is 35 to 45. While watching these levels as well as the display for TB-2, we tried different approaches. Within a few seconds of decreasing the amplitude of the in-breath (breathing less deeply), $ETCO_2$ levels rose sharply to 35. Appropriate ANS levels were observed on the monitor, light-headedness disappeared, the peak heart-rate variability increased from 0.09 to the more desirable 0.10, and he reported feeling calmer and more relaxed than ever. Since then, other clients have had similar responses to TB-3.

F-06: Summary of TB-1, TB-2 and TB-3

Usually in the initial training session clients experience relaxation and inner calmness after using these breathing techniques. Moreover, they are able to see positive changes in ANS functioning on the computer monitor; for example, slowed respiration rate, increased respiration amplitude, decreased heart rate, increased heart rate variation, decreased skin conductance, decreased blood pressure, increased peripheral blood flow, increased RMSSD numbers, peak power at 0.10 Hz, and $ETCO_2$ levels between 35 and 40. Relatedly, a 2006 study by Hopwood and Bryant indicated that an opposite and dysfunctional breathing method known as hyperventilation served to increase the number of intrusive thoughts among those with acute stress disorder, a PTSD-like response to trauma that can occur in the first month after exposure to trauma. Hyperventilation decreases RSA and both hyperventilation and lower RSA are very common among the clients I see with PTSD prior to learning tactical breathing.

In-session training and then home practice of the techniques several times a day allows clients (a) to live in a calmer state than before and (b) to achieve that calm state quickly when needed. Moreover, tactical breathing is easily extended into daily life such as when in wait mode, such as while standing in line at a bank, during television commercials, while on the telephone, while 9-1-1 runs the licence plates after a suspect's car has been stopped, during meetings, and so on. With practice, these techniques become more and more automatic, like learning to drive a car with a standard transmission. Moreover, as my one-time consultant, Dr. Tony Hughes, reminded me, "You have to breathe anyway, John. So why not do it right all the time?"

F-07: Tactical Breathing-4 (TB-4)

One of the characteristics of PTSD is that symptoms of anxiety or anger arise rapidly when the individual is faced with reminders of the trauma, and they also arise at other times when it is difficult to determine what set off the reaction. For these situations, I recommend the following approach, which I have labelled Tactical Breathing-4 (TB-4).

First, notice and label what is happening; for example, "I had a thought about the incident and my emotions just got activated." Or if the emotional trigger is unknown, then notice something like, "My emotions just got activated by something."

Second, think about it realistically: for example, "Nothing of a threatening nature is happening right now. It is just a memory or a reaction to something."

Third, tell yourself what you need to do: for example, "So what I need to do right now is to begin tactical breathing (1, 2, and 3), which will decrease my reactions." Then do it.

Fourth, continue the tactical breathing until symptoms subside. The immediate effect of this approach is to decrease the arousal/anxiety/anger response of PTSD as it occurs.

Fifth, once symptoms subside, shift your attention immediately to something that is pleasant and relaxing. Keep a list in your pocket of things you can think about (such as remembering a pleasant time in your life), things you can do (like going for a walk with your IPod playing calming music), people you can call or hang out with, and so on. This will help to minimize a re-triggering.

Sixth, repeat the above sequence each and every time the reactions are activated. Over many and consistent repetitions, the ultimate effect is to uncouple the reactions from whatever is setting them off. This approach is effective even if you do not know consciously what the trigger is.

I use additional methods of biofeedback for regulation of the ANS as required. For example, there is good research indicating that learning to increase finger temperature at will can be helpful in establishing a state of calmness, in lowering blood pressure, in increasing circulation, and in decreasing migraine headaches. Also, sweat gland activity is one of the quickest ANS responders when traumatic cues are present even if the client is unaware that the cue is a traumatic one; hence, a client's psycho-physiological response can help to identify traumatic cues as well as determining when any particular cue no longer has a negative impact.

F-08: Other ANS Interventions

Other measures of autonomic nervous system (ANS) functioning are taken during the assessment of PTSD.

One is *electrodermal activity (EDA)* as assessed by levels of skin conductance. When the ANS becomes aroused, the activity of sweat glands in activated very rapidly and this can be assessed via sensors on the surface of the palm of the hand, where there are hundreds of glands per square inch. The sensor measures sweat gland activity 32 times a second, amplifies the signal, processes it through the software, and displays an average on the computer monitor every one second. According to Andreassi (2007), EDA is a complex reaction involving multiple centres in the central nervous system including the pre-motor cortex, the sensory-motor cortex, the hypothalamus, the anterior cingulate gyrus, and the reticular formation.

A second is *finger temperature (FT)* taken by a very sensitive thermometer (called a thermistor) taped to the end of one finger. This is an indirect but fairly responsive measure of blood flow or circulation. When a person is relaxed, a variety of processes allow the blood vessels to expand (vaso-dilation), thus there is an increase in blood flow, and consequently the temperature at the end of the finger increases. Under conditions of stress, the blood vessels become narrower (vaso-constriction), so there is a decrease in blood flow, and the finger temperature decreases. Additionally, via stress mechanisms in both central and autonomic nervous systems, blood is directed from areas where it is not needed (for example, fingers, kidney, and lower abdominal area) to areas where it is needed (such as large muscles and heart) so that the individual may fight or flee. Thus, under such circumstances, finger temperature lowers further. On my system, finger temperature is measured 32 times a second, the signal is amplified, it proceeds through the software, and the second-by-second average is displayed on the computer monitor. Nash (2009) concluded that "skin temperature has everything to do with complex decisions being made   in   the brain stem and midbrain regarding the need for blood flow to the skin and digestive organs versus the skeletal muscle system" (p. 22).

Since most of our biological systems are bi-directional, this means that altering skin temperature via standard methods of biofeedback can have a positive effect on our central nervous system and in particular the brainstem and midbrain.

A third is *blood pressure (BP)*. During assessment, this is recorded periodically on my Dynapulse system. BP levels and additional information are displayed on the computer monitor. Under conditions of psychological stress, systolic blood pressure (SBP) tends to increase.

All three of these components are included in the psycho-physiological assessment I complete on all clients with PTSD. Moreover, the data from each are part of what I use to determine the effectiveness of the various forms of tactical breathing noted above. Usually, the tactical breathing results in a decrease in EDA, and increase in FT, and a decrease in SBP both over baseline levels and following trauma reminders.

Sometimes, one or more of these indicators of ANS arousal do not decrease sufficiently once clients have mastered tactical breathing. This can occur because each is a partly independent sub-system of the ANS. Sometimes the various components function together in synchrony and at other times one or more operate independently of the others. Regardless, because reduction of ANS functions is so critical in my treatment of PTSD, when any sub-system remains elevated after tactical breathing, specific biofeedback training is initiated for that sub-system. Thus, depending on results with the client, relevant components can be added such as biofeedback to decrease EDA, to increase FT, and to decrease BP.

Also, various of these components can be included to address conditions that can co-occur with PTSD. For example, FT biofeedback has been demonstrated to be successful for a condition known as Raynaud's condition, wherein finger temperature can be so low that the tips are white or blue, and in the treatment of migraine headache. Direct biofeedback can decrease blood pressure by 10 to 15 points among those with essential hypertension.

F-09: Decreasing Muscle Tension and Reactivity
In a review of the literature in 2007, Pole found evidence that those with PTSD had different muscle tension and reactivity than those without PTSD; both frontalis and corrugator muscle tension was elevated while recalling the traumatic experience. In addition to these muscles, many of my clients have shown elevated tension levels while resting during baseline measurements of the masseter, cervical paraspinal, and upper trapezius muscles. This was supported by the 2008 research reported by Afari and colleagues wherein there was a significant association between the re-experiencing and avoidance symptoms of PTSD and co-occurring temporomandibular disorders and consequential pain. Relatedly, (a) many with elevated muscle levels also suffer with tension-type

headaches that decrease by 40% to 60% as they learn methods to relax relevant muscles and (b) some are able to interrupt the development of migraine headaches through relaxation of relevant muscles. This clinical observation is consistent with the 2008 review of research by Nestoriuc and associates.

Also, there is evidence linking facial muscle tension and hyper-arousal/hyper-reactivity of the ANS. For example, increased activity of the sympathetic nervous system can lead to excessive muscle readings as well as high heart rate, low HRV, elevated skin conductance, decreased temperature at the extremities, and respiration that is both rapid and shallow. I have found such a combination to be common among those with PTSD with or without accompanying depression. Additionally, there is some research suggesting that higher than normal levels of or an increase in *muscle tension* for any reason may lead to hyper-arousal/hyper-reactivity of the ANS directly or via activation of the threat sensors in the brain such as the amygdala, which will be discussed later. For example, in his excellent 2007 book on psycho-physiology, Andreassi (2007) noted that increasing muscle tension resulted in corresponding increases in a number of measures including EEG, heart rate, muscle activity, and breathing rate. In addition, he cited some studies indicating that decreases in the levels of activity in one set of muscles resulted in decreased tension levels in other muscles; however, other studies did not obtain such results.

Finally, as discussed in the 2009 book edited by Budzynski and others, Price and Budzynski cited studies indicating that (a) level of muscle tension can be reduced via biofeedback of muscle activity, (b) reduction of muscle tension in the frontalis had a positive effect on the ANS, and (c) decreasing frontalis muscle tension led also to an increase in the amplitude of EEG waves in the alpha and theta frequency range. The importance of the latter relates to my finding discussed later that many with PTSD have insufficient brain activity in the theta range. Also, as discussed in the stabilization strategies, there is evidence that decreasing muscle tension results in a decrease in symptoms of PTSD.

Nash (2009) summarized his experience and knowledge of the published research by noting that the measurement of muscle levels "gives the most exact, localized, and precise measure of activity within the motor cortex and secondarily within the premotor or "motor planning" regions. Further, this control is bidirectional" (p. 23). Therefore, he concludes that getting muscles relaxed will promote brain wave rhythms that lead to a calm and relaxed state.

Years before the above information was available, in light of the post-graduate training I had completed, the electronic/computerized instruments I had, and my supervised clinical practice with the late Dr. Jeff Cram, I added the assessment and treatment of frontalis, masseter (TMJ), upper trapezius, and cervical paraspinal muscle tension to my practice. The field is referred to as sEMG (surface electromyography). Attention to the frontalis muscle is helpful in stress assessment as well as in learning general relaxation. Applications with the masseter muscle are useful for stress assessment and decreasing anxiety as well as helping with teeth grinding (bruxism) and TMJ pain. Focus on the upper trapezius has value in stress assessment, general relaxation, and headache management. The importance of the cervical paraspinals has to do with neck muscle activity, stress assessment, general relaxation, and headache management.

The rationale for sEMG is discussed in the 2009 manual *Basics of Surface Electromyography* available from Thought Technology.

> To innervate a muscle fibre (stimulate it to contract), an electrical signal from the central nervous system must first reach an alpha motor neuron. These neurons are responsible for initiating muscle contractions. As the contraction signal spreads from the alpha motor neuron across the muscle fibre, a series of electrophysiological and electrochemical processes take place. This produces an electrically measurable depolarization and repolarisation event known as the action potential. sEMG looks at the action potential signals from a number of innervatedmuscle fibres located near the pickup electrodes. . . .
> Contraction intensity is controlled by how often the nerve impulse arrives and enervates the muscle fibres. Each action potential generates a certain amount of energy in the sEMG signal. So as the action potentials arrive more often, the muscle contracts harder and the sEMG signal level increases. (Thought Technology, n.d.)

Levels are determined by placing an sEMG sensor on the skin over the body of the muscle of concern. The sensor measures the raw signal continually (analog recording), amplifies the signal, and then filters out some of the artifacts in the signal. Then, the raw signal is sampled 32 times a second and converted to digital data by an analog/digital converter (known as an encoder). The resulting information then goes to the software program, which calculates an average each second. This information then is displayed on the computer monitor and saved in the program.

During the assessment phase, recording of muscle activity is made during baseline conditions and while attending to trauma cues. If these results and/or other information indicate that muscle tension is elevated, then, via biofeedback technology (BFT-M) and gentle stretching, clients learn methods (a) to lower overall (baseline) muscle tension levels, (b) to be able to sense sooner than previously when muscle tension is increasing, and then (c) to use effective methods to decrease muscle tension as needed. Again, clients are encouraged to practice effective methods at home several times a day and additionally as required for life. I have found this approach to be more effective than other methods I have used over the years and clients report that the muscle-tension reducing exercises can be implemented easily during a variety of other activities throughout the day.

Treatment Step G: Establishing Restful and Refreshing Sleep

As indicated below, restful/refreshing sleep is critical to recovery from PTSD.

Contents

G-01: Introduction

With regards sleep, folks with PTSD often differ from those without PTSD in a number of ways, such as initiating and maintaining sleep, nightmares, panic attacks during sleep, various sleep disorders, and sleep architecture. However, while clients often report concerns about falling and staying asleep, subsequent results of assessment in sleep laboratories have been inconsistent. The reason for this might have to do with where the sleep study is conducted. A 2006 pilot study by Germain and associates among crime victims with PTSD confirmed clients' reports of sleep problems and EEG differences when data were recorded while they were at home as compared with when they attended a clinic.

There is reason to believe also that sleep quality has an impact on the severity of PTSD symptoms. For example, Belleville and co-investigators (2009) reported on about 100 persons with PTSD. Even without the influence of confounding variables (such as alcohol use, medications, and co-occurring conditions), results indicated that poor sleep had a negative impact on the severity of PTSD symptoms.

Most of us have no idea of what happens during the roughly one-third of our lives that we spend in sleep. This is true in spite of the considerable amount of research-based information that is now available on topics such as why we need sleep, what happens to us during sleep, what we need to do in order to obtain a restful and refreshing sleep, what the common sleep disorders are (there are 78 of them), and how sleep disorders can be treated effectively. Moreover, although 10% to 60% of patients seeing their physician have a sleep problem, only 30% raise the subject.

Therefore, some basic information is provided to clients as needed with the recommendation that they at least read the most applicable sections. For those wishing additional information, I suggest the book *The Promise of Sleep* by Dr. Dement, who has been instrumental in developing our understanding about sleep. Armed with such information, clients are able to start putting things in place to sleep well.

My continuing study of sleep began because there is a strong relationship between sleep problems and psychological reactions to traumatic events including post-traumatic stress disorder and depression. In fact, the co-occurrence is so marked that an expert in sleep disorders recommended at a recent conference I attended that when sleep disorder and PTSD were both present in an individual each should receive therapeutic attention as if they constituted a separate psychological condition. Also, the relationship is bi-directional; PTSD and depression following trauma can worsen sleep problems and sleep problems can increase symptoms of PTSD and depression.

G-02: Basic Information about Sleep
From Dr. Dement's work as well as the research-based findings of others in the field, the following points are supported at this time.

a. A sufficient amount of restful sleep is essential to health. Far from being idle, parts of the brain such as the thalamus and hypothalamus have increased activity during sleep, body tissue is built/repaired, EEG tracings alter, and hormone levels change throughout the night, as does body temperature, muscle tone, and blood pressure. Thus, insufficient sleep over long periods as compared to adequate amounts of sleep can have the following effects:
*less alertness;
*greater daytime fatigue and sleepiness;
*less ability to concentrate, remember, and learn;
*slower reaction time;
*less energy and stamina;
*greater tension and anxiety
*up to three times more likelihood of developing depression;
*fewer positive feelings and greater moodiness;
*greater likelihood of developing problems with substance abuse;
*greater irritability;
*greater health problems (high blood pressure, coronary heart disease, stroke, and diabetes);
*greater susceptibility to infections including the common cold;
*more stomach upset;
*lower levels of TNF and NK cells in the immune system;
*less natural repair of cells;
*increased sensitivity to pain and other bodily problems;
*decreased ability to ignore bothersome stimuli;
*increased utilization of medical care;
*more absenteeism;
*increased motor vehicle and other accidents;
*increased incidence of falls among the elderly; and
*shorter life span.

In fact, after only one night's sleep loss, the activity of the natural killer cells of the immune system can be lowered by as much as 30%.

b. Sleep studies reveal some frightening findings. As reported in the September 2007 internet article from *Force Science News*, one Harvard study of over 5,000 law enforcement organizations indicated that about 40% of officers suffer from sleep abnormalities such as sleep apnea, restless leg syndrome, and narcolepsy with temporary paralysis. Taken directly from that publication with some minor changes, it was noted that:

> *after 20 hours of wakefulness, neurobehavioural functions are impaired equivalent to a blood alcohol level of 0.10 and noticeable impairment sets in well before that. Generally, motorists are considered to be impaired if the reading is 0.08 or above;
> *the ability to maintain speed and road position on a driving simulator is significantly reduced when the normal awake period is prolonged by just 3 hours;
> *after 24 hours of sustained wakefulness, the brain's metabolic activity can decrease by up to 65% in total and up to 11% in specific areas of the brain involved in judgment, attention, and visual functions;
> *as people try to fight through periods of fatigue, the human body, in an effort to rest, goes into microsleeps during which people actually fall asleep from 2 to 10 seconds at a time, often without being aware of what has just happened; and as a consequence
> *fatigue is four times more likely to cause workplace impairment than alcohol or other drugs.

The 2009 study by the Schnyer research team found that various regions in the anterior prefrontal cortex of the brain were activated during complex decision-making tasks when the individual was rested. However, after 24 hours of sleeplessness there was a breakdown in this integrated neural activity with a resulting decline in making integrative decisions compared with the rested condition.

c. Sleep is a very complex process involving brain chemistry and various neurological pathways. Thus, the need to sleep is controlled in part by a number of homeostatic processes, self-regulatory processes involved in balancing bodily functioning.

First, from the individual's viewpoint, it is as if the hours we are awake must be paid back in some ratio by time spent asleep. Sometimes, the term *sleep debt* is used to describe this process; we accumulate sleep debt during waking hours and pay it back by the right number of hours of sleep. Thus, as sleep debt increases during the day, we begin to feel tired; sleep debt then is one of the mechanisms that alerts us that we are tired and thus promotes sleep. However, if we get less sleep than we require, we begin the next day with sleep debt. The amount of this debt can accumulate night after night. As this debt reaches 20 hours or more, one consequence is that people fall asleep for brief periods throughout the day without even knowing about it (micro-sleeps). Research indicates that once your eyelids start feeling heavy, you are only a few seconds away from sleep. Consequently, there is good reason to believe that micro-sleeps resulting from sleep debt are the cause of at least 200,000 annual motor vehicle accidents in the U.S., many of which are fatal. Thus, getting sufficient refreshing sleep is a matter of life and death as well as a health concern. Also, it is true that even a small amount of alcohol combined with a large sleep debt increases even further the chances of micro-sleeps. So regardless of where you are, if you are driving or operating other potentially dangerous machinery and your eyelids become heavy, stop immediately and have the nap you require. You may have only a few seconds to stop before experiencing micro-sleeps.

Second, an independent process related to circadian rhythm is involved in determining when we sleep and when we are awake. Known as the biological clock and the sleep-wake cycle, it is a process whereby we normally feel awake during the daylight hours and become sleepy as darkness arrives. Basically, two pin-sized clusters of nerve cells (called the suprachiasmatic nuclei or SCN) located directly above the optic nerve detect the amount of light surrounding us and when that amount of light falls below a specific threshold, they send a chemical message to relevant parts of our brain to shut off the powerful alerting mechanisms and to engage in those promoting sleep. Thus, if we are bathed in light when the darkness outside should trigger sleep processes, then the action of this clock is altered and we continue to be awake even if we have some sense of sleepiness because of accumulated sleep debt. Additionally, for many of us the biological clock slacks off somewhat in the early afternoon, resulting in some drowsiness, at which time a nap may overtake us if we have accumulated too much sleep debt. In addition to governing the timing of sleep onset and sleep offset, the SCM plays an important role in the distribution of rapid eye movement (REM) during sleep as well as sleep spindle activity.

Third, levels of arousal are governed in part by a group of neurons located in the brain stem referred to collectively as the reticular activating system (RAS). When their firing rate decreases, we feel relaxed enough to fall asleep. However, when these neurons are activated, such as by other brain areas involved in the symptoms of PTSD, they keep the central nervous system in arousal mode, making it impossible to achieve restful sleep.

Fourth, the thermo-regulatory system is involved. Failure of this system to promote a rapid decrease in body temperature interferes with the ability to sleep. That is why for some people, going into a hot tub two or so hours before bedtime can result in sleep as the body temperature drops.

d. Although the needs are individualized and may depend on other circumstances, the average adult requires somewhere between 7 and 9 hours of sleep every 24 hours. However, recent studies indicate that only about 33% of us get 8 hours and 50% get less than 7 hours. Many seem to be cutting back on sleep deliberately so as to have more time for work, commuting, household-related duties, hobbies, and entertainment. This of course results in an ever-increasing accumulation of sleep debt, such that the symptoms described in #01 above begin to emerge. The point is made eloquently by Dr. Dement:

The more we learn about the science of sleep, the more apparent it becomes how radically the course of modern life has diverted us from our bodies' natural rhythms. In virtually every aspect of contemporary living–from electric lights to all-night television to split shifts at work–we are literally punching the clock that maintains the synchronicity of our mind and body. In just a few decades of technological innovation we have managed totally to overthrow our magnificently evolved biological clocks and the complex biorhythms they regulate. . . . Over millions of years our bodies and minds have evolved to use sunlight as a Universal Standard Time, as the infallible index against which we reset out internal clock. Our daily activity. . . . [is] driven by hours of darkness and light. Our whole pattern of existence is based on a genetic code formed by the cumulative experience of millions of human ancestors who knew only the light of the sun and whatever dim illumination could be rendered from burning fat. Edison's bright electric lights gave us supernaturally long days. . . . Societal pressures to work more and more at odd hours...have reduced our sleep time over the past century by 20%. In the three decades since 1969, working Americans have added 158 hours annually–nearly a full month of work–to their workday (including the commute). Many of these extra hours are being carved out of sleep time. (p. 98)

I understand that Dr. Dement rarely deviates from his regular bedtime of 2100 hrs (9 p.m.) and I have followed suit for decades.

e. Throughout the night we have cycles involving various stages of sleep as measured by changes in brain waves. As noted by Dr. Ancoli-Israel (1997) in the book *Understanding Sleep* (pp. 178-179):

*The first stage of sleep, *Stage 1*, is transitional sleep, the point where most people find themselves drifting off, realizing that they are not fully asleep but no longer fully alert. Stage 1 sleep is defined by a decrease in alpha EEG activity (8-12 cycles per second) and an increase in theta brain waves (4-7 cycles per second). There is high activity in the chin muscles and there may be some slow rolling eye movements. As reported in the book by Andreassi (2007), increased amplitude of theta led to the poorest performance while monitoring a radar simulator while training to suppress theta amplitude in the occipital area of the brain led to improved monitoring.

*Stage 2 sleep is most often considered the official onset of sleep. It is characterized by sleep spindles (brief periods of activity in the 12-14 cycles per second range), with low muscle tension and increased amounts of slow-rolling eye movements.

*Stages 3 and 4 are progressively deeper, with stage 4 being the deepest level of sleep. Stages 3 and 4 are sometimes combined and called deep sleep or slow-wave sleep. Both stages 3 and 4 are characterized by delta waves (1-4 cycles per second). In stage 3 sleep, 20% to 50% of the time is spent in delta wave activity.

*The first state of sleep, *REM* sleep, is defined by low voltage, fast frequency EEG and rapid eye movements. The EMG (measuring the amount of muscle tension) during REM sleep is essentially flat, because except for the eyes and respiration, the body is paralysed during this stage of sleep. Because 85% of dreams take place during REM sleep, this paralysis is believed to be a protective mechanism that keeps us from acting out our dreams. Failure of this paralysis may explain the aggressive behaviours during sleep exhibited by some with PTSD.

*About 25% of the night is spent in REM sleep and 75% in NREM [non-REM] sleep. Among adults, NREM sleep is distributed with 5% in stage 1, 45% in stage 2, and 25% in each of stages 3 and 4. As cited by Andreassi (2007), 75% of NREM awakenings produced recall of mental activity and 54% resulted in accounts of what could be called dreams. As the night progresses, people cycle in and out of different stages of sleep. This cycling is called sleep architecture.

*Sleep begins in stage 1 and progresses through stage 2 to stages 3 and 4. It then goes back through stage 2 into the first REM period, generally 90 to 100 minutes after sleep onset. One continues to cycle in and out of the different stages throughout the night. Most of deep sleep occurs during the first third of the night and most REM sleep occurs in then last third of the night, in the early morning hours.

f. Generally, our sleep-wake cycle is related to a variety of other processes that together are referred to as the circadian rhythm. For example, the release of growth hormone is typically highest during the first 90 minutes of sleep. Cortisol secretions tend to increase during the early morning hours.

g. Our sleep needs and patterns change through the life-span including how much we sleep, when we sleep, and the ratio of REM to NREM sleep.
First, those about to be born spend 60% to 80% of their sleep in REM; this decreases to 50% of sleep time for the typical newborn, decreases further to about 25% of sleep time in mid-life, and drops to 15% to 20% of sleep time among seniors.

Second, there is a change in the biological clock during the teenage years such that teens get a troublesome kick in alertness so they are not biologically ready for bed until an hour or so after others in the house are getting sleepy and going to bed. Correspondingly, they are not yet biologically ready for getting up until much later than the rest of the household. The combination can lead to much conflict in the family.

Third, the most recent data indicate that the amount of sleep we need decreases as we get older so that the average is about 7 hours a night for ages 40 to 65 as contrasted with 9 to 10 for older teens and those in their twenties.

Fourth, as Dr. Dement notes:
The release rate of growth hormone after the night's first period of stage 4 sleep grows smaller and smaller as the years go on–as does the amount of stages 3 and 4 sleep. Without further need to make us taller, growth hormone mostly takes on the important work of repairing and renewing cells and tissues . . . many people in the 60s and beyond have very little stage 4 sleep left at all, and comparatively little growth hormone is being released at night. (p. 269)

h. Many factors can disrupt sleep:
*illegal drugs;
*medications (such as asthma drugs, prednisone, beta-blockers, and oral decongestants);
*surgery;
*alcohol (especially if taken within 4 hours of bedtime);
*nicotine (especially if you smoke more than 20 cigarettes per day or before bedtime);
*caffeine (especially if taken anywhere from 3 to 7 hours before going to bed);
*pain (including headache, rheumatoid arthritis, and fibromyalgia);
*heat;
*noise;
*things we eat;
*medical conditions:
chronic obstructive pulmonary disease such as emphysema and asthma;
kidney disorders such as uremia and urinary tract infection;
endocrine disorders such as hyperthyroidism and hypothyroidism;
heart disorders such as angina pain at night;
waking due to shortness of breath caused by impaired heart functions;
acid reflux;
AIDS;
dermatologic disorders;
severe anemia;
diabetes and
peptic ulcers.
*lifestyle;
*false ideas about sleep;
*being overweight;
*bad sleep habits;
*sex hormones
progesterone may shorten the time to sleep onset and reduce wakefulness after sleep onset while estrogen decreases the time between sleep onset and the beginning of REM sleep;

*depression and post-traumatic stress disorder; and
*stressors.

i. For most people, just one bad night will result in few major negative effects, although the sleep architecture the following night will be different from the normal pattern. For others, even the loss of 1-2 hours of sleep will result in an increase in both tension and tiredness the next day.

j. Those who experience significant sleep difficulties for a year or more without successful treatment have a high risk of developing depression.

k. Significant sleep problems can develop at any age, although difficulties with falling asleep are more likely to begin in the early thirties while issues of remaining asleep throughout the night are more likely to begin in the mid fifties.

G-03: Some Common Types of Sleep Disorders

a. Insomnia
One of the most common complaints of those with PTSD is insomnia: taking a very long time to fall asleep (for example, hours instead of the normal 15 or so minutes), waking frequently during the night, and then either taking a long time to fall back asleep or not returning to sleep at all. Together, this is known as difficulties in initiating and maintaining sleep (DIMS) and people with such difficulties are said to have insomnia.

Studies show that about 70% of military veterans with PTSD indicate problems with falling asleep and this is a common finding in my practice. By comparison, only 6% of combat veterans without PTSD report difficulties getting to sleep. As far as maintaining sleep is concerned, the evidence is that for combat veterans, 91% of those with PTSD report problem with night waking compared to 63% without PTSD. Likely related to this is the finding from psycho-physiological monitoring of those with PTSD showing increased arousal of the sympathetic nervous system during sleep often associated with demonstrated dysfunctions in the HPA axis. Also, EEG and neuro-imaging studies of those with PTSD reveal high activation of brain systems involved in the fear/threat response such as (a) the limbic system in general and the amygdala in particular, which has connections to cholinergic nuclei in the area of the brain stem involved in sleep and (b) activation changes in the pre-frontal cortex, especially the rostral portion of the anterior cingulate. These brain structures are discussed later. It is not yet clear to what extent the issues of insomnia were present before exposure to the trauma and to what extent they began after the trauma. On the other hand, sleep studies performed in the clinic do not report much of an objective difference between those with and those without PTSD; those with PTSD take about 9 minutes longer to fall asleep, have a 5% decrease in sleep efficiency, and spend about 3% more time in deep sleep. One interpretation is that those with PTSD are mis-perceiving sleep problems, another is that their sleep is improved in the laboratory compared to being at home because they feel safer, and another is that the results may depend on whether or not the person with PTSD has a co-occurring depression.

b. Nightmares

Although only 2% to 6% of the general adult population report one or more nightmares a week, between 52% and 71% of combat veterans report frequent, disturbing, trauma-related nightmares throughout the night, some so bad that upon wakening, the individual does not want to return to sleep in case the nightmare replays. For some, the content of nightmares is a pretty exact replication of a recalled traumatic event and these nightmare tend to repeat. For most, the content is not similar to any actual event but the theme continues to be one of personal threat although the details vary from one nightmare to another.

All of the preceding are common complaints among my police clients. Interestingly, 57% of posttraumatic nightmares occur outside of REM sleep, a state usually associated with dreaming. A small percentage of combat veterans engage in self-defence or attack actions during nightmares that pose a risk to bed partners. The occurrence of night-time disturbances certainly disrupts sleep, can result in various forms of insomnia, and leads to decreased restedness upon morning waking with the attendant consequences.

c. Night-time Panic Attack

Some with PTSD wake abruptly during the night with symptoms of a panic attack (palpitations or accelerated heart rate, sweating, trembling, shortness of breath, chest pain, nausea or abdominal distress, feeling dizzy, feelings of unreality, feelings of being detached from onself, fear of losing control or dying). A percentage of these folks do not have panic attacks when awake. Until recently, the thinking among clinicians was that this was related to waking from a nightmare even though clients could not recall any dreams even when awakened in a laboratory immediately after symptoms were noticed. Studies of night-time panic attacks using EEG technology indicated that most of these events occurred outside of the REM stage of sleep associated typically with dreaming. Rather, they occured mainly in NREM sleep between stage 2 sleep and the transition to slow-wave sleep. Thus, at the present time it appears that the panic attacks represent an episode of autonomic and central nervous system hyper-arousal. Perhaps related are reports of night terrors wherein the person wakes abruptly from slow-wave sleep with intense fear and panic but without any recall of a terrifying nightmare and with no recall the next day of having awakened at all. Among combat veterans with PTSD, 54% reported night terrors while the rate for those without PTSD is less than 10%. Following my use of BFT-C with a WWII veteran who had been also in a POW camp, the frequency of his night-time panic attacks decreased from 1 to 2 a week to about 3 a year. Similar results have been reported by other clients.

d. Sleep-disordered Breathing

Studies show that a high percentage of combat veterans with PTSD (40% to 54%) suffer from sleep-disordered breathing including obstructive sleep apnea (OSA). During OSA, breathing may stop for significant periods (at least 10 seconds) up to hundreds of times throughout the night with an accompanying decrease in the level of blood oxygenation to lower than biological requirements. For example, in one study, Krakow and associates reported that 52% of combat personnel with PTSD also had OSA, and in their 2001 study of victims of crime, 22 of the 44 participants had OSA while another 18 suffered from upper airway resistance syndrome. Also, about 40% of the elderly have OSA (Celle and associates, 2008).

After a stoppage in breathing (an apneic event) automatic processes are triggered and bed partners often can hear a gasping sound as breathing re-starts. EEG analysis reveals that apneic events are accompanied by micro-arousals from sleep states lasting up to one minute, although the client is not aware of having awakened. Thus, when one wakes frequently throughout the night, sleep is fragmented at best. As a consequence of these two facets, those with OSA wake tired and unrefreshed in the morning with no memory of having awakened throughout the night.

Indications of OSA can include loud snoring, feeling sleepy or fatigued each day for no apparent reason, waking with a headache or a sore throat in the morning, and symptoms such as diminished performance, forgetfulness, poor concentration, and low libido.

The attention of the family physician is critical in the assessment and treatment of OSA for a number of reasons.

First, the possibility that the person may have any of a number of the medical conditions that can cause OSA needs to be investigated. These conditions include COPD, gastroesophageal reflux, renal failure, and diabetes. Also, disordered night-time breathing can result from other factors that require assessment. These include a naturally narrowed throat, enlarged tonsils or adenoids, a family history of OSA, using substances that cause relaxation of throat muscles (such as alcohol, sedatives, and tranquillizers), smoking, and chronic nasal congestion. In addition, the condition is often worse among those who are substantially overweight, in males over the age of 40, and among women in the menopausal stage of life.

Second, OSA can cause any of a number of conditions that can lead to premature death. For example, Khyat and co-authors (2008) reported OSA as a factor in hypertension, left ventricular dysfunction, coronary artery disease, and stroke. In addition, the likelihood of traffic and occupational accidents increases due to micro-sleep episodes during the day. Moreover, a series of studies by Morrell and Twigg (2006) demonstrated that the intermittent oxygen decrease among the general population with sleep-disordered breathing resulted in neuro-degeneration in the hippocampus and pre-frontal cortex, areas associated with cognitive processes; presumably this is worse yet among those with persistent OSA. Furthermore, a 2007 study by Eskelinen and her group indicated that those with OSA often suffer functional disturbances in the pre-frontal lobes; for example, they experience less slow-wave sleep than those without OSA. Eskelinen and associates' paper refers also to a proton magnetic resonance spectroscopy study finding of axonal loss or dysfunction and myelin metabolism impairment in the frontal white matter. In their 2008 study, Sarchiellli and co-authors documented undesirable changes in various brain chemicals with OSA, which might relate to breakdown of cell membranes. Other consequences can include cardiovascular problems (such as arrhythmia, atrial fibrillation, congestive heart failure, and myocardial infarction), metabolic complications (such as leptin and insulin resistance), depression, anxiety, impaired neuropsychological functioning (including those associated with vigilance, executive skills, and coordination), and a shortened life span. Untreated, it can lead also to sympathetic nervous system hyper-arousal both during the night and during the day.

I am not aware of any studies indicating clearly that shift-work can cause OSA. The 2005 study by Klawe and associates found no differences between police officers on shift work and those who were not. On the other hand, a number of studies have reported a relationship. A 2007 study of police officers by Laudencka and co-investigators indicated that many of those who were already diagnosed with OSA experienced a worsening of their condition after night shift but not after day shift. The 2005 study by Tafil-Klawe and colleagues had previously demonstrated that shift work among police officers increased instances of sleep-disordered breathing as compared to normal breathing following day shift. The 2008 study by Paim and colleagues of shift workers in a nuclear power plant revealed that 9% had OSA and 5.5% had periodic limb movement disorder. Relatedly, the 2007 study by Takil-Klawe and associates of shift-workers with and without OSA indicated that those with OSA had higher blood pressure readings.

As cited in Andreassi (2007), one of the documented characteristics of OSA is a decrease or complete disappearance of activity in the muscles of the throat and/or upper air passage during attempts to inhale, even after medical conditions have been ruled out. Thus, air cannot be delivered to the lungs. Certainly for these folks, the most common and effective intervention is nightly use of a continuous positive airway pressure (CPAP) device, which provides a flow of air from a compressor, thereby maintaining relevant muscles in the open position. The client can choose between a full-face mask, a mask over the nose only, or two nasal cannula. With regular use of a CPAP system at least four hours a night, micro-arousals lessen significantly, restful sleep increases, and improvements are noted in sustained attention, spatial memory, and motor agility. The 2000 study by Krakow and associates of clients with both PTSD and OSA after completing treatment for OSA reported that sleep improved (93% versus 33% for a control group); 9 of the 15 with PTSD receiving treatment for OSA reported a significant decrease in PTSD symptoms while the 6 who did not receive treatment reported a worsening of about 43%. Badr (2008) reported that CPAP may be helpful also among clients with central sleep apnea; one cause of this form of sleep apnea is significant lowering of arterial partial pressure of carbon dioxide (pCO_2). However, many clients have considerable difficulty in adjusting to the use of CPAP. Sometimes, a gradual program of desensitizing to the mask is helpful. For those unable to tolerate the CPAP even after a program of desensitization, radio frequency ablation has shown success among about 31% whose OSA was not the most serious (Farrar and others, 2008) and upper airway surgery helped 8 of 11 clients to use the CPAP device afterwards (Chandrashkaria and associates, 2008).

Common additional recommendations in managing OSA are to bring weight down to normal levels and to avoid alcohol, nicotine, and sedatives. Mild sleep apnea may be lessened by sleeping on your side rather than on your back; sewing a pocket on the back of your pyjamas between your shoulder blades and placing a tennis ball in the pocket may help to train side sleeping. Frequent practising of good regulation of the ANS via tactical breathing may help to decrease OSA.

e. Periodic Limb Movement Disorder
Some people have repetitive jerky movements throughout sleep that qualify for a diagnosis of periodic limb movement disorder (PLMD). Those with PLMB are unaware of these movements. A 2008 population study of Detroit residents by Scofield and colleagues reported that 7.6% had 15 or more such episodes per night. In contrast, one study found that up to 76% of combat veterans with PTSD had PLMD. Other studies of combat veterans show movement rates between 11 to 38 per hour compared with 0 to 1 for controls. About one-third of PLMD movements have been observed to result in micro-awakenings up to 9 seconds in duration and thus have the same effects as OSA due to multiple micro-arousals throughout the night. Thus, some of the reported insomnia is attributable to PLMD. The medication Clonazepam taken at bedtime has been shown to decrease PLMB.

f. Restless Leg Syndrome
Folks including those with PTSD can suffer from restless leg syndrome (RLS), characterized by uncomfortable sensations and an irresistible and frequent conscious urge to move the limbs (usually legs). This leads to sleep disturbances including significantly decreased sleep time, lower sleep efficiency, and increased night-time waking. The 2008 study in Sweden by Mallon and co-workers indicated a rate of 10.3% in the Swedish population, and RLS also was associated with shorter periods of sleep at night, several health problems, and depression in the group studied as well as increased mortality among females but not males. Dopaminergic dysfunction, problems with iron insufficiency (serum ferritin levels), renal failure, and abnormalities of supraspinal inhibition are considered to be the primary causes. About 80% of those with RLS will have PLMD also. Typically those with RLS are heavy users of nicotine and caffeine and get little regular exercise. Currently, ropirinole is the only medication approved for treating RLS and has been found effective in two large double-blind, placebo-controlled, randomized control studies although oxycodone, propoxyphene, and codeine are prescribed sometimes.

g. Narcolepsy
This is a condition in which the person has symptoms such as excessive daytime sleepiness (EDS), thereby struggling all the time to remain awake, falling asleep when relaxed or sedentary, being unrefreshed after daytime naps, often experiencing cataplexy (a sudden loss of muscle tone with such results as a mild

head or jaw drop, buckling of the knees, or falling to the ground, after which the person usually falls asleep suddenly), sleep paralysis (being in an awake state but unable to move any muscles including those required for speaking), vivid hypnagogic hallucinations (seeing things or hearing voices as one falls asleep), and disturbed night sleep. Also, the condition is often associated with double vision, memory problems, balance disturbances, and changes in personality. For most folks, the attacks come without warning, while for a few people, a narcoleptic attack will be triggered by an emotional situation such as an argument with a spouse or laughing at a joke. Although little is known about the pathophysiology of narcolepsy, for some patients symptoms begin following viral illnesses, hepatitis, mononucleosis, and viral pneumonia. Those suffering from narcolepsy should be seen by sleep disorder clinics or specialists. Pervasive daytime sleepiness has been treated successfully with stimulant medication (such as amphetamines, methylphenidate, and pemoline) while low dosages of anti-depressants have been used effectively to treat cataplexy. Also, the medication alertec (Modafinil) is approved for treating narcolepsy. Among psychological methods, there is research support for sleep hygiene, structuring daytime naps, and avoiding shift work.

h. Circadian Rhythm Sleep Disorders

This group includes a number of specific conditions in which there is disruption in the way the 24- hour biological clock functions. Many of us have experienced this as jet lag; the disagreement between the body's internal clock and the light/darkness levels associated with the local time. As a result, we are alert when local time suggests we should be preparing for sleep and we are sleepy when local people are awake. Additional common symptoms include fatigue, difficulty concentrating, gastrointestinal distress, impaired coordination, reduced cognitive skills, and alterations of mood. Some of these symptoms may be due to changes in other aspects of the circadian rhythm such as melatonin concentration, core body temperature, and cortisol levels. Generally, at the rate of one day to adjust to each time-zone change, the circadian rhythm eventually becomes synchronized with the new environment.

While some of the symptoms are the same, folks with either of three types of circadian rhythm sleep disorders do not achieve a desirable sleep-wake pattern naturally even if they have never left their time zone. Fortunately, depending on the cause, typically these patterns can be changed with consistent use of the approach noted below over a 12- to 25-day period. However, difficulties in changing the pattern may be encountered in those in whom the brain section known as the SCN is not functioning properly and in some folks taking beta-blockers (since they alter melatonin secretion).

> * *Delayed sleep-phase syndrome (DSPS)* is the condition in which people are so alert at night that they cannot go to sleep and are so sleepy in daylight hours that they have great difficulty waking/getting up in the morning. This is common during the adolescent years and is not the fault of the young person. For example, one common pattern is difficulty falling asleep before 2 a.m. to 6 a.m. and difficulty waking up before 10 a.m. to 1 p.m.

Folks with this pattern have a high prevalence of other psychological conditions. DSPS becomes problematic if the person has to fall in line with typical morning start time for work. Research to date supports the strategy of increasing darkness levels in the evening (for example, lights off, use of blinds on windows, and wearing dark glasses) and exposure to bright light levels for an hour or two after morning waking (for example, through use of a SAD light at brightness levels of 10,000 lux). Another effective strategy known as chronotherapy involves delaying bedtime by one hour each day and getting up one hour earlier than the previous day until the desired bedtime and wake time comes around. I tend to use both strategies together. Also, there is some support for taking melatonin in the early evening. Dr. Dement takes the position that low doses (up to 0.5 milligrams for the average adult) after the sun sets appear to be helpful for some people. However, he also notes, "When patients ask me if they should be experimenting with melatonin, I first remind them that the hormone is less carefully regulated for purity and dosage than are prescription medications, and that much more research into its actions remains to be conducted" (p. 116).

*Advanced sleep-phase syndrome (ASPS) is the condition in which the person wakes early in the morning and cannot resist the urge to fall asleep early in the evening. There is evidence indicating the effectiveness of using a bright light (such as a SAD light) from 7 p.m. to 9 p.m. consistently in conjunction with increasing room darkness in the morning to help correct this pattern. Also, the opposite strategy to that used in DSPS is often helpful; each evening the person forces him-/herself to stay awake one hour later than the previous evening until the desired sleep-wake cycle is in place. I tend to use both strategies together. Dr. Dement suggests that taking melatonin in the morning may be helpful for some.

*Shift Work Type (SWT) of sleep-phase syndrome is characterized by (a) excessive sleepiness while working during normal sleep periods when the circadian alerting mechanisms are low and (b) insomnia when trying to sleep during the usual wake period when circadian alerting mechanisms are high. Consequences of SWT can include decreased hours of sleep, non-refreshing sleep, sleepiness while at work, and both diminished work performance and increased accident rate. Depression is higher among those with SWT even after retirement from work. Exposure to bright light for 3 to 6 hours during the night shift terminating 2 hours before the end of the shift, some aerobic activity at night when sleepiness increases, minimal exposure to bright light in the morning before going to bed, creating a very dark room for daytime sleep, and attention to reduce both noise and bedroom temperature are among the strategies found to be most helpful.

i. REM Behaviour Disorder

This is a condition in which the muscles that should be paralysed during REM sleep are not. Thus, the body is able to act out the brain's dreams by movement of arms and legs in bed or even getting up and running around. This can be dangerous if the dreams are violent or frightening, as can happen to persons with combat or certain police experiences. Interestingly, during episodes the autonomic nervous system remains calm and unaffected. Those with this condition need to be seen at a sleep disorder clinic.

G-04: Sleep Architecture

Even though we are asleep, our brain and bodies are still active throughout the night. Some of the studies have assessed brain wave activity during sleep (sleep architecture) and have demonstrated differences among those with PTSD as compared to those without as follows:

-more stage 1 sleep;
-less time in slow-wave (delta) sleep during which the body self-repairs; sufficient time in delta sleep is associated also with waking refreshed;
-increased brain wave activity in the beta frequency during sleep likely related to increased levels of arousal;
-more rapid eye movements (REM) (typically associated with dreaming) throughout the night;
-greater REM density of eye movements, that is, greater number of eye movements per unit of time;
-more frequent transitions from REM sleep to stage 1 sleep;
-disproportionately more awaking during REM sleep; and
-reduced sleep movement time particularly among those with nightmares or co-morbid panic attacks (as if they were in "freeze" mode).

Disruptions in REM sleep have been reported to be associated with centrally measured hyper-arousal. Some of the differences between those with and those without PTSD may be due to elevations in both brain hormones and the neurotransmitter CRF associated with the response to threat. There is some reason to believe that deviations from normal sleep architecture are greater among males exposed to military trauma than among others. However, it is not yet clear from research whether the differences above were present before traumatic experiences and thus perhaps predictive of who might develop PTSD or whether they represent a consequence of exposure to trauma.

However, a complicating factor is that in many of the studies published to date, the clients suffered from both PTSD and signs of major depression. Thus, the relative contributions of PTSD and depression to sleep architecture anomalies remain in doubt. Also, some of the anomalies noted above do not appear among those who are taking SSRI medications and it is known that such medications can affect the sleep architecture of those with major depression.

G-05: Putting It All Together into a Plan of Action
Many experts believe that significant sleep problems such as those described above must be treated as a separate or co-occurring disorders rather than hoping that they will improve with the direct treatment of PTSD. Since I have been persuaded by this viewpoint and in view of the consequences of sleep loss, I take an aggressive approach to resolving sleep problems.

Initially, I assess sleep characteristics by asking the client to complete the Maimonides Sleep Quiz (sleeptreatment.com/QSQ.htm), followed by careful questioning, analysis of information provided by this and by the client's bed partner, and by other methods as required. Then, depending on overall findings, one or a number of the following well-researched interventions are implemented until sleep is normalized

First, if there are signs of a sleep disorder such as sleep-disordered breathing or periodic limb movement disorder, I encourage the physician to complete an appropriate medical evaluation and to consider referral to a clinic for a complete investigation of sleep, known as polysomnography (PSG). Then appropriate interventions are implemented if the PSG reveals a specific disorder; for example, use of a CPAP device for sleep-disordered breathing.

Second, daily practice of the stabilization strategies is encouraged.
Third, if there are problems with adjusting the sleep-wake cycle to time of day (for example, not being tired until 2 a.m.), then use of proven strategies such as light therapy and perhaps the hormone melatonin can be considered.

Fourth, clients are encouraged to maintain tactical breathing (as long as there is no evidence that it worsens sleep-disordered breathing) and muscle relaxation strategies throughout the day and especially at bedtime. Many of my clients have reported that doing so is very effective in decreasing the time to sleep onset as well as promoting a return to sleep after night waking.

Fifth, nightly use of an AVED (Audio-Visual Entrainment Device) to calm the mind is recommended. The unit I use is made by MindAlive in Edmonton, Alberta. The AVED works by using light and/or sound at specific frequencies to stimulate similar brain wave frequencies; slower brain wave frequencies are associated with sleep.

Sixth, use of Krakow's imagery rehearsal therapy (IRT) is implemented for those whose nightmares do not abate with the above interventions. IRT involves initially practicing pleasant imagery during the day, then a week or so later writing a different script to the nightmare (for example, changing the ending), and finally rehearsing the new script frequently before bedtime with as much imagery as possible. A number of studies and one large randomized control study demonstrated significant reductions in both nightmares and insomnia with this therapeutic technique. Some studies report relatively consistent decrease in other PTSD symptoms also while some studies have not found this to be the case. Many of my police and military clients have found IRT to be helpful with the result that either a change in the upsetting details of the nightmare takes place or nightmares stop all together.

With about 70% of my clients with PTSD, the combination above has been successful in restoring normal sleep. If the above is not completely successful in re-establishing restful sleep, then two additional approaches are considered.

Seventh, if analysis of a qEEG indicates anomalies, then I consider BFT-C since there is evidence that it can serve to improve sleep. Although I proceed only following analysis of a qEEG, some colleagues following the pioneering work by Dr. Sterman have reported success by increasing the amplitude of the EEG signal in the 12-to 15 Hz. Additionally, significant increases in memory performance were reported.

Eighth, seeing a physician for sleep medication (known as hypnotics) is an option. For example, although both are approved in the treatment of PTSD, the SSRI Paxil has been shown to decrease sleep disturbances among some clients while insomnia is a frequent side-effect of the SSRI Zoloft. If sleep still does not respond to a sedating SSRI, then a couple of open-label studies have shown that adding Trazodone or an intermediate to long-lasting benzodiazepine such as temazepam can be helpful. As indicated in the section above on medication, several small and one large study has demonstrated that prazosin has resulted in large reductions of both insomnia and nightmares; one study found that sleep disturbances returned once the medication was withdrawn. Yet other small studies have found that the sleep aid zolpidem tended to improve sleep. (However, in 2007, federal regulatory bodies issued a warning of possible significant safety issues with prescription sleep medication including Ambien, Butisol, Dalmane, Doral, Halcion, Lunesta/Immovant, Placidyl, Prosom, Restoril, Rozerem, Seconal, and Sonata/Starnoc). Sleepiness is a common side-effect of the anti-depressant mirtazepine and this medication was found in one open-label study to reduce the frequency and intensity of nightmares among refugees. In addition, other small studies have suggested that buspirone, and gabapentin tend to improve sleep. An open-label study adding respiridone to an SSRI prescribed to combat veterans published by Dr. Davis and her colleagues in 2006 indicated decreases in the frequency of night awakenings and trauma-related dreams. Similar findings were reported in the 2006 open-label trial of adjunctive respiridone among combat veterans by Dr. David and associates. A 2008 publication by Neubauer reported the value of ramelteon for sleep onset; this medication does not have sedating effects but influences sleep regulatory mechanisms within the SCN. A physician colleague in a service for returning U.S. veterans with blast injuries (some of whom had PTSD as well) has found that the medications trazodone and Seroquel are effective; otherwise the over-the-counter medication benedryl can be employed to advantage. However, large randomized placebo-controlled studies are needed to assess the findings above.

If daytime fatigue is the problem, then a trial of modafinil or Ritalin can be considered. It needs to be recalled, however, that any medications can result in unwanted side effects, mostly benign but sometimes having more impact, and this needs to be considered as well as being monitored carefully.

Treatment Step H: Attention to Memory

Memory processes and disorders of memory are central to PTSD. In fact, some have argued that PTSD is basically a disorder of memory since those with the condition are unable to forget the trauma both consciously and unconsciously. Also, most of those with PTSD experience from minor to major difficulties in remembering details of everyday life, such as which items to obtain at the store.

Contents

H-01: Memory and Re-Experiencing

The most obvious way that memory is affected has to do with the core PTSD symptom of re-experiencing. The memory of the trauma occurs unbidden, is vivid, is emotion-laden, is most often visualized, and involves a reliving of past traumatic events in the present. Clients describe it as if someone pushed the re-play button on a VCR. Then the scene replays over and over again. Research to date indicates that a brain structure known as the locus coeruleus located in the brain stem is significantly involved in traumatic memories. This structure has abundant projections to other parts of the brain involved in memory formation including the amygdala, the hippocampus, and the prefrontal cortex. Research to date suggests that the locus coeruleus releases the brain chemical norepinephrine, which ultimately inhibits the prefrontal cortex from extinguishing or suppressing memories. Other stress hormones are involved also, which results in over-consolidation of the traumatic memory/memories. In addition, some have argued that any reactivation of the memory leads to another round of over-consolidation, which suggests that revisiting the trauma repeatedly in treatment is contra-indicated for this reason as well. My clinical experience to date with police and military clients is that with treatment methods such as BFT-A and BFT-M described in Step F, for most clients the traumatic memories no longer evoke an anxiety reaction (the link between memory and reaction is weakened) and ultimately the frequency of these memories decreases. Among those whose traumatic memories continue, my clinical experience to date after using additional methods based on applied neuroscience is that (a) traumatic memories occur rarely or typically are not re-triggered by exposure to cues and that (b) during those rare times when they occur, their intensity, duration, and anxiety-arousing properties are so reduced that clients are not bothered by them.

H-02: Verbally Accessible Memory

Verbal recall of the details of the trauma is sometimes referred to as conscious memory or as *verbally accessible or autobiographical memory*, a term used by Dr. Brewin. When they remember consciously the details of the trauma, many of those with PTSD re-experience the trauma and demonstrate signs of distress accordingly. Sometimes the memories are triggered by a reminder (such as a question from another or news coverage of a similar incident). The person can verbalize both the trigger and the resulting experience. Hence, the memory is verbally accessible. The reason for this experience was investigated in a 2007 study by Piefke and colleagues, in which an fMRI was completed on surgical patients with acute PTSD. They found that compared to non-traumatic memories, traumatic ones increased neural activity in a number of brain regions implicated in memory and emotion such as the amygdala, lateral temporal area, retrosplenial area, hippocampus, and anterior cingulate.

However, it is true also that some victims cannot remember all of the relevant details of their traumatic experiences. This may be because (a) so much information was incoming that it could not all be encoded in verbal memory, (b) the focus of the person's attention was automatically restricted to only those aspects related directly and immediately to survival, and (c) the activity of the hippocampus (one brain structure known to be involved in short-term memory) was disrupted (by sleep loss or the release of excess stress hormones in the brain) so that portions of the event never made it to long-term storage. In relation to this, many studies and my own clinical experiences indicate that this type of memory impairment is clearly evident in some with PTSD. This was particularly poignant in my clinical work with police officers who had been forced by the actions of the suspects to shoot such that the suspects died of their wounds. The police officers had taken these actions in order to avoid their own deaths or grievous bodily injuries. Typically the events took place within a very short time span. To make matters worse, authorities both internal and external to the police organization often did not give officers sufficient time to recover from physiological and psychological states so as to increase the chances that some of the details would be recalled when questioned later. At other times, investigators suggested possible details of the shooting, which officers agreed could have been the case even though they had no conscious memory of these possibilities. At other times, failure to remember was taken to mean that officers were faking memory impairment because they knew that they had made an error of commission or omission.

H-03: Situationally Accessible Memory

Another type of memory is unconscious memory or, as labelled by Dr. Brewin, *situationally accessible memory*. In this type, information about which the person was not consciously aware (including sights, sounds, feelings, smells, and so on) becomes associated with the trauma. This non-verbalized information and trauma association is stored in part so that a more rapid and automatic response can occur in the future if the same stimuli should be present again. Unfortunately, this means that exposure to such stimuli can provoke a PTSD response and the person will be unaware of what the trigger was. Without specialized psychological interventions, this information cannot be recalled voluntarily since it is below the level of conscious awareness and not under verbal control. Such memory is a sub-cortical process that we share with our animal ancestors that allowed them to survive a hostile environment. This is why even police officers who are trained to notice small details sometimes do not remember such details after exposure to a significant traumatic incident such as the shooting of a suspect. Also, it explains why those with PTSD often experience sudden and extreme distress, like the symptoms of panic disorder, without knowing why; it often transpires that a stimulus that was present during the initial trauma reappeared outside of conscious awareness but nevertheless triggered the amygdala and related automatic structures to respond as if the emergency was happening again. It is known that trauma-related pictures presented too fast for people to be aware lead to activation of the visual association areas in the occipital lobe.

In their 2008 publication, Peres and associates refer to the dual representation theory with one aspect primarily associated with vivid re-experiencing and the second as autobiographical memories of trauma. Also, they consider neuroimaging studies that suggest different brain functioning associated with each kind of memory. Moreover, they note that the two systems are not static over time, nor is the expression of traumatic memories. The recall of the situation as well as the triggers of reactivity change over time. Moreover, they indicate that multiple memory systems can be activated simultaneously.

From the perspective of clinical intervention, I have noticed one or more of three things happening with situationally accessible memory. First, with the interventions in previous sections, generally there is a reduction in the frequency of reactivity. Second, if an unidentified cue triggers a reaction, clients can still use tactical breathing, which usually reduces psycho-physiological hyper-reactivity; over time, this seems to lead to a decrease in connection between a situationally accessible memory and client reactivity. Third, by keeping a record of the circumstances in which they find themselves when symptoms emerge, clients can begin to make a connection between the cue and their traumatic event. In the few instances where significant reduction or complete remission of hyper-reactivity did not occur, usually such an objective was achieved ollowing BFT-C.

H-04: Working Memory
Subsequent to the trauma, some with trauma-related conditions such as PTSD report significant difficulty with aspects of *working memory*; for example, they have problems in learning or retaining new information. Also, even simple things like where they placed the keys, whether they locked the door, and the three things they were supposed to get at the store are hard to remember whereas before the trauma no such memory difficulties were noted. Research on memory has supported such observations by clients with PTSD: for example, a number of studies have documented PTSD-related deficits on one or more measures of memory and learning. The 2008 study of 50 Dutch peacekeepers (25 with PTSD and 25 without) by Geuze and colleagues found that those with PTSD had memory deficits even though intelligence levels were similar for both groups. The 2008 research reported by Dickie and her group found that greater memory impairment was found among those with higher versus lower severity of PTSD. Other studies have not found such evidence; for example, Demakis and co-investigators in their 2008 study found no relationship between severity of PTSD symptoms and cognitive ability. Adding to the complication are a series of 2008

findings: Bonne and collaborators found smaller hippocampal volume among those with PTSD compared to controls; Whalley and co-workers found that those with PTSD had a substantially intact neural system; Wild and Gur found that those with PTSD who did not respond to cognitive behavioural therapy had poor performance on measures of verbal memory; the Werner team found that PTSD has an effect on memory- related brain function even though memory functioning was intact; and the Galletly group reported that those with PTSD differed from both normal control subjects and patients with schizophrenia in aspects of N2 and P3 on a target detection task dependent on auditory memory. Clearly, although the bulk of the research to date indicates impairment in working memory among those with PTSD, much more research is required to sort this all out.

Also, at this time it is not clear why those with PTSD might suffer from memory problems apparently unrelated to details of the trauma. A number of possibilities have been mentioned including disruption of attention such that the information is never encoded, dis-coordination among the multiple cortical regions involved in working memory, decrease in brain activity related to reduced cerebral blood flow, altered functioning of key neurotransmitters such as GABA and norepinephrine, the negative effect on memory of over-consumption of alcohol, the negative impact of sleep problems, the negative impact of co-occurring depression among some with PTSD, and the fact that some with PTSD may have received a traumatic brain injury. For example, the 2008 fMRI study of Iraq veterans with PTSD by Morey and associates indicated that the presentation of combat-related stimuli led to decreased activation in a number of brain areas associated with executive functioning including the dorsal anterior cingulate, the middle frontal gyrus, and the inferior parietal lobule. In their chapter of the 2008 book edited by Squire and associates, Rapp and Bachevalier note that damage to the front part of the brain leads to problems in such "executive" functions as the strategic use and manipulation of remembered information, the ability to identify the source from which information was acquired, and the order in which it was presented

A major focus of research on impaired working memory among those with PTSD has been the hippocampus, a key brain area involved in short-term memory. Although some studies have not supported them, among the fairly consistent findings are:

> *decreased bilateral hippocampal volume among those with PTSD compared to controls. This is true also among those with severe PTSD as contrasted with those exposed to trauma who did not develop PTSD;
>
> *a relationship between amount of hippocampal volume and degree of memory deficit;
>
> *decreased hippocampal activity shown by PET studies during memory tasks; and
>
> *a reduction in neuronal integrity among those with chronic PTSD.

Also, studies have shown a 5% increase in hippocampal volume with a 35% increase in memory after one year on the SSRI medication Paxil.

As noted in a 2006 review of the topic by Dr. Bremner, researchers have reported that

prolonged stress affects the hippocampus as a result of any of a number of processes including the impact of elevated levels of gluocorticoids released during stress, stress-related inhibition of brain-derived neurotrophic factor (BDNF), changes in serotonergic function, or inhibition of neurogenesis (the growth of new neurons in the hippocampus).

A 2006 review of evidence by Alfarez and associates proposed that stress and the associated elevated levels of corticosteroid hormones has a negative effect on the excitatory amino acids that mediate synaptic transmission and synaptic plasticity, which are considered to underlie learning and memory processes in the hippocampus.

Budzynski and colleagues (in Evans, 2008) note that chronic and repeated stressors can produce atrophy of the neuronal dendrites that allow brain cells to connect with each other as required. This state can be reversed. Also, they cite a study demonstrating that perceived stress can alter telemere length, which is a DNA measure indicative of cellular aging.

Bremner (2006) notes also that

it is unclear at the current time whether these changes are specific to PTSD, whether certain common environmental events (e.g., stress) in different disorders lead to similar brain changes, or whether common genetic traits lead to similar outcomes. (p. 84)

A growing functional imaging literature suggests that PTSD is also associated with alterations in neural activation during memory tasks such as under-activation of the left hemisphere of the brain. In addition, a 2006 fMRI study of those with PTSD by Moores and her colleagues observed a pervasive reduction of brain activity during a test of working memory and specifically during those periods when updating that memory was required.

A 2008 study by Schuff and co-workers assessed N-acetylasparate (NAA) levels in the hippocampus of those with PTSD and those exposed to trauma who did not develop the condition. NAA measures the integrity and functioning of neurons. The authors found reduced NAA levels in the hippocampus among those with PTSD but not among control subjects. They did not find significant reductions of hippocampal volume in either group.

Interestingly, in addition to the role it plays in respect to memory, studies have shown that the hippocampus has an important role to play in regulation of stress by its effect on such brain chemicals as glucocorticoids, CRF, and BDNF as well as its capacity to alter serotonergic receptor activity–all of which are of importance in PTSD.

Regardless of mechanism, if working memory does not appear to be restored among those whose PTSD symptoms are in remission with the intervention methods discussed to date, then the computer-administered standardized

MicroCog test is repeated. This test assesses a range of cognitive activities such as attention, memory, and information processing. It provides an objective evaluation from which decisions can be made about both intervention and fitness to return to work with the safety of the person and the public in mind. Any continuing differences between the re-administered MicroCog and expected levels are presumably due to some factor other than PTSD symptoms.

H-05: Interventions For Working Memory
In the absence of other possible contributing causes to memory problems (for example, untreated hypothroidism) and depending on assessment results, clients have a number of options. The list below is often one I follow in sequence.

One, to decrease worry about the significance of any memory problems. It is helpful to be aware (a) that memory is often affected in PTSD, (b) that this is due to brain processes, (c) that it does not mean that the person is going crazy, (d) that it usually improves, and (e) that it can be managed until improvements are noticed.

Two, until memory improves, clients are encouraged to use aids to memory; for example
_____*make and consult lists,
_____*place post-it notes in strategic places,
_____*do one thing at a time,
_____*organize tasks into established routines, and
_____*use timer/reminder functions from an electronic planner.

Three, it is important to follow faithfully all of the stabilization strategies relevant to memory. For example, keep stressors within manageable limits, continue aerobic exercise, decrease or eliminate alcohol, and eat a healthy diet with the addition as required of supplements containing vitamins, minerals, and omega-3. Also, research as summarized in the 2006 review by Walker and Stickgold and again in the 2008 review by Walker, points to the absolute importance of sufficient restful sleep due to its proven and substantial role in memory. Comparing those who slept well versus those who were awakened during various stages of sleep, studies have shown that disturbed sleep leads to more forgetting generally, more memory loss when competing information is presented, more remembering of negative versus positive stimuli, less ability to learn material the next day, and less speed and accuracy the next day on motor skills learned before going to sleep. From all this, it is clear that good sleep is critical to learning, remembering, and consolidation of memory. Most clients with PTSD have significant sleep problems.

Four, since it is one of only two medications recommended for the treatment of PTSD anyway, and given results of improved memory with its use noted by Bremner, physicians have the option of prescribing Paxil. Furthermore, there is evidence that Paxil promotes the growth of new neurons in the hippocampal area. On the other hand, studies indicate that this SSRI is less effective with military folk (and thus is likely to be less effective for police also) than for female victims of a single trauma. Also, some cannot tolerate the common side effects of a significant decrease in sexual interest and/or performance, night sweats, and, for women, weight gain.

Five, clients are encouraged to continue with my basic PTSD treatment approach. Following its completion, upwards of 70% of police and military clients find that they no longer have the symptoms of PTSD and most discover also that memory has improved.

Six, procedures to reduce monkey-mind as described in section "L" can be implemented. The net effect is an increase in attentional processes. Sometimes memory problems are not due to memory process but rather are due to problems in attention/concentration. If one's mind is elsewhere, then information does not get to memory.

Seven if memory has not improved with the approaches above, my next step is to recommend a qEEG. Subsequently, any dysregulation in either amplitude of the EEG signal and/or reduced coherence between pairs of cells can be addressed with BFT-C if there is reason to believe that such anomalies are associated with memory processes.

Alternatively, some colleagues recommend the option of using the near-infrared spectroscopy (NIRS) form of hemoencephalography (HEG) to increase blood flow to one of more areas of the brain involved in working memory. Clinical experience to date has shown that such treatment has positive effects on memory. For working memory, generally prefrontal locations (Fp1, Fp2, and Fpz) are chosen as the site. A velco strap is placed around the head with the sensor on it positioned in front of the desired location. The blood flow level at that site is measured continuously and presented on a computer screen, providing feedback to the client as to whether it is increasing or not. Clients are encouraged to use that feedback in the attempt to increase the strength of the signal. This encourages an increase in blood flow, thereby bringing glucose and oxygen to that area of the brain. Ultimately this results in an increase in the number and/or size of blood vessels in that area of the brain. With increased blood flow and vascular growth, memory functions are increased. I am not aware of any negative side effects of this approach.

Eight, based on a study by Dr. McFarlane that showed positive effects on memory among those with PTSD, a trial of the medication donepezil (5 to 10 mg) can be considered by the physician. Donepezil is a central acting selective acetylcholinesterase inhibitor. Various case reports such as by Foster and Spiegel (2008) indicate that it has been used to treat disturbances of attention, arousal, memory, and executive functions in folks with either dementia or head injury. It is presumed to have a positive effect by preventing further neuronal atrophy in the hippocampus. As demonstrated in a 2006 publication by Chen and associates donepezil (5 to 10 mg/day) administered to older adults with mild cognitive deficits may be effective since cerebral blood flow in the left frontal and temporal region was not reduced over time, as it was among those given a placebo. However, caution is required since there are some reports of traumatic memories being re-triggered among some for whom donepezil was prescribed. Also, many physicians seem reluctant to prescribe this medication.

Treatment Step I: Elements of Cognitive Behavioral Therapy

The cognitive portion of Cognitive Behaviour Therapy (CBT) has a long association in both the understanding and treatment of PTSD. The behaviour aspect of CBT is included in what I refer to as *Stabilization Strategies* as presented in Step E.

Contents

I-01: Introduction and Assessment

There is research supporting the position that beliefs and thinking processes in some folks with PTSD are not as helpful as they could be. Dysfunctional beliefs and problematic information processing induce stress that can lead to maladaptive emotional responses. Attention to these aspects is known as cognitive therapy (CT). Correcting cognitive issues with CT has proven to be effective in treatment outcome among those exposed to a single traumatic experience. Thus, studies have demonstrated that the intensity of PTSD symptoms can be reduced with CT. Furthermore, other studies have shown positive changes in the electrical activity of the brain following CT. For example, a 2008 experiment published by Rabe and associates divided those with PTSD following a severe motor vehicle accident into two groups, only one of which received CT. At pre-treatment, activation of the right side of the brain resulted when clients were exposed to a trauma-related picture. At post-treatment, the group who had received CT showed a greater reduction in right-side activation than the control group when exposed to the trauma cue. Also, the more the decrease in right-sided activation, the greater the reduction in PTSD symptoms. For me, the jury is still out as to the overall relevance of CT to police and military victims of PTSD or its impact on their treatment outcome. Also, at this time it is not clear to me as to the extent to which some of the noted thinking issues were present before trauma or are a result of trauma.

I assess beliefs and thinking informally. Sometimes, I ask questions as needed (for example, "Do you believe you are responsible for what happened?"). I pay careful attention to what clients say (for example, when they talk about why they believe things are happening to them). I look for patterns in how they deal with information (for example, if everything is either black or white for them). I check on the methods they use to moderate emotional arousal. Sometimes, I administer tests of beliefs.

Then, when the timing seems to be appropriate, (a) I relay my observations, (b) I recommend readings , (c) we discuss cognitive interventions, (d) the client selects an approach s/he believes will be effective, (e) we try out this approach, and finally, (f) we evaluate results and modify the method as needed. Generally I have found that for most clients, concerted attention to beliefs, distortions, and goals is most successful after completion of the treatment components above. However, up to 30% of clients to date did not resolve cognitive issues completely until after abnormal patterns of cortical activation were brought under control through BFT-C, as discussed in step "K." For these clients, often the cognitive anomalies resolved without any further intervention.

I-02: Inaccurate Thoughts about Self
It is common for those with PTSD to have inaccurate thoughts about themselves: for example, that they are damaged for life, that they are responsible for what others did, that they should have known better or acted more wisely, that they have failed to live up to their self-image (such as being completely in control at all times), that their sense of their own basic goodness was shattered by the act of killing another, that they should live with shame and guilt because they were helpless during a traumatic event, and so on. I have found that de-mystifying PTSD in part by providing psycho-education to the client helps in this regard. Additionally, as inaccurate thoughts about self come up in the course of the treatment elements above, they are submitted to reality testing and subsequently are changed to more accurate ones by the frequent rehearsal of replacement thoughts.

The importance of a well-regulated sense of "self" has emerged in some preliminary and exciting recent findings in neuroscience. It began with what has been referred to as the default brain, the state wherein the thinking part is not engaged in any specific tasks. As summarized by Dr. Thatcher in his 28 July 2008 message to a listserv I receive:

> The default brain was discovered by PET and fMRI researchers when they noticed that the baseline oxygen, blood flow and energy consumption of the brain decreases when one is engaged in a task in comparison with simply resting. . . . It turns out that there is a system of the brain that includes the posterior cingulate gyrus, the precuneus, the bilateral temporal lobes and the midline prefrontal cortex that is involved in maintaining the "self" or creating the "self narrative" or the "autobiographical self.' This system of "self homeostasis" or "self-maintenance" is a very energy expensive system and consumes large amounts of oxygen and blood flow within the most interconnected parts of the brain. Subjects that have a better regulated or homeostatic "self" have more resources to allocate for tasks. When the metabolically expensive "self-referential" system is out of balance, then there is a high price to pay in diminished functioning or efficiency in other brain regions. (personal communication)

Thus, it is reasonable to conclude that any psychological interventions leading to a better regulated sense of self will result in an increase in the central nervous system resources needed to manage PTSD. Accordingly, the methods discussed to this point (including stabilization strategies, applied psychophysiology, and restful sleep) as well as affective neuroscience and stress management presented in steps "K" and "L" below assume added importance.

Also, for those who exhibit problems with self-view, I recommend working through the most recent version of the book *Self-Esteem* by M. McKay and P. Fanning (New Harbinger Publications, Oakland, CA). ISBN for the 3rd edition I have is I-57244-198-5. The book covers critical topics such as disarming the critic within yourself, how to do an accurate assessment of yourself, how to replace cognitive distortions, and how to handle mistakes and criticism.

I-03: Beliefs about the World

The beliefs or assumptions held about the world provide a foundation for making sense about things, a guide as to what to expect, and a basis for our actions. They too can be shattered following trauma. Commonly, such beliefs can include the view that there is fairness in the universe, that other people generally are good, that one is usually safe, and that there is meaning in what happens to us. With the shattering of such beliefs, some folks with PTSD are left questioning everything they once thought to be true, not knowing what can be counted on, and unable to predict what will happen. The consequence is increased anxiety. The options of honestly acknowledging what has been shattered, developing alternate views about reality, testing them out, and finally rehearsing them until they become automatic are considered as these issues arise during the treatment process above. Given the seedy side of life to which police officers and military veterans are constantly exposed, I have seen them often develop strong but still erroneous beliefs that are the opposite of the above; there is <u>no</u> fairness anywhere, life is basically capricious, most people are or could easily become evil, and most things are virtually meaningless. Examining these views, developing alternate ones, testing them, and incorporating them become part of the intervention as the issues arise naturally during the course of treatment.

To begin the process of developing accurate world views, often I recommend the self-help book *Don't Believe It for a Minute: Forty Toxic Ideas That Are Driving You Crazy* by A. A. Lazarus, C. N. Lazarus, and A. Fay (1993, Impact Publishers, ISBN0-915166-80-2). A brief chapter is devoted to each of 40 self-defeating beliefs, with ways to change it. As a result, a format for considering other personal beliefs is developed.

Impact of retirement, reduced physical capacity, concurrent ph

Treatment Step K: Adding Methods Based on Applied Neuroscience

In summary, about 70% of my combined group of police and military clients achieved remission with my basic approach; following treatment, they no longer met the diagnostic criteria for PTSD. Most of the remaining 30% achieved a meaningful reduction in symptoms but not enough to bring the PTSD into remission.

For those who do not reach remission with my basic approach, I review with them the elements and rationale for a subsequent intervention. I refer to this step as BFT-C, direct biofeedback of the central nervous system with the EEG as the signal of interest. Once they have all the information they need, clients can then choose to complete an informed consent to proceed with this additional method of treatment. This consent is found in Appendix III. The basis and results of BFT-C are presented in the following sections.

Contents
K-01:   Affective Neuroscience
K-02:   The Fear Circuitry and the Role of the Amygdala
K-03:   The Role of the Prefrontal Cortex
K-04:   Affective Neuroscience the Amygdala-PFC Interactions
K-05:   Electro-Physiology Abnormalities
K-06:   Other Brain Research
K-07:   Reports on the Use of BFT-C
K-08:   Implementing BFT-C
K-09:   Results of My Clinical Series
K-10:   Event Related Potentials
K-11:   Closing Comments
K-12:   Further Information

K-01: Affective Neuroscience

Beginning in about 2000, I began to study a then relatively new area of investigation for me, affective neuroscience, which attempts to integrate findings about emotional activity at the level of brain functioning. Affective neuroscience is based on research in clinical psychology, neuropsychology, neuropsychiatry, neurology, neuro-imaging, EEG, neurobiochemistry, psychopharmacology, genetics, molecular biology, and neuro-endocrinology.

My involvement was a natural outgrowth of my experience with BFT-A and BFT-M as well as my long-term interests in neuropsychology, the relationship between brain and cognitive functioning. This study of the field of affective neuroscience included attendance at many conferences with experts in the field, informal post-doctoral course work, and thousands of hours of reading. Consistent with that reached by other psychologists, my conclusion was that there was good reason to believe that BFT-C described below might be helpful for some of my clients.

Some of the information leading me to that conclusion is presented below.

Also, developments are occurring at a breathtaking rate. For example, at one time, amplitude of the client's raw brain wave signal was the only variable for

BFT-C. Then came raw coherence statistics, which measure the communication degree between two areas of the brain. Recently, z- score training referenced to the NeuroGuide and LORETA software has become available. Compared to raw score training, the expectation is that the NeuroGuide modification may result in more rapid training while using LORETA z-scores may permit also more precise localization of the BFT-C training. The utility of other statistics derived from the qEEG such as phase shift, phase lock, and phase reset are under investigation. Also, it is becoming increasing clear that the brain's electrical activity may be concentrated in a limited number of "hubs" and work is proceeding on this front. According to Dr. Thatcher, "brain function is the coordinated interplay of five to six massive 'Hubs' or 'Modules' and three basic pacemakers" (personal communication, NeuroGuide listserv of 16 May 2009).

Analyses of my use of BFT-C indicated that the vast majority of police officers who still met the diagnosis of PTSD after completion of my basic approach were able to achieve remission after a course of BFT-C based on raw amplitude of the EEG signal. Attention to both amplitude and coherence among subsequent clients indicates that the pattern of success is still evident.

K-02: The Fear Circuitry and the Role of the Amygdala
Incoming sensory stimuli such as sights and sounds are processed by both of two routes. The *slow route* developed in our most recent evolution as a species. In this route, sensations from vision, hearing, and other inputs travel to the thalamus of the brain, which then directs the signal to cortical areas of the brain involved in that sensory experience (for example, what we see goes to the visual association areas at the back of our head). The processed information from these primary sensory regions then is forwarded to the frontal lobes of the brain and is associated with conscious awareness of the sensory signal a half-second or so later. Subsequently, information can then be transmitted to the amygdala. However, because the frontal lobes of the brain (including the anterior cingulate gyrus and orbital frontal cortex) have the capacity to inhibit emotional responses, the amygdala does not necessarily respond to the sensory information via the slow route. For example, the frontal lobes may identify an object as a twig and therefore inhibit a fear response from the amygdala; this process would be quite different if the object were identified as a poisonous snake. Also, there is evidence that CBT strengthens the brain's frontal areas, thereby helping to inhibit amygdala over-activation.

The *fast route* is a left over from our primordial motivation-emotion system whereby crucial features of the external environment travel an ancient thalamic-amygdala shortcut to the limbic system and arrive significantly sooner than the signals that travel to the sensory primary association areas via the slow route. Since the fast route was developed before our species had language, we do not have conscious awareness of what is happening within us and survival reactions occur before pre-frontal areas of our brain are aware or can exert any control. Rudimentary rather than refined or verbally comprehended sensory stimuli go within one-quarter of a second from the thalamus to sub-cortical brain structures in the limbic system centred on sections of the basolateral amygdala (including the stria terminalis or extended amygdala). The function of the amygdala and connected structures then includes (a) maintaining vigilance, (b) orienting /paying selective attention (c) evaluating very rapidly the incoming stimuli for harm-potential even before we are aware of the stimuli, (d) reacting, and (e) influencing other brain regions so as to produce an integrated response to any stimuli that might be arousing, ambiguous, or threatening to us. Additionally, as we evolved as a species, internallygenerated stimuli such as feedback from the cortical, autonomic nervous, neuro-endocrine, and skeletal-motor systems also reached the amygdala and resulted in the same responses. The fast route probably accounts in part for why trauma cues of which we are not consciously aware, internallygenerated memories of trauma, activation of the autonomic nervous system by any stressor, and muscle tension for any reason have the capacity to switch on the amygdala and other parts of the fear circuitry, resulting in an increase of PTSD symptoms. The central nucleus of the amygdala sends output to the hypothalamus and brain stem region.

Whatever the route, when a stimulus is perceived as threatening (which is likely to happen with an overactive amygdala), the resulting unconscious response from the amygdala is accompanied by activation of other areas and processes of the brain:

* the lateral nucleus of the hypothalamus subsequently releases the peptide hormone CRF. The hypothalamus is part of the HPA axis;
* the release of CRF acts on the medulla and the locus coeruleus in the brain stem;
* the resulting release of norepinephrine prevents sleep, increases amygdala activity, reduces the inhibiting capacity of the pre-frontal cortex, increases attention and vigilance, and increases autonomic nervous system arousal leading to such symptoms as increased blood pressure and heart rate;
* there is a co-release of norepinephrine with neuropeptide Y, a neurotransmitter found in the sympathetic nervous system and in multiple stress-responsive brain regions such as the locus coeruleus, amygdala, and hippocampus. This is released only with intense stress and potentiates the effects of norepinephrine;
* the periaqueductal grey is activated, leading to defensive behaviours and postural freezing;
* the parabrachial nucleus is activated, which increases respiration rate;
* the nucleus pontine reticularis, related to the startle response, is activated;
* the paraventricular nucleus of the hypothalamus is activated, which stimulates the anterior pituitary gland to synthesize and release ACTH;
* there is a resulting increased release of adrencorticoids, which takes longer to return to baseline levels as we age;
* functioning of the pre-frontal cortex is impaired by increasing catecholamine levels (epinephrine and nor-epinephrine). Also, these brain chemicals are toxic to the hippocampus, which shrinks when exposed to them, thereby disrupting its functioning, lowering the chances of the growth of new neurons, inhibiting the development of new connections between brain cells, and

causing cells to burst and release the brain chemical glutamate, which stimulates surrounding cells;

* functioning of the hippocampus decreases, which (a) impairs memory, (b) decreases its capacity to inhibit overactivity of the HPA axis, (c) may decrease the moment-to moment integration of thoughts, feelings, and sensations, perhaps leading to symptoms of dissociation from which some with PTSD suffer, and (d) perhaps turning off processes that may be involved in the extinction of fear conditioning;

* glutamate is released, which decreases the neurotransmitter GABA's capacity to inhibit the amygdala. Normally, GABA allows the thalamus and amygdala to filter out sensory information;

* serotonin levels are reduced, which lowers the firing threshold of the amgdala (increased activation of the amygdala);

* hormones, neurotransmitters, and neuropeptides such as those above are released and interact with each other in a complex fashion, ultimately leading to an increase in hyper-arousal and hyper-vigilance;

* systems involved in skeletal and muscle systems are mobilized; and

* the breakdown of stored fats and glycogen is increased to produce usable energy such as required for the movement of large muscles.

The threat characteristics and their context are then stored in unconscious memory (involving the hippocampus and amygdala) so that an even more rapid response can occur when these stimuli are present again and "recognized" as potentially threatening. This process does not occur among those whose amygdala is not functioning, nor does it happen when those with PTSD are exposed to non-threatening situations. The encoding and consolidation of fear memories seems to be related to the increase of noradrenergic activity during a traumatic event, which has implications in the prevention of PTSD as discussed towards the end of this section. The extinction of fear-learning seems to involve both the amygdala and the ventro-medial pre-frontal cortex in ways that I do not yet understand.

Most of the human research conducted since about 1997 in regards the generation of many of the symptoms of PTSD has focussed on the role of the amygdala, a complicated, small, paired, almond-shaped collection of neuron clusters buried in the anterior medial temporal lobe of the brain. A part of the amygdala receives sensory information (except smell). It is considered part of the limbic system of the brain, has abundant receptors that are modulated by serotonergic neurons, and is a key component of the circuitry involved in processing emotionally significant stimuli, detecting threat or potential punishment, generating negative emotions such as fear and anxiety (in part by orchestrating endocrine and autonomic components of the fear response), the storing of fear associations, and preparing the person for a response. In addition, studies suggest that the amygdala is associated with processing emotional expression, making judgements about the trustworthiness of unknown people, and the attribution of internal states, beliefs, and desires to other people, all of which relate directly to those with PTSD. Moreover, dysregulation of the amygdala as well as other brain structures can result in symptoms associated with PTSD such as increased rage, violence, explosive temper, hostility, and irritability, as well as increases in placidity, passivity, and apathy.

Many of the early studies of the amygdala were focussed on its size among those with PTSD as compared with those without the condition. As discussed in the 2009 paper by Woon and Hedges, a meta-analysis by Karl and associates found that a mixed sample of children and adults with PTSD had a smaller left amygdala than those without PTSD. However, in their 2009 meta-analysis of careful studies to date with adults only, they did not find differences in amygdala volume between those with and those free from PTSD.

Furthermore, research indicates that the presence of one genetic variation (5-HTTLPR S allele) showed a five-fold increase in activation of the amygdala (relative failure of regulation of the amygdala) compared to those with a different genetic variation (L allele), perhaps offering an explanation of why some people may be more reactive than others. Relatedly, the reported study of Schardt and associates indicated that the s-allele carriers showed significantly more activation of the amygdala in reaction to aversive stimuli than the comparison group. Also, some differences in responding may be related to the use of reappraisal in everyday life (for example, identifying an object as a twig and not as a snake). The 2008 fMRI study of healthy women indicated that those who used appraisal in daily life showed decreased amygdala activity and increased pre-frontal and parietal activity when shown emotionally negative faces.

While much of the recent research and clinical focus has been on the amygdala, it is clear that this structure has many connections to numerous cortical and sub-cortical systems that participate in sensing and evaluating the emotional properties of stimuli. So it is simplistic to consider the amygdala as the only part of the fear circuitry. Still, it gives us an idea of how things happen.

A substantial number of neuro-imaging studies of those with PTSD have demonstrated consistently increased and rapid over-reactivity in the amygdala when threat/fear stimuli are presented, for example combat-related sounds for military veterans with PTSD. This reactivity seems not to occur among folks without PTSD nor among those with PTSD when cognitive demands unrelated to the trauma are presented. Many of the studies were summarized in the 2008 literature review by Liberzon and Sripada. Also, high PTSD scores on the CAPS, a measure of severity of PTSD, have been linked with increasing activation of the amygdala. Other investigatory approaches have been consistent, including studies of those whose amygdala was not functioning and studies wherein drugs were administered that impacted on the functioning of the amygdala; among such clients, fear and anxiety reactions were absent. A 2008 study by Koenigs and colleagues of veterans whose exposure was traumatic and resulted also in brain injury reported that PTSD rates were reduced among those with damage to the amygdala compared with those whose amygdala remained intact. The same was true if there was damage to the ventral medial pre-frontal cortex.

Moreover, those with PTSD have high concentrations of cortisol-releasing factor receptors in the amygdala. Furthermore, as noted in a 2005 review by Dr. Bremner, combat veterans with PTSD had increased blood flow to the amygdala and decreased blood flow to in the middle temporal and left inferior frontal cortex when exposed to combat imagery (but not to neutral images), while those vets without PTSD showed increased blood flow to the orbitofrontal and medial prefrontal cortex. A 2008 study led by Felmingham found increased amygdala activation to non-conscious fearful stimuli among accident or physical assault survivors with PTSD who dissociated as compared to those who did not dissociate.

The evidence of amygdala activation among those with PTSD is so strong that Dr. Evian Gordon indicated at the 2007 ISNR conference that reactivity of the amygdala as measured by the fMRI is a marker for PTSD. At the same conference, Dr. Sitaran noted that amygdala functioning can be modified using neurofeedback from fMRI imaging, although it is still unknown as to whether such changes lead to a decrease in PTSD symptoms. However, outside of large and well-funded research institutions, fMRI options are not widely available and PTSD can be treated successfully with other methods such as BFT-A and BFT-C.

K-03: The Role of the Pre-Frontal Cortex (PFC)

The ability to master difficult circumstances cognitively, to gain control over emotional arousal, to regulate the sympathetic and parasympathetic branches of the autonomic nervous system, and to increase control over attention resides to a great extent in the frontal cortex of our brain. For example, those with sufficient activation in the frontal cortex are particularly skilled in the down-regulation of negative emotions. The frontal cortex makes up about one-half of our brain and is pretty much connected to all other parts of our brain. Commonly, it is understood as being sub-divided into four regions: the medial frontal cortex (MFC) (which includes the anterior cingulate cortex [ACC]), the orbital frontal cortex (OFC), the dorsolateral prefrontal cortex (DLPFC), and ventromedial prefrontal cortex (VMPFC).

The front (rostral) portion of the anterior cingulate cortex (rACC) has received the most attention in recent research. This is important for a number of reasons. *First*, the rACC has reciprocal connections to such other key brain areas as the amygdala and other parts of the affective (limbic) system, the anterior nucleus of the thalamus, pre-motor areas, spinal cord, red nucleus, and locus coeruleus. *Second*, the rACC along with other areas of the PFC has been associated with such PTSD-relevant functions as priorizing where attention should be directed, decreasing hormonal responses to stress, suppressing reactivity of the HPA axis, inhibiting or extinguishing fear-related processes by regulating emotional reactivity through adjusting cognitive controls, and reducing the generalization of fearful behaviour. Thus, the rACC can be said to play a vital role in regulating our emotional states. *Third*, the rACC appears to function best between 4 Hz and 7 Hz, most easily measured between brain location Fz and Cz. The importance of this is discussed later.

Many lines of evidence point to dysfunctioning of the ACC among those with PTSD.

First, a large number of studies (see the 2008 review by Liberzon and Sripada) comparing those with PTSD and without PTSD have assessed the functioning of the ACC or other frontal areas with fMRI and related neuro-imagining following exposure to traumatic stimuli. Generally, these studies have demonstrated reduced functioning of such brain regions among those with PTSD (for example, Bryant et al., 2008; Falconer et al., 2008; Kim et al., 2008). Typical is the Liberzon and Sripada study in which combat veterans showed more deactivation of the ACC area following emotionally evocative stimuli than matched non-combat controls. In many of the reported studies, evocation procedures have led to increased activation of the ACC among those without PTSD, presumably as a successful way to control emotional reactivity. An interesting 2007 study by Lanius and co-authors examined fMRI data after provoking symptoms in those with PTSD. Those with PTSD with or without accompanying major depression showed decreased brain activation in the anterior cingulate. However, those with PTSD without depression showed even less activation in both left and right anterior cingulate and in both left and right posterior cingulate than those with both PTSD and depression. Relatedly, other studies as reviewed by Liberson and Sripada have provided evidence of reduced cerebral blood flow in the ACC areas among those with PTSD in response to sounds and images of their traumatic experience. Generally, reduced blood flow has been taken to mean reduced functioning of those areas. Moreover, a 2007 study by Liberzon and co-authors of combat veterans with PTSD found that the release of cortisol co-varies with regional blood flow in the rACC as assessed by PET scans while no such relationship was found among combat veterans without PTSD

Second, the importance of the ACC has been indicated by a variety of other studies. Neuro-imaging studies among adult humans as summarized for example by Karl and colleagues in 2006 have often shown that those with PTSD have smaller bilateral ACC volumes as well as a smaller ACC when compared with trauma-exposed controls without PTSD. One indication of the importance of this is the 2008 study by Bryant and associates, which showed that among clients with PTSD who received cognitive behavioural therapy, symptom reduction was associated with larger initial rACC volumes; that is, those with larger ACC volume had better outcomes than those with smaller volume. Relatedly, the 2008 study by Woodward and associates found that a smaller ACC was associated with less parasympathetic regulation of heart rate among combat veterans when under stress. This is consistent with decreased autonomic nervous system regulation among those with PTSD.

Third, and related to the preceding, is the information presented at the 2005 ISTSS conference by a member of Dr. Shin's research group. Those with chronic symptoms of PTSD but not controls show long-term changes in the rACC; for example, decreased activation/responsiveness, decreased volume, a change in shape, and a decrease in neural integrity. This means that the rACC becomes compromised in the long term among those with chronic PTSD and therefore the rACC has much less of an inhibitory role on the amygdala.

Consequently, without normal inhibitory controls, the amygdala is easily activated and threat processes can continue for a long time.

Fourth, studies of the biochemistry of the neurons themselves point to dysfunctioning of the ACC among those with PTSD. For example, Schuff and co-workers published a study in 2008 that used magnetic resonance spectroscopic imaging (MRSI) to investigate N-acetylasparate (NAA) levels in the ACC of those with and without PTSD. NAA is a marker of the integrity and functioning of neurons. Results indicated reduced NAA concentrations in the ACC among those with PTSD but not among those trauma-exposed people without PTSD; all had normal NAA levels globally in other areas of the brain. This suggests a selective abnormality in the ACC. Consistent with this finding was the previous 2007 study by Ham and associates, who reported that, following exposure to a subway fire in South Korea, those with PTSD but not those without had decreased levels of NAA in both ACC and hippocampus; also, the lower these levels were, the more frequent were their symptoms of trauma re-experiencing.

Fifth, ACC dysfunction among those with PTSD is indicated in the clever 2008 study by Felmingham and associates based on the knowledge that those with PTSD who dissociate in the presence of trauma or trauma cues show fewer symptoms of hyper-arousal than those who do not. Their research presented conscious and non-conscious trauma cues to both groups of clients who had been in a single traumatic incident. The dissociation group who had less arousal had more pre-frontal activation to conscious cues than the non-dissociaters.

Evidence that the trauma itself is the cause of differences among those with and those without PTSD was provided in the 2008 study by Kasai and collaborators. As measured by MRIs, combat veterans with PTSD showed significant grey matter density and reductions in right hippocampus, pregenual anterior cingulate cortex, and both left and right insulae as compared with their monozygotic twins who had not been exposed to combat. Also, the rACC has rich connections to other parts of the brain including the anterior insula, nucleus accumbens, hippocampus, hypothalamus, and brain stem regions, thereby affecting autonomic, visceromotor, and endocrine systems.

However, it is clear that other parts of the prefrontal cortex, both singly and via inter-connections with other regions, have an important role to play in PTSD symptoms.

(a) *The MPC* governs emotional, autonomic, and attentional processes and its dysfunction can lead to such symptoms common in PTSD as decreased curiosity, less interest in learning, reduced social interest, diminished affection, and reduced starting of tasks.

(b) *The OFC* has a role to play in filtering emotional stimuli so as to decrease unnecessary arousal, regulating autonomic nervous system, managing the startle response, determining resistance to extinction, inhibiting irritability, and maximizing our participation in those things that are rewarding.

(c) *The DLPFC* is involved in working memory, which is often disturbed in those with PTSD, in victims of brain injury, and in those with depression.

(d) *The VMPFC* plays a role in inhibiting activation of the amygdala.

In summary, with the automatic activation of the fear/threat response described in the section on the amygdala above, sometimes the rACC and other structures of the pre-frontal cortex involved in the filtering, processing, and evaluating of emotional information essentially go off line. Such decreased activation has been found in many studies to occur in those with PTSD but not in others exposed to the same provocations. Without the inhibitory influence provided by the ACC, the amygdala activation is not adjusted downward and therefore both threat reactions and processes continue. This is desirable if the perceived stimulus represents a real threat to the person's survival in that moment but is simply not functional and distressing when there is no actual threat present. Also, with decreased pre-frontal activation, the HPA axis is less well regulated, as demonstrated by Kern and colleagues (2008). As well, as supported by the 2008 research by Falconer and co-authors, those with PTSD generally commit more errors than controls in non-threatening executive tasks requiring inhibition of responding.

Fortunately, a growing number of neuro-imaging studies among non-clinical populations have shown that the functioning of DLPFC, VMPFC, ACC, and OFC can be activated by cognitive reappraising of negative stimuli (for example, when the object is recognized as a twig and not a snake). Moreover, the 2008 study by Bryant and associates and the 2007 report by Felmingham and colleagues demonstrated that cognitive behavioural therapy was successful in both reducing symptoms of PTSD following a single trauma as well as both increasing ACC and decreasing amygdala functioning. Furthermore, studies have shown that biofeedback training in relaxation (e.g., BFT-A) serves to increase activation of the ACC, other parts of the MFC, and the OFC. Finally, indirect effects of increased ACC were found in my study below in that BFT-C led to both increased levels of frontal theta and decreased symptoms of PTSD among police officers exposed to multiple trauma.

K-04: The Amygdala and PFC Interactions
Typically, once the threat and/or the emotionally charged stimuli are no longer present, physiological systems return gradually to pre-threat levels; for example, amygdala and related activation decreases and the rACC/frontal cortex switches on again. If this process did not occur, then the person would be in constant threat mode, medical conditions would develop, and ultimately death would result.

However, in those with PTSD, (a) these instinctive changes of fear-system activation continue as if the threat were still present even after the actual threat has long passed, (b) further activation of the threat system is very easily re-triggered by even incidental reminders of the traumatic event, by presentation of emotional stimuli, and by exposure to any stressors, (c) the amygdala and hippocampus are sensitized so that less intense or more remote stimuli acquire the capacity to activate the fear systems, and (d) some respond to trauma reminders by automatically engaging in physically aggressive actions that may have been appropriate at the time of the trauma but are no longer so.

Furthermore, among those with PTSD following duty-related accumulated trauma (DRAT), a stimuli or intrusive memory of one traumatic incident can easily lead to the automatic recollection of other traumatic experiences such that neuro-processes and physiological reactivity become even more marked and more easily triggered than otherwise.

In a 2006 article summarizing the implications of affective neuroscience research in PTSD, one of the pioneers in PTSD, Dr. Van Der Kolk, noted that "traumatized individuals are prone to experience the present with physical sensations and emotions associated with the past" (p. 4).

In her 2006 publication, Dr. Shin noted that while the vast majority of studies have supported the findings above, some investigators have reported different findings. Also, it is unknown whether the differences in brain functioning were present before the trauma or were caused by the trauma. Finally, it is not yet known if the differences above are true for PTSD only or are present with other psychological conditions.

K-05: Electro-Physiological Abnormalities
In addition to differences as measured by various neuro-imaging techniques (fMRI, PET, and SPECT), many studies indicate that those with PTSD have what is known as electrophysiological abnormalities. These differences from the norm are determined by statistical analysis of single- or multiple-site recording of brain waves (EEG) and by qEEG (quantitative EEG, as discussed below). These differences are not to be confused with major neurological conditions such as brain tumours, aneurisms, and neurological diseases, which neurologists attempt to detect in part by visual inspection of the EEG tracing, nor are they life-threatening.

In the above section on memory, volume decreases and other abnormalities in the hippocampus were associated with those with PTSD. From these findings, it is reasonable to wonder if deficits in cortical activation will appear on the qEEG and, if so, if changes achieved through BFT-C would improve memory. Positive changes in memory have been reported following BFT-C among those with minimal traumatic brain injury. Also, most of my clients with memory problems have noted memory improvement with BFT-C.

Compared to those without PTSD, those with the condition have demonstrated consistently differences in the amygdala, as discussed above. Although the amygdala cannot be accessed directly by EEG scalp sensors, qEEG results from some of my clients with PTSD due to DRAT have demonstrated amplitude abnormalities in nearby areas, sometimes excesses and sometimes deficiencies. Moreover, PTSD symptoms were reduced when BFT-C was demonstrated to alter such abnormalities. Furthermore, there is reason to believe that, since frontal lobes inhibit amygdala activity, BFT-C at sites such as Fp1, Fp2, F7, F8, F9, and F10 would be relevant (Dr. D. C. Hammond, personal communication, 2007).

Similarly, as discussed above, among those with PTSD, both acute and chronic deactivation as well as decreased functioning of the (rACC) have been noted.

For those interested, the corresponding Broadman areas are 24 and 32 and the 10-20 sites include Fz, Fcz, F1 and F2 (Dr. D. C. Hammond, personal communication, 2007). Thus, it is reasonable to wonder if qEEG statistical differences from the norm in scalp sites proximate to the location of the rACC could be identified and thus be amenable to BFT-C. QEEG results from about 80% of my police clients with PTSD due to DRAT have shown power insufficiencies in the 2- to 7-Hz frequencies in such frontal cortex locations as Fp1, Fp2, F7, F3, Fz, F4, and F8.

Before proceeding further, a word about terminology. The most common EEG element measured in studies to date has been the *amplitude* of the wave form from the crest of the wave to its valley in the frequency of interest. When compared with a norm, we can then consider amplitude power to be low (insufficient) or high (excessive). Another term is *cortical arousal*, which refers to the frequency of the wave form measured in the number of cycles per second, also known as Hz. Brain waves in the slow frequencies such as 0.5 Hz to 7.5 Hz are considered as indicators of low cortical arousal while faster frequencies such as 22 Hz to 30 Hz are considered to indicate high cortical arousal. Note, though, that the term *arousal* is not the same as behavioural arousal, such as high or low rates of activation, nor is it the same as arousal of the autonomic nervous system, such as that associated with high blood pressure. For example, many of my clients with PTSD (with symptoms including high rates of agitation) have lower amplitude than normal in the slow frequencies at frontal locations of the brain and a common successful intervention for them is to increase (reinforce) the amplitude of such mplitudes.

Rabe and co-workers (2006) used the EEG to compare victims of a motor vehicle accident (MVA) with PTSD, victims without PTSD, and those who had not experienced an MVA. Only those with PTSD showed increased levels of alpha waves in both the right anterior (F4, F8, T8) and right posterior (Cp6, P4, P8) areas of the brain when exposed to trauma-related material. No differences between groups occurred when they were exposed to neutral or positive material. The authors note a small 1993 study by McCaffrey and associates showing right-sided activation among Vietnam vets when exposed to smells related to combat. Some of my police/military clients have demonstrated elevated right anterior and/or right parietal alpha, but most did not.

In the 2001 study of combat veterans from Zagreb with and without PTSD, Begic and associates obtained baseline levels of EEG activity. In comparison with those without PTSD, veterans with PTSD showed decreased levels in the 7.5 to 9.5 Hz frequencies in a variety of areas (F3, F7, C3, C4, T3, T4, T5, T6, O1, and 02) as well as increased levels in the 13.5 to 18 Hz frequency (Fp1, Fp2, F3, F7, F8, C3, C4,T3, and T4). Their 2003 study (Jokic-Begic & Begic) had found that compared with veterans without PTSD, those with PTSD showed increased 3.5 - 7 Hz activity at C3 and C4, increased 13.5 - 18 Hz levels over F3, F4, C3, C4, and O1, and increased 18.5 - 30 Hz activity at F3 and F4. Some of my police/military clients had EEGs similar to those in the 2001 study above but others did not. Also, some of my clients had decreases in the 7.5 to 9 Hz frequency as in the 2003 study but most did not.

A 2004 EEG study by Dr. Metzger and associates comparing female Vietnam War nurse veterans with and without PTSD determined that arousal symptoms as measured by CAPS were associated with higher amplitude in the 8- to13-Hz band at scalp location P4 (right occipital area) than at P3 (left occipital area). Amplitude in the 13- to 20-Hz frequency in the occipital areas was not significantly related to CAPS, nor was the amplitude power in the 8- to 13-Hz band at the other scalp locations that were considered (F3, F4, T3, T4). The abnormalities exhibited by some of my police/military clients have included high beta frequency amplitudes at P4 and some had similar results at PZ and P3.

Chale and associates (2004) reported that those with PTSD had increased activity in the 13.5- to to 18-Hz range over frontal, central, and left occipital areas and more 18.5- to 30-Hz activity in frontal regions. A couple of my clients showed some of these aspects, but most did not.

Rabe and colleagues (2006) compared victims of motor vehicle accidents with and without resulting symptoms of PTSD. There were no between-group differences at baseline. However, compared to both accident victims without PTSD and healthy controls, those with full or partial symptoms of PTSD displayed a pattern of increased right-sided over left-sided activation (lower alpha amplitude on the right side) in anterior and posterior regions upon exposure to a trauma-related picture. This is consistent with other studies demonstrating right posterior activation during anxiety-provoking situations. Also, these findings are consistent with functional neuro-imaging studies showing that emotionally laden visual stimuli elicit increased activation in posterior cortical regions. In a 2008 follow-up study, Rabe and associates found CBT to be effective in reducing right-side frontal cortical activation among those whose PTSD was the result of a motor vehicle accident. Also, a decrease in PTSD symptoms was associated with a reduction in right-side activation in the frontal cortex.

In a 2009 review of studies to date, Clark and associates concluded that what most anxieity disorders had in common was the finding of cortical arousal instability as measured by quantitative electroencephalography (qEEG). Also, they noted two specific patterns in the qEEG of those with PTSD. One was high beta and/or low alpha in frontal regions of the brain. The other was increased asymmetry in alpha on the right side of the brain.

In a previous study by many of the same authors (Veltmeyer et al., 2006), those with PTSD stemming from a variety of traumatic incidents exhibited increased cortical arousal as noted in resting qEEG levels of reduced alpha , reduced theta, and reduced theta/beta ratio. Moreover, such findings were found to be related to clinical and physiological measures of arousal. Furthermore, PTSD was associated with abnormalities in measures of working memory and attention.

The findings to date in regards PTSD are consistent with a much larger body of research that has been published on other conditions (see the review at www.isnr.org). For example, three different patterns of brain dysregulation have been found to be associated with attention deficit/hyperactivity disorder. Also, a 2008 literature review published in the *Journal of Neuropsychiatry* indicated that 25% to 30% of those with panic disorder had demonstrable EEG abnormalities.

Similar abnormalities have been discovered over a wide range of conditions including addictions, seizures, reading disabilities, anxiety, and traumatic brain injury. Moreover, other studies have demonstrated that correcting the appropriate EEG dysregulation leads to a decrease in symptoms including seizures, anxiety, reading disabilities, seizures, and cognitive impairment subsequent to traumatic brain injury.

K-06. Other Brain Research
The bulk of brain research to date has focussed on the both the limbic system in general and the amygdala in particular as well as the pre-frontal cortex with much attention to the rACC. However, various well-designed research studies have implicated additional areas of the brain in PTSD.

For example, fMRI techniques have documented areas of the temporal cortex, the fusiform gyrus (FG), and the superior temporal sulcus (STS) that detect features such as movement, voice, and facial expression of others to determine the level of personal risk. If no danger is detected, then adaptive defences mediated by areas of the limbic system are subdued so that social engagement can occur. On the other hand, if danger is detected, then adapted defences are not inhibited and, via connections to other brain structures and processes, the body prepares physiologically to respond to the challenge. In the 2007 study by Hedges and colleagues, grey and white brain matter regional brain volumes were assessed by structural magnetic resonance imaging in those with and without PTSD. Significant white matter volume differences between the groups was observed in the superior temporal gyrus, fusiform gyrus, parahippocampal gyrus, white-matter stem, middle temporal gyrus, and inferior temporal gyrus. A similar 2008 study by Geuze and co-workers compared cortical thickness between combat veterans with and without PTSD and discovered reduced thickness among the former in the bilateral superior and middle frontal gyri, the left inferior frontal gyrus, and the left superior temporal gyrus.

The insular cortex (Broadman area 13) is an area of the brain located in the centre of the cerebral hemisphere and is involved in processing many functions including visceral sensory, visceral motor, vestibular, attention, chronic pain, emotion, and motor information. It is involved also in the regulation of the sympathetic and parasympathetic nervous systems; for example, when stress becomes uncontrollable, the left insula shuts down the vagal nerve activity involved in the parasympathetic response and only the right insular sympathetic activity prevails. In addition, the insula has connections to a number of other brain regions such as the anterior cingulate, the amygdala, the prefrontal cortex, the superior temporal gyrus, the temporal pole, the hippocampus, and the orbitofrontal cortex related to PTSD. PET and fMRI scans of the insula have indicated its involvement in conditions such as PTSD and mood disorders as discussed in the 2007 review by Nagai and associates. Using fMRI, a 2007 study by Lanius and associates reported less significant activation of the left insula among those with PTSD without depression than among those with both PTSD and major depression. Simmons and collaborators reported a 2008 study of 15 women with PTSD as a result of partner violence and 15 without such experiences; when presented with cues to negative events, those with PTSD showed greater increased activation of the bilateral anterior insula than control subjects. Currently, the electrophysiology of the insula can be detected only by the LORETA software programme. At this time, it is not possible to alter the functioning of the insula directly via methods such as BFT-C, but altering the electrophysiology of its connections to other structures in the brain (for example, the ACC) may have an impact on it.

Also, there has been much interest in EEG asymmetry between the left and right frontal cortex and its relationship with depression. Moreover, some but not all studies have reported improvements in depression symptoms by altering the asymmetry via BFT-C. A large first-class study reported by Beauregard at the 2008 ISNR conference provided strong evidence that individualized BFT-C based on qEEG resulted in remission in depression. In regards resting asymmetry, the 2008 study by Shankman and co-investigators determined that individuals with PTSD did not differ from controls.

Freeman and co-investigators published a study in 2009 on prosody among combat veterans with PTSD as compared with a group of healthy people, a group with known left-hemisphere brain damage, and a group with known right-hemisphere brain damage. Measures of prosody evaluate one's ability to comprehend the emotional aspects of interpersonal communication and prosody is relevant to PTSD since emotional restriction is one of the symptoms of the condition. Previous studies had demonstrated that the capacity for prosody is related to the functioning of an area of the right hemisphere analogous to Wernicke's language area of the left hemisphere. Results indicated that (a) those with PTSD had significant deficiencies in the comprehension and discrimination of the emotional components of communication, (b) the healthy subjects and those with left-hemisphere brain damage had no such difficulties, (c) those with right-hemisphere brain damage displayed significant impairment in prosody, and (d) the deficiencies among those with PTSD were similar in severity and performance to those with right-hemisphere brain damage.

In a 2008 study by Sailer and co-investigators, those with and without PTSD had the opportunity to learn a decision-making task while brain areas were monitored by fMRI. Consistent with findings reported by others, controls learned the correct response faster than those with PTSD. Also, those with PTSD showed lower activation of brain areas associated with the reward pathway (nucleus accumbens and mesial pre-frontal cortex) than those without the condition. The authors surmised that this finding may reflect decreasing motivation as the task progressed.

Based on the notion that deficits among those with PTSD may reflect hypervigilance and deficient inhibitory control, in 2008 Falconer and associates reported a relevant study. FMRI was used during a go/no-go inhibition task. Results indicated that those with PTSD made more inhibition-related errors than individuals without trauma exposure. Also, compared to control subjects, those with PTSD showed a reduction in right cortical activation as well as increased activation of striatal and somatosensory brain regions.

K-07: BFT-C Reports on PTSD

In the early 1990s, a study by Dr. Penniston and colleagues demonstrated that, as compared with combat veterans with both PTSD and alcohol addiction who received only traditional treatment, similar combat veterans who received BFT-C achieved better results. The BFT-C group first completed 5 to 6 sessions of relaxation training using temperature biofeedback and autogenics. Then, they learned to increase the amplitude of alpha/theta frequency brain waves at scalp location O1 over 15 sessions of 30 minutes each. At 30 months post-treatment, all of the 14 who received traditional treatment had relapsed and been hospitalized, while this was true for only three of those receiving the BFT-C training. Also, the entire BFT-C group who had been on medication had been able to decrease medications, while among the control group only one had decreased medications, two reported no change, and 10 required an increase in medication. With respect to scores on the accepted Millon standardized psychological test, overall scores on the 10 clinical scales improved significantly over baseline among the BFT-C group while there were no significant before-to-after differences among those receiving traditional treatment. Some of my clients have exhibited low amplitudes in the alpha and/or theta frequencies at O1, but most have not.

However, since measures of PTSD appear not to have been taken after the relaxation training and autogenics, there is no way of knowing what if any degree of improvement was due to this treatment component and not to the BFT-C. As noted in Step J above, a sizeable majority of my military and police clients achieved remission without the addition of BFT-C. Also, we now know that training in applied psycho-physiology training can lead to changes in brain waves.

During informal meetings at professional conferences I have attended, BFT-C professionals have reported positive effects of BFT-C among some clients with PTSD symptoms. These reports must be considered cautiously since they are anecdotal in nature. Also, generally, these colleagues have not relied on a prior qEEG assessment and due to the informal nature of these sessions, I do not know the types of clients they were seeing nor do I know how PTSD symptoms and severity were determined.

a) Some reported success by increasing the amplitude of alpha or SMR brain waves while others decreased the amplitude of high beta brain waves at various single sites (Fz, Fcz, Cz, Pz, and F4).

b) One colleague reported success by decreasing beta for 7-15 minutes among those with elevations at sites F8 and F4 or increasing alpha at Pz/Oz.

c) Others reported positive results from conducting the training above but at two brain locations at the same time (C3-C4, T3-T4, and, P3-P4).

d) One clinician reported good effects by increasing the amplitude of alpha/theta brain waves at location FPO2 and another by increasing alpha and decreasing theta at that location.

e) Another found it helpful to decrease 8-12 Hz at either FPO1 or FPO2.

f) Using locations T3 or T4 as a common reference, one group reported efficacy by training clients to increase amplitude in a frequency comfortable to them (for some as low as 0.1 Hz) while at the same decreasing amplitude in all other frequencies between 1 Hz and 30 Hz. According to their review of their approach (2009), physical calming is achieved by targeting the right parietal region; emotional calming and stability are targeted by right pre-frontal training. Overall system stability is promoted with the interhemispheric bipolar placement T3-T4. In the second phase of their approach, two-channel alpha/theta sum training is begun at P3 and P4 with theta at about 7 Hz and alpha at about 10 Hz.

f) One neurologist proposed that excess amplitude in 13-20 Hz or 21-30 Hz, or insufficient amplitude 1-10 Hz at locations T3, T4, were associated with PTSD.

Based on the information in this section, in 2008 I reported qEEG data based on NeuroGuide in regards to my police clients who had not reached full remission of their PTSD following implementation of my basic treatment approach. Results included the following:

*80% of my clients showed insufficient cortical activation in 4 Hz to 7 Hz range either frontally or over much of the cortex including frontal locations. Studies have demonstrated that the septal region of the brain contains pacemaker neurons that regulate such frequencies in the cingulate (Dr. Robert Thatcher, personal communication, 2009. Relatedly, Kropotov (2009) presented evidence that (a) low frontal theta activity during rest is linked with low cerebral metabolism in the medial frontal area of the brain as well as the rACC and (b) low frontal midline theta is associated with high anxiety scores. This raises the question as to whether my clients' demonstrated low amplitude in slow waves is related to under-functioning of the rACC. Certainly, my outcome data indicated decreased PTSD symptoms following BFT-C to increase the amplitude of slow waves and this would fit nicely with what is already known about the inhibitory role of the rACC on the amygdala.

*On the other hand, 17% did not show insufficiencies but rather showed frontal excesses in the 4HZ to 7 Hz frequency range. It may be that excess amplitude in this area of the brain indicates dysfunction of the rACC in the same way that lower than normal amplitude is associated with PTSD symptoms.

*About 50% had excess cortical activation in the high beta ranges at central and/or posterior areas of the brain.

*40% exhibited both frontal low frequency insufficiencies and excess central/posterior beta.

*Smaller percentages of my clients produced various other differences from the normative sample; excess left frontal SMR, insufficient SMR at various locations, excess alpha or alpha/theta in various locations, insufficient alpha or alpha/theta in various locations, excess temporal alpha and/or beta, insufficient alpha or beta at most sites, and so on.

As presented later on, BFT-C directed to relevant sites as determined by NeuroGuide resulted in a very high rate of symptom remission. At the same time, subsequent repeat qEEGs indicated improvement in the targeted frequencies of specific brain locations.

Walker (2009) described a clinical series of 23 patients who had been identified by some unreported process as having PTSD. He indicated excess amplitude in the 21- to 30-Hz range in a wide variety of brain sites. Down training proceeded subsequently in the sites identified and he documented resulting impressive decreases in self-reported levels of anxiety. The effect on the overall and specific diagnostic symptoms of PTSD was not reported.

Aroche and colleagues published an informal report in 2009 on their work in Australia with trauma refugees with PTSD. They noted that conventional approaches often are not sufficient to achieve remission. However, providing BFT-C was found to be a valuable tool in reducing symptoms of over-arousal, thereby enabling clients to benefit from psychotherapy. Success rates around 85% were noted including among those with chronic, complex presentations.

Price and Budzynski in the 2009 book edited by Budzynski and others reported that a number of studies described a decrease in anxiety (including for those with PTSD) by increasing alpha waves. Also, they cite studies showing increased anxiety levels (not specific to PTSD) to be associated with activity in the basal ganglia area of the brain, or with more activity in the 12- to 18-Hz range in the right hemisphere, or increased 4- to 6-Hz activity and 12- to 18-Hz activity in the parietal-temporal area.

The presence of different qEEG among those with PTSD should not come as a big surprise. *First*, many different combinations of symptoms are still classified as PTSD such that it is reasonable to conclude that different sets of symptoms may involve different brain processes. *Second*, with the longer history of neuroscience research for such conditions as attention deficit/hyperactivity disorder (ADHD), a number of different brain wave patterns have been demonstrated. *Third*, it is becoming increasingly clear that any cognitive or affective process requires the integrated activity of multiple areas of the brain. For example, 13 or more areas of the brain are involved in vigilance, one form of attention. Multiple areas of the brain are involved in affect regulation, including the rACC and other parts of the frontal cortex, the amygdala and other parts of the limbic system, the basal ganglia, and areas of the cerebellum. Thus, dysfunction in any one area of the brain required for completion of a task can result in functional problems. Moreover, depending on the location of the "weak link," different patterns would be expected in a qEEG analysis. *Fourth*, single or groups of brain cells are connected to other neurons or groups of neurons. Some of the connections are between cells at the same vertical level of brain organization, some of the connections occur within and between any of the remaining five levels of cells in the neocortex, some connect to each other via long fibers (such as the fasciculi linking directly the very back of the head to the front), some provide connections between the neo-cortex and the sub-cortex, and some link the two hemispheres of the brain. Thus, disruptions in inter-connections can have the same affect as the non-functioning of one brain area as noted above. *Fifth*, it is known that changes by BFT-C in one area of the brain are likely to create changes in other areas.

All of the above research points again to the vast complexity of the human brain. Congedo (2009) interviewed one of the giants of the neuroscience field, Dr. E. Roy John, who stated:

> We have found in the electrical activity of the brain the fact that there exists a number of parallel systems, operating in parallel, maybe five or six parallel pathways, sharing structures, sharing neuro-anatomical regions performing multitasking, integrating this. (p. 15)

K-08: Implementing BFT-C

Consequently, on the basis of research to date, information from attendance at specialized professional conferences involving applied neuroscience practitioners and scientists, and my own clinical experiences, I now employ BFT-C in the treatment of PTSD as required. As noted earlier, BFT-C simply refers to biofeedback training, but in this instance the target organ is the central nervous system and the signal of interest is the EEG. Also, as noted previously, tactical breathing exerts a direct positive effect on the autonomic nervous system as well as an indirect effect on the central nervous system. Conversely, BFT-C exerts a direct effect on brain functioning with an associated indirect effect on the autonomic nervous system.

I discuss the possibility of BFT-C with clients only (a) if my basic approach has been implemented but has not resulted in complete remission of the PTSD, or (b) if symptoms are too easily re-triggered after the basic treatment has concluded, or (c) if impaired cognitive processes are not resolved. I inform clients that there is good evidence that qEEG-based BFT-C is effective with very few side-effects in the treatment of a variety of psychological conditions including seizures, attention deficit/hyperactivity disorder, and depression. In addition, I mention the research to date with PTSD, which now includes my own clinical case series. Finally, I refer clients to web sites of professional associations; for example www.isnr.org.

If the client gives informed consent for a trial of BFT-C (see Appendix III), standardized psychological tests are re-administered to give us a measure of symptom intensity before starting the procedure; this is repeated periodically to assess changes after BFT-C is started.

Then a qEEG is completed since I have the post-graduate training, equipment, and case consultation to do so. In this painless and non-embarrassing procedure, I place a cap on the client's head that contains sensors for EEG recording. The EEG, as measured on the scalp, records electrical fields that are due to post-synaptic potentials of neocortical cells and pyramidal cells in particular. About 50% of the power of the EEG arises from directly under the sensor and 95% of it within 6 centimetres of the sensor (Dr. Thatcher, personal communication, 2008). The cap is attached to an electronic biomedical processing device. For both hardware and software, I use the Truscan-32 by DeyMed Diagnostic with a sample rate of 128 per second, a sensitivity of 70 microvolts, and filters set between 0.5 Hz and 70 Hz. In this way, brain wave activity is measured and recorded simultaneously at 19 different scalp locations according to the accepted 10-20 site map for sensor placement. This results in about 73,000 bits of information per second that are recorded digitally and then stored on my computer.

The following information about the qEEG is from a 2009 article in *NeuroConnections* by Jonathan Walker, a board-certified neurologist practising in Texas and an ISNR member. *Firstly*, he points out that hundreds of papers have been published in peer-reviewed professional journals, all of which consider the qEEG to be valid and reliable. *Secondly*, he notes that its sensitivity and specificity is "equivalent to or greater than clinical standards for the MRI, sonograms, blood analysis, and other common clinical diagnostic procedures" (p. 4). *Thirdly*, he notes that a number of established agencies of the U.S. government consider the qEEG to represent the standard of care for the diagnosis of such conditions as traumatic brain injury. Also, it is true that the reliability rates of qEEG analysis far exceed the average of about 44% from visual inspection of the EEG by trained neurologists.

Then, for every client the qEEG information I obtain is web-mailed to Dr. Robert Thatcher, a well-known and well-respected researcher who is director of the Applied Neuroscience Laboratory in Saint Petersburg, Florida. He is the author of six books, numerous book chapters, and hundreds of research articles in respect to qEEG. In addition, Dr. Thatcher is certified as an EEG specialist (ECNS) and is the developer of the NeuroGuide software that I and many colleagues use. He receives my digitized qEEG, examines it visually, remove artifacts (such as muscle movement) electronically, and processes the result through the NeuroGuide normative software he developed and continues to revise. As well, he processes the data through LORETA, a software program that gives three-dimensional localization of the relevant cortical generators. In his 18 June 2009 listserv communication, Dr. Thatcher noted that

> there is a vast qEEG literature showing good correspondence between the surface EEG and the location of strokes, tumors and lesions as well as abnormal activity in "modules" associated with clinical problems... There are over 500 qEEG LORETA studies and the vast majority of these studies show good correspondence with fMRI and PET and MEG . . . related to localization of function, or actually "functional systems" or modules.

The resulting "*brain maps*" from NeuroGuide and LORETA are returned to me via web mail. This information highlights any brain abnormalities with respect to such aspects as amplitude, asymmetry, coherence, and phase. In addition, Dr. Thatcher suggests a possible BFT-C plan based on the analyses in relation to the normative groups, the symptoms needing alteration that have been already assessed, and the current state of knowledge in regards the relationship between these symptoms and brain locations/functions.

After explanation to clients, a BFT-C plan is then implemented using the Truscan-32 Deymed system. Clients see a visual and hear an auditory representation of their brain wave activity in relevant sites and in real time, from which desired changes are made. "Real time" means that brain wave activity appears on the monitor within 0.10 seconds of its occurrence. Once clients achieve the pre-set standard over 50% of the time in two consecutive minutes, the reinforcement criteria is made more difficult gradually. Sessions are held once a week and during each session fifteen consecutive minutes of training is devoted to each protocol. Typically, the training plan involves 3 different protocols as identified in the qEEG analysis. Commonly, training sessions continue from between 20 and 40 weeks. Many experts have noted that BFT-C allows the brain to normalize itself into a more stable and functional pattern of neuronal activity than previously.

Before and after each session, clients are quizzed on any effects or side-effects of BFT-C. Additionally, on a regular basis, appropriate standardized psychological tests are re-administered to determine progress. Also, after 15 to 20 sessions of BFT-C, usually the qEEG is repeated and again is sent to Dr. Thatcher for analysis and suggestions. The key matter is the extent to which desired changes in PTSD symptoms have taken place. If remission has not been achieved, then information from the repeat qEEG and other sources may determine any modifications to the BFT-C approach. The importance of continuing BFT-C until symptoms are diminished sufficiently is because otherwise the brain may find other other patterns of dysregulation.

Interestingly, often as PTSD symptoms decrease, clients report improvement in aspects of their functioning that had not been targeted specifically. For example, while targeting PTSD symptoms, I have been fascinated by client reports of deciding to engage in a new hobby, treating their spouses more lovingly, re-establishing contact with friends from the past, making therapeutic changes to their jobs, reporting of a desireable increase in libido, and taking the risk of revealing their condition to other people in their acquaintance who they believe might have symptoms of PTSD.

By the time symptom remission is reached, usually I am able to see clear changes in the brain maps in the direction of increased similarity to the normative sample. During my own exposure to BFT-C, I noted that very small changes in the amplitude of my EEG signal was associated with significant changes in the mental calmness I was trying to achieve. However, it is true also that the person can experience dramatic changes in symptoms with no obvious difference between the pre- and post-qEEG. For example, Arnes (personal communication, ISNR listserv, 2008) noted this in his study with those experiencing depression. This does not mean necessarily that no change in brain functioning took place, since it undoubtedly did as the symptom changes had to be the result of something happening. For example, changes could have occurred in areas of the cortex not measured, or changes in neurochemistry could have taken place, or changes in electrophysiology in the sub-cortex could have occurred. Regardless, as I remind my clients frequently, the primary objective is changing symptoms, not changing the brain maps. Or, as my physician once said, "I treat people, not lab results."

K-09: My BFT-C Results to March 2009

In March 2009, I examined BFT-C results on 12 consecutive male police officers with PTSD due to DRAT on whom I had complete data. They had completed an average of 24 sessions of BFT-C after my basic approach had not been successful in achieving symptom remission. Their average age was 48 and all but one had completed a minimum of 20 years of front-line policing.

Results indicate that:

    a. eleven (92%) achieved symptom remission; they no longer met the diagnostic criteria for PTSD;

    b. the one still meeting all the diagnostic criteria for PTSD continued to be exposed to significant reminders of trauma by virtue of ourt proceedings not yet completed 10 years after the major traumatic event;

    c. whereas a score of 64 or above on the TSI is consistent with the diagnosis of PTSD, the average PTSD summary score immediately before BFT-C (but after all the basic approach had yielded some positive effect) for the 12 was 73.1 and after BFT-C was 61.8. This is a statistically significant finding and indicative of remission;

    d. the mean intrusive re-experiencing score on the TSI decreased from 79.2 to 65.0, again a statistically significant difference;

    e. the average defensive avoidance score on the TSI decreased from 72.5 to 63.0, again a statistically significant change. The final levels were in the remission range;

    f. the average anxious arousal score on the TSI decreased from 74.5 to 59.4, also statistically significant and consistent with remission of that symptom;

    g. although the average dissociation score on the TSI decreased from 68.9 to 56.9, the difference was not judged to be statistically significant. However, remission was achieved;

    h. eight were taking medication at the beginning of BFT-E; four of them were able to stop their medication, and no one started medication;

    i. of the 12, three were retired before we started BFT-C, seven opted for a medical discharge during treatment (generally they did not want to run the risk of attending traumatic scenes that might lead to a return of symptoms); two returned to full operational duties; and

    j six months later, I was able to follow-up on four of the 12 who had completed BFT-C. All four were still in remission. Moreover, the average PTSD score on the TSI for them had continued to decrease from 58 after BFT-C to 50 at follow-up.

The overall results of using BFT-C described above seem to be echoed among both police and military clients I have seen since I worked with the twelve police officers described. BFT-C has not been required with any of the civilians I have seen. Also, four female 911/dispatch personnel who did not reach optimal levels following my basic approach then completed BFT-C; all four achieved remission of PTSD symptoms.

The International Society for Neurofeedback and Research (ISNR) to which I belong is a group of about 800 professionals world-wide, most of whom are Ph.D. psychologists and some of whom are neurologists. Members of this group use BFT-C in their work with a variety of clients and/or conduct research in this area. We publish a professional journal (*Journal of Neurotherapy*), a newsletter (*NeuroConnections*), maintain a listserv so we can inform each other of new developments, produce an annual 5-7 day conference, and are associated with a number of other groups. One well-respected member of our group on the medical faculty of the University of Utah, Dr. Cory Hammond, keeps up on all publications related to the use of BFT-C and updates it regularly on a section of our website (www.isnr.org). Evidence that BFT-C has positive results is overwhelming, with studies demonstrating effectiveness with conditions such as anxiety, depression, seizures, attention deficit/hyperactivity disorder, sleep problems, addictions, reading disabilities, some autistic behaviours, and traumatic brain injury.

In a chapter written by Dr. Evans and Dr. Rubi in the 2008 book edited by Dr. Evans, possible reasons for the success of BFT-C in treating a variety of conditions include the fact that it requires clients to participate actively in treatment; other than medications, it alone targets the electro-chemical functioning of the central nervous system; it deals with recognized and reliably measurable electromagnetic energy; it alone can address any abnormalities of neural timing and connectivity; it is supported by scientific research; and it has few if any negative side-effects. Additionally, it seems to me that control of treatment is left in the hands of the client who is able to see progress. This appeals to police and military clients for whom control and relying on their own judgement are viewed as critical to their own survival.

K-10 Event Related Potentials (ERP)
I am still learning about ERP and have yet to include this in my clinical practice. Still, knowledge about this remains instructive as it provides information from yet another branch of neuroscience regarding the automatic operation of our brains outside of our conscious awareness.

There is ample evidence that basic, automatic, and very rapid attentional processes occur among humans within .25 to .75 of a second after a stimulus is present. This is assessed by analysis of brain activity in a format known as event related potentials (ERP). Following a stimulus such as a light or a sound, usually in the first milliseconds the ERP components known as P-1 and N-1 can be seen, which reflects an early stage of brain activity wherein the stimulus is noticed/perceived. At about 300 milliseconds after the onset of stimulus, the P-300 wave is seen, which seems to be related to selective perceptual processes used in identifying stimulus relevance and assigning the amount of attentional resources devoted to stimulus evaluation.

Most of the studies with those experiencing symptoms of PTSD report ERP differences compared to those without the condition. For example, about 20

ERP studies to 2005 were completed and as summarized in the book *Neuropsychology of PTSD*: "This work has revealed evidence consistent with impaired sensory gating, increased sensitivity to stimulus change, heightened orienting response, impaired attention to neutral stimuli, and heightened attention to trauma-related stimuli in PTSD" (Vasterling & Brewin, 2005, p. 83).

A 2006 summary of research to date by Karl and associates concluded that, compared with those
without PTSD, those with the condition showed:

-responses to auditory sensory gating (P-50) indicating they were less able to ignore or get used to continuing stimuli associated with the trauma;

-an augmented P-200 response to sounds of increasing intensity, suggestive of a sensory system on the lookout for stimuli,

-overall reduced P-300b amplitude to neutral stimuli;

-significantly longer P-300b latencies at central and parietal areas of the brain but not at frontal locations; and

-significantly higher P-300 amplitudes to trauma-related distractors at scalp sites Fz, Cz, and Pz, meaning that they reacted more strongly to cues related to their traumas.

At the 2007 ISNR conference, Sokhadze indicated that those with PTSD demonstrated increased N-200 and N-450 responses compared to those without PTSD. He interpreted this to mean decreased capacity to monitor responses.

In regards to the P-300 at frontal brain sites, the 2008 study of combat veterans with PTSD by Shucard and colleagues found that those with high amplitude and low latency had higher symptoms of hyper-arousal and re-experiencing than controls. The authors concluded that attentional problems are related to slowed attentional processing when response inhibition is required and to an impaired ability to screen irrelevant information. They reported also that heightened arousal appears to worsen attentional dysregulation among those with PTSD.

A 2008 study by Kiss and Eimer indicated that fearful stimuli resulted in the same ERP characteristics whether the duration of stimuli was 200 milliseconds or 8 milliseconds. This in spite of the fact that subjects had no conscious awareness of the content of the shorter exposure.

In 2008, Metzger and co-investigators presented evidence (a) that the P2 amplitude intensity slope of the ERP among combat veterans was elevated among those with PTSD compared to their monozygotic twin brothers without combat experience, and (b) that there was an association between slope and PTSD symptom severity. Also, they concluded that results were the result of combat experiences rather than pre-combat differences between the twins.

In a 2008 study by Galletly and co-authors, various measures of ERP were contrasted between patients with PTSD and those with schizophrenia while doing an auditory task of target detection. Task performance was impaired in both groups with some differences between them suggestive of specific automatic patterns of cognitive dysfunction for each condition. N1 amplitude reduction was found only among those with schizophrenia,

> an increase in N2 amplitude and latency was found only among those with PTSD, . . . both groups showed a reduction in amplitude of the non-target and target P3 . . . and a reduction in non-target parietal P3 amplitude in the schizophrenia group and a reduction in target P3 amplitude over the left posterior parietal region in the PTSD group. (p. 201)

A 2008 study by Falconer and co-investigators used a variety of measures including ERP to determine if they could reliably differentiate between civilians with PTSD and a control group who had not been exposed to trauma. They reported successful discrimination between the two groups based on such automatic characteristics as changes in levels of arousal, time to switch attention, and reaction time.

Studies were reviewed in the 2009 publication by Clark and his group. Using ERP, one study of combat veterans indicated that those with PTSD as compared with veterans without the condition showed a number of differences in brain activity indicating that they were unable to avoid paying attention to stimuli reminiscent of traumatic experiences. In other words, without conscious attention or effort, their brains automatically directed their attention. This finding is consistent with many studies using what is known as a modified Stroop procedure, which show that those with PTSD have impaired attention to required elements of a task when trauma-related cues are present.

At the 2009 ECNS conference, Boutros and Amirsadri reported results of their review of studies to date and reaffirmed P200 and P300 differences to trauma-related stimuli among those with PTSD.

The 2009 study of combat veterans by Metzger and colleagues noted that ERP deficits such as increased N2 latency and P3b amplitude were determined to be the result of traumatic exposure among those with PTSD rather than prior characteristics of those who go on to develop PTSD.

Findings such as those reported above again illustrate the operation of critical and automatic basic brain functions in those with PTSD. By their very nature, they are outside of conscious awareness and volitional control but still exert a significant and negative effect.

Similar ERP finding have been reported in folks with other neurological and psychological disorders and in normal aging. For example, very much relevant to PTSD, studies of people with specific fears or phobias (such as to snakes or spiders) show automatic shifting of their attention to the feared object when it was present, thereby demonstrating difficulties in disengaging attention to such objects.

In their 2009 review of electrophysiological studies, Clark and co-authors concluded that also, common to most of the anxiety disorders are condition-specific difficulties with sensory gating and the allocation and deployment of attention. These are clearly evident from evoked potential and event-related potential (ERP) electrical measures of information processing in obsessive compulsive disorder (OCD), post-traumatic stress disorder (PTSD), panic disorder (PD), generalized anxiety disorder (GAD), and the phobias. (p. 92)

A few studies have reported that undesirable ERP characteristics can be altered substantially. For example, psychological medications have produced such an effect. Also, at the 2007 ISNR conference, Kropotov indicated that ERPs had been changed towards normal reactivity with the use of BFT-C and his specialized software. Additionally, at the 2008 SPR conference, Bostanov and co-researchers reported that clients with depression exhibited positive changes in some ERP variables following mindfulness-based cognitive therapy while a control group showed no such changes. Thus, there is evidence that ERPs can be changed, although I am not aware of any studies attempting to do so among those with PTSD.

K-11: Closing Comments

1. In spite of the evidence above, some professionals and organizations still consider BFT-C to be an experimental approach. The view of others is that it has more evidence of usefulness than is indicated in the research on anti-depressant medication, which is used widely. My own view is that, while it is certainly not yet mainstream, when implemented appropriately and with the safeguards I maintain, BFT-C has proven to be an effective component in the treatment of many psychological conditions including PTSD.

2. Even so, BFT-C is not for everyone. To date, I have discontinued BFT-C with about a few clients who appeared to develop headaches from this component. Also, based on the information I provided, a number of other clients declined the opportunity to begin BFT-C.

3. Hopefully, the field of affective neuroscience will continue to develop and new findings will emerge that will help those who do not experience remission with BFT-C. I keep abreast of the newest research and theory in this regard.

First, exciting work is continuing with another aspect of the qEEG known as coherence, which indicates the degree of connectivity between different populations of neurons. Areas of the brain can fire together too frequently (hyper-coherence) or not often enough (hypo-coherence). Generally, increased coherence is required between two or more areas of the brain as the task becomes more complicated. Existing research has shown coherence-training to be helpful among those with autism spectrum disorders, learning disabilities, and traumatic brain injury. Just recently I have included treatment protocols for hypocoherence based on qEEG findings. BFT-C data indicate clearly significant increases in targeted hypocoherence. Also, PTSD remission rates appear to be as robust as previously. However, since the average client's treatment plan includes both amplitude and coherence training, I have not yet been able to determine the relative contribution of each to outcome effects.

Second, some colleagues have reported success in their practices with the strategy of using BFT-C initially to decrease overall brain dysregulation as determined from a qEEG and then adding further training to increase power in the alpha/theta frequency range for those who do not achieve remission. Sometimes independent sound feedback at both 4-8 Hz and 8-13 Hz are provided in an eyes-closed condition. Sites such as O1, O2, Cz, Pz, and P4 have been used. In what is known as the Scott-Kaiser modification, clients completed an average of 13 BFT-C sessions before about 30 alpha-theta sessions. Many of the studies using these approaches were treating alcoholism. Interestingly, before qEEG was widely available, the first published study of BFT-C in the treatment of PTSD included alpha/theta training at site O1.

Third, the developers of software (such as Dr. Thatcher) continue to find other measurable aspects of qEEG (for example, burst metrics and aspects of phase) that might provide additional evidence-based strategies for intervention at some future time. In the 2009 interview article with by Gismondi, Thatcher mentioned the EEG statistic of phase reset, noting that it

> is made up of two components, phase lock and phase shift. Phase lock is positively correlated with present measures of coherence, which is a measure of stability of phase differences over time. Phase shift, the disengagement from phase lock between sites, is negatively correlated to coherence. Phase shift is related to the functional circuit's ability to recruit momentary neural resources and processing pathways. Phase lock is the binding together of the recruited resources for a brief period as a commitment to a particular information processing task OR compensation, which is just another way to speak about function . . . phase reset measures are consistently at or near the top of our predictors for not only pathology, but intellectual ability as well . . . . (p. 39)

My summary is that there is ample evidence to indicate that the functioning of the central nervous system (CNS) can be altered by "peripheral" methods such as BFT-A and also by BFT-C, which targets the CNS directly. Moreover, there is ample evidence both that (a) the CNS is dysregulated with respect to PTSD and other psychological conditions and that (b) directing therapeutic attention to the CNS can lead to positive changes in psychological functioning. Moreover, given the continuing new developments in the field, I find increased optimism in the field of applied neuroscience.

K-12: Further Information
For those interested in learning more about neuroscience and/or its application to psychological conditions, below are some of the books I have profited from in the past few years.

Andreassi, L. L. (2007). *Psychophysiology: Human behavior and physiological response* (5th ed.). Mahwah, New Jersey: Erlbaum Associates.

Budzynski, T. H., Budzynski, H. K., Evans, J. R., & Abarbanel, A. (Eds.). (2009). *Introduction to quantitative EEG and neurofeedback: Advances and applications* (2nd ed). New York: Academic Press.

Evans, J. R. (Ed.). (2007). *Handbook of neurofeedback: Dynamics and clinical applications.* New York: Haworth.

Evans, J. R., & Abarbanal, A. (Eds.). (1999). *Introduction to quantitative EEG and neurofeedback.* New York: Academic Press.

Kropotov, J. D. (2009). *Quantitative EEG, event-related potentials and neurotherapy.* New York: Academic Press.

Miller, B. L., & Cummings, J. L. (Eds.). (2007). *The human frontal lobes: Functions and disorders* (2nd ed). New York: Guilford.

Schwartz, M. S., & Andrasik, F. (2003). *Biofeedback: A practitioners guide* (3rd ed.) New York: Guilford.

Squire, L. R., Berg, D., Bloom, F. E., du Lac, S., Ghosh, A., & Spitzer, N. C. (Eds.). (2008). *Fundamental neuroscience* (3rd edition). San Diego, CA: Academic Press.

Thompson, M., & Thompson, L. (2003). *The neurofeedback book: An introduction to basic concepts in applied psychophysiology.* Wheat Ridge, CO: Association of Applied Psychophysiology and Biofeedback.

Whalen, P. J., & Phelps, E. A. (Eds.). (2009). *The human amygdala.* New York: Guilford.

Also, for the interested person, a number of legitimate, professional, and accurate web sites can be considered.

1. www.appliedneuroscience.com. This is Dr. Thatcher's web site, which includes a list of some of his 200+ publications to date.

2. http://psyphz.psych.wisc.edu/web/index/html: This is the website of a prolific affective neuroscience researcher, Dr. R. Davidson, and colleagues, who are currently at the University of Wisconsin. Using the EEG and neuro-imaging methods such as fMRI, Dr. Davidson was one of the first to identify one of the brain EEG patterns associated with major depression.

3. www.isnr.org: This is the official website of the International Society for Neurofeedback and Research. Neurofeedback is the term they use for what I refer to as BFT-C. This group is comprised mostly of Ph.D. scientists and practitioners in clinical psychology and neuropsychology along with others such as neurologists. We have such members from all over the world. I am on the annual conference committee and was elected to the Board of Directors for the 2007-2009 term. On the website, you will find a very complete definition of neurofeedback (BFT-C), which I was instrumental in developing.

Treatment Step L: Minimizing Relapse

Contents

L-01: Introduction

Studies have demonstrated that even when PTSD symptoms seem to have decreased, they can be re-triggered/re-activated by such things as anniversary dates, media reports of the event or similar events, preparing documents for court, watching the news or movies involving police action, and involvement in another traumatic event, as is common among police officers. Results are similar with military personnel in regards to media reports, anniversary dates and especially Remembrance Day, sounds of or similar to rifle fire, overhead helicopters, and war movies. I can still recall the tale told by an 80 year-old WWII veteran. His regular barber changed the razor for his haircut and as soon as it was turned on, the veteran dove to the ground. Later he realized that rasor's sound was similar to a dive bomber. Additionally, symptoms can be re-triggered by current stressors unrelated to the traumatic experiences or they may appear for reasons unknown.

Under circumstances such as those noted above, studies demonstrate psycho-physiological reactivity such as elevated heart rate, increases in skin conductance, and changes in bio- and neuro-chemistry consistent with re-activation of the threat system. In fact, even after successful traditional treatment, research demonstrates that the most common pattern, especially for those who suffered from PTSD as a consequence of DRAT, is for the symptoms to come and go throughout the life cycle. In other words, with the methods used today by most treaters of PTSD clients (which do not include BFT-A, BFT-M, or BFT-C), the symptoms have tended to follow a chronic course for many with PTSD. However, studies reporting these results have typically employed only one or two of the methods described above while most clinicians I know take a multi-component approach. The only direct information I now have is three-fold. *First*, there have been few reports of significant relapse among the police officers or military veterans who completed my basic treatment sequence successfully. I have seen very few of them and those who have returned have usually done so because a new stressor emerged. *Second*, because I did not know what to expect, I provided several monthly follow-up meetings to those who completed the course of BFT-C successfully. There was no evidence of relapse during these sessions, often standardized psychological testing indicated further improvements in symptom levels, and EEG statistics continued to change in the direction suggested by analysis of the qEEG. *Third*, six-month reviews of some after follow-up sessions were terminated demonstrated that all were still in remission and in fact their PTSD scores had continued to decline.

I do not know whether the follow-up sessions contributed to the maintenance of gains or whether they would have continued without these sessions after acute treatment ended. However, being very cautious in temperament and in view of reports of others, once a client's PTSD is in remission, I do encourage follow-up visits. Typically, they are scheduled once a month for the the first several months, then quarterly for the next year, and every 6 to 12 months thereafter. Maintenance of gains is the criterion for decreasing the frequency of sessions. Depending on the client, a number of objectives can be included in the follow-up sessions.

One objective is to encourage the client to keep all of the intervention plans in place, which decreases the likelihood or intensity of a relapse. Most commonly, when clients suffer a relapse, one of the reasons is that they have discontinued the recommendations that brought about remission. Clients need to be reminded that they are feeling better more because they are maintaining recommendations and less because they have been cured.

A *second* goal is to problem-solve with him or her when stressors arise, since any new stressor or combination of stressors can lead to an increase in symptoms even if the stressor is unrelated to the original traumatic experience(s). More about this is found below in L-03: On Stress and Stress Management.

A *third* purpose is to prepare for those times when we know that the chances of re-triggering are high (for example, anniversary dates and court appearances).

A *fourth* aspect involves booster sessions of BFT-C.

A fifth option is providing clients the opportunity to learn how best to manage monkey-mind. Reports in the areas of both anxiety and depression indicate that so doing decreases the incidence of relapse. Information about this is found below under section L-02: Reducing Monkey-mind.

A *sixth* element of follow-up is addressing as needed the Level 2 stabilization strategies found at the end of this section.

L-02: Reducing Monkey-mind
If we pause for long enough and just become aware of our thinking, many of us humans will notice that our mind is on the go constantly; now a memory, now a plan, now a self-criticism, and so on. For those with PTSD, thoughts related to traumatic experiences are part of this. Sometimes this is referred to as "monkey-mind" and one can picture the mind as a monkey swinging from one tree branch to another. Monkey-mind can be exhausting, the negative self-statements can contribute to depression, remembering trauma can maintain uncomfortable states of hyper-arousal, and it can be difficult to maintain control over attentional focus, thus decreasing capacity for remembering and other aspects of cognitive functioning.

Often, the BFT-A methods noted in section F above are successful in curbing monkey-mind so the client is able to achieve mental calmness and focus. If clients are not completely successful in gaining control over monkey-mind to this point, then a form of meditation is discussed which, with other methods, has been shown to have a positive effect on stress and depression. For example, Waelde and collaborators (2008) used only a manualized meditation intervention for 20 mental health workers who chose to attend following the assistance they provided to victims of Hurricane Katrina in the U.S.A. The intervention comprised a 4-hour workshop followed by an 8-week home-study program. Although the drop-out rate was 25%, results indicated that those who completed the intervention decreased PTSD symptoms significantly. Moreover, the number of total minutes of daily meditation was correlated with such improvements. Other studies have noted the lessening of relapse of both anxiety and depression when monkey-mind is addressed successfully. Also, this and other forms of meditation have been demonstrated to reduce the stress of cancer and high-stress work environments as well as being helpful for managing panic and anger.

Moreover, the meditative approach to monkey-mind has been used by millions within Eastern traditions such as Buddhism. Furthermore, this is something I have found to be helpful personally in keeping my own monkey-mind under better control; along with additional meditative practices, it has played a major role in allowing my mind to remain calm, clear, and responsive.

The initial sequenced steps I recommend for meditation/controlling monkey-mind are as follows and are fairly typical.

One, sit comfortably in a quiet place. Plan to do this initially for 5 to 10 minutes at a time and gradually extend this. I am up to about 30 minutes in the morning, which works well for me. Setting a timer allows me to practice without needing to check on the time, thus decreasing one distraction. Also, I am now able to meditate anywhere regardless of noise and other distractions, but it took years to get to this stage.

Two, begin tactical breathing (1, 2, and 3) and continue this for a few minutes. At this point and throughout the session, I keep my eyes lightly closed. For those with an initial tendency to doze off, keeping eyes open and focussed on a distant object often is successful.

Three, pay full attention to one part of your body at a time (while breathing); note whether it is tense or relaxed. If it is tense, do some stretching exercises for that part. Then, when it is relaxed, move on to another part. Continue this until all parts of your body are relaxed. This often is referred to as a body scan. Generally, I meditate after my morning exercise program, which includes running, gentle stretching of many muscles, and weight training. So my muscle systems are fairly relaxed by the time I get to meditation time.

Four, continue breathing and become aware of any sensory sensations such as sounds, smells, parts of your body touching the chair, and so on. Simply note each sensation while continuing tactical breathing. Note that sensations simply rise up in consciousness and then fall away. This pattern in sensations, feelings, and thoughts is often referred to as impermanence.

Five, become aware of your breath: the temperature during intake, the temperature on exhalation, the parts of your body that move with it, what the breath feels like, and so on.

Six, count the breaths ("breathing in one," "breathing out one," "breathing in two," "breathing out two," and so on) while focussing all of your attention on the breath and physical sensations associated with it.

Seven, continue the counting as long as no thoughts intrude.

Eight, if a thought does intrude, simply, calmly, and matter-of-factly label it ("ah, a memory" or "ah a plan for the future" or "there's another thought," etc.) without evaluation, anger, or frustration. Don't push it out or dwell on it. Just observe it objectively and label it without judgment. Note the impermanence of thoughts, they simply come and go.

Nine, immediately after labelling, instruct yourself gently and calmly as to what to do next. "Okay. Time now to return focus on my breathing."

Ten, start the breath-counting over again.

Eleven, continue this day in and day out until monkey-mind is under control. When I began this decades ago, it took well over six months before I could get up to a count of five. It is truly amazing how frequently thoughts arise, how rapidly they can pass through mind, how negative and self-destructive they can be, and how compelling they can be.

Twelve, once monkey-mind is under control, you can drop the counting and consider adding other elements to your meditation practice. For me, this now includes getting in touch with exactly what I am feeling in the moment, reviewing/remembering how best to increase mental calmness and clarity throughout the day, and often some time spent on an objective observation of thoughts or images that occur with significant frequency. I have done this for decades and would be pleased to help start you on this path.

One of the consequences of this approach as well as some reading for me has been the realization that thoughts are simply what is generated by the mind. Generally, they don't mean much and are rather like an itch or bowel gas. They do not necessarily have any reality base and they certainly do not have to be believed. Moreover, the do not have to be acted upon.

Just like other aspects of treatment for PTSD, for some clients the steps above to decrease monkey-mind are not successful until after completion of BFT-C. Presumably, this is so because the basic electro-physiological brain functioning must first be within some normal range.

Other methods of meditation have demonstrated their positive effect on psychological health. For example, the 2007 review reported by Rainforth's group demonstrated that Transendental Meditation resulted in significant reductions in blood pressure as compared to biofeedback, relaxation-assisted biofeedback, progressive muscle relaxation, and stress management training.

L-03: On Stress and Stress Management

Introduction
In this outline, I will use the term *stress reaction* to refer to our response to stress and the term *stressor* to stand for that which causes a stress reaction.

Basically, a stressor is a demand placed on a person and stressors can be grouped into a number of types. As discussed in the 2007 book edited by Lehrer and colleagues, stressors can be acute time-limited (such as preparing for a medical procedure or an examination), involve a sequence of events (such as a natural disaster or loss of a loved one), be chronic intermittent (such as repetitive evaluations in competitive sports, recurrent headaches, or police or military duty), or be chronic continual (such as a debilitating medical or psychological condition or marital discord).

A stress reaction is the natural and automatic response we as humans have to a stressor. The reaction involves many physical and psychological processes and ultimately energizes us and prepares us for action such as to fight or to flee or to freeze in the face of a stressor. Stress reactions allowed our early ancestors to survive in a hostile world. Our reaction to fight, flee, or freeze is not always within our conscious control; frequently it depends on which hormones our body produces at the time. Also, unless they are under our conscious control, the automatic processes are triggered whether the stressor is a life-threatening one or a psychological one.

Common Misconceptions about Stress
A number of myths or unexamined but false ideas about stress are found throughout our society.

One myth is that an appropriate goal is to reduce one's stress level to zero. The truth is that if there were no stressors or demands, then there would be nothing done and humans would be bored, have low energy, and achieve no results. However, it is also true that if the stressors are too great, then the results are as bad, including overstimulation, psychic pain, lack of productivity, and being unfocused. The most sensible goal, then, is to learn how to find the best levels of stress for you so that your life can be enjoyable and productive.

A second misconception is that what is a stressor for one person will be a stressor for all. That is clearly false. The impact of a stressor as well as the magnitude of a stress reaction varies greatly from one person to another. For example, if three people are trapped for a time in an elevator, it is possible that one will shown signs of a stress reaction early on, another will be fine during the "crisis" but perhaps avoid elevators from that day forward, while another will be unfazed by the experience. Our reactions are very individualized and depend on such factors as our physiology, our history of past stressors, our stress management skills, how we perceive the situation, how extreme the stressors are, and the number of other stressors we face at that time.

A third false belief is that all of us humans start with the same basic levels of reactivity. In fact, we can be cured of that myth by spending just a little time in the hospital nursery, where newborns show significant variation in basic physiological levels as well as what it takes to evoke a reaction.

A fourth myth is that having a stress reaction is a sign of weakness, or unworthiness, or unmanliness, or inadequacy, etc. That is patently nonsense. Given sufficient stressors, all of us humans will react; no one is immune. Stress reactions have occurred since humans were on this planet, among both genders, all races, all religious groups, and those with and without disabilities, across the whole socioeconomic spectrum, and independent of sexual orientation. Simply put, being susceptible to stress reactions is part of how we are made. Besides, it is less than intelligent to ignore what is happening to us, since by coming to grips with stressors and stress reactions we will be able to deal effectively with them and get on with a meaningful life.

A fifth erroneous thought is that the consequences of stressors are always bad. The fact of the matter is that facing stressors in one's life can lead to positive outcomes such as a better prioritizing of life goals, an increase in self-esteem and growth (the sense that "I did it"), the learning of new coping skills, and the strengthening of social relationships (the notion that "we did it").

Can There Be Too Much Stress?
You bet there can. The stressors and stress reaction can be so great for any one of us at any time that our coping skills are insufficient to prevent our system from breaking down. In fact, some researchers have identified four stages that a person might go through, depending on such factors as the number, intensity, and duration of stressors. In *stage one*, one experiences mild anxiety or distress but generally feels quite motivated and energized. In *stage two*, the feeling grows that one is going under and barely able to keep head above water. Performance is impaired and one feels significant distress. In *stage three*, frustration increases and both physical and psychological symptoms appear. Physical symptoms can include tension headaches, sleep problems, digestion problems, heart palpitations, and other conditions. Psychological symptoms can include irritability, inability to concentrate, problems with memory, tension/inability to relax, thoughts that won't stop, and withdrawal from others. In *stage four*, one has the sense of burnout. Experiencing emotional exhaustion and an inability to make oneself hang in there is common. Depending on other characteristics of the reaction, sometimes this stage is called depression.

What Is Stress?
I use a definition of stress pieced together from what I have read, experienced, and learned from my clients. The definition is divided into a number of parts and each part is discussed in turn below.   Stress is (a) a threat/demand/constraint (a stressor), (b) perceived by a person, (c) having some traits, (d) a history, (e) a current status, and (f) a perceived range of resources, (g) who evaluates all of the preceding, (h) decides if the stressors overtax resources or not, (i) and then reacts physiologically, (j) cognitively, (k) affectively (emotionally), and (l) behaviourally.

A. A Stressor

As humans, we can face a bewildering number of stressors. As noted previously, operational police officers and many military personnel are continually confronted with both the death and injury of others as well as threats to their own safety. In addition, for some folks organizational and administrative matters serve to increase the stressor load. These can include expectations that are unclear or constantly changing, not having the expected knowledge or skills to do the job, inadequate supervision or supervisors, unenlightened personnel policies, insufficient resources, job instability, a perceived sense of being treated unfairly, and the worry that if something goes amiss that the organization will not be supportive.

(B) Perceived and Interpreted by a Person

Generally, a person needs to both notice the situation and be affected by it before it becomes a stressor. For example, if you are home alone at night and you are awakened by a sudden noise at 0200 hours, you will experience stress if you think it was caused by an intruder but you will just feel some annoyance if you think it was caused by your cat. However, some people (1) do not notice symptoms until they cannot be ignored and are severe enough to be consistent with an established psychological diagnosis, (2) often cannot tie the symptoms to any precipitating event(s), and (3) often have to check out any sound in the night for fear that it is being caused by an intruder whose intent is to harm the person or other members of the family. I remember several police clients who, when awakened by a sound in the night, would grab a loaded gun and cautiously explore the perimeter of the house. Many others, including both police and military veterans, sleep with a loaded firearm nearby.

(C) Who Has Some Traits

We all bring some personal characteristics to any stressor. Some characteristics are mostly learned, some are mostly inherited, and some are a combination of both. If we have not learned any effective strategies for managing or coping with stressors, then this trait will make stressors worse for us compared to a person who has the skills to deal with stressors. Our natural inclination (likely inherited) to fight, flee, or freeze in face of the stressor is a trait that can cause a positive or negative outcome. Also, characteristics (probably a combination of inherited and learned factors) such as resting level of physiological arousal, tendencies to suppress negative events and psychological symptoms, the need to appear strong and invincible, fragility or ease of becoming overwhelmed, and sense of optimism or pessimism are aspects we bring to each stressful situation that can have an impact on outcome.

(D) A History

Experiences that a person has during life have an impact on how that person will fare when stressors mount. For example, for some folks the negative things that happened to them during childhood (such as physical or sexual abuse or the loss of a parent) can increase the chances of crashing under stressors as an adult. There is reason to believe that this happens because such early experiences alter brain functioning. So the slogan of "Deal with past trauma or it will deal with you" has a lot of truth to it. On the other hand, having positive experiences as a child within the context of a supportive and well-functioning family is the kind of early history associated with an increased ability to manage or cope with stressors later in life.

(E) A Current Status

What else is happening at the time the stressors mount has an impact also on how successful the person will be in handling them. One area is physical and mental health; a person who enjoys good health and does not suffer a psychological condition will fare better than one who has either a physical or psychological condition. A second area is the number of other stressors present at the same time; things will be much worse if the person is facing four other stressors rather than facing no additional ones. A third area is psychological needs. All of us humans require such things as affection, a sense of belonging, adequate stimulation, and so on, in order to function well. If these basic requirements to our psychological health are absent, then we will have a more difficult time dealing with the stressors that come our way. A fourth area is our stage of development; for example, losing a job at age 45 is likely to be worse than at age 65.

(F) A Perceived Range of Resources

The more resources the person has, the greater the chance s/he will be able to manage a stressor. The fewer personal resources a person has to draw upon, the greater the impact of a stressor will be. Some of the resources of importance are healthy living (for example, having a history of making healthy decisions, a balanced diet with alcohol intake low and no drugs, and a helpful pattern of physical activity and fitness), healthy feelings (such as ability to manage negative emotions and to stay calm in the face of stressors), healthy thinking (including having reality-based beliefs instead of delusions, perceiving reality accurately, putting things in proper perspective rather than using cognitive distortions, having a good and accurate sense of self, and being able to notice and enjoy the positives in life), healthy skills (for example, having a successful strategy for problem-solving what is one's control and being able to stay calm and cope well with things outside of one's control), and healthy relating (such as being able to see things from another's perspective and to experience empathy, using well-intentioned and effective communication skills, enjoying the company of positive people, being able to resolve conflicts with a "win-win" approach and having friends and relatives who are supportive).

(G) Who Evaluates all of the Preceding

When the stressor(s) come up, most of us consciously or automatically consider all of the things I have noted and essentially answer a series of questions quite quickly such as the ones I have listed below. Sometimes this is referred to as a "decision tree."

1. Is this important or not?
   If no, then it can be ignored at this time?
   If yes, then go to # 2.
2. Can something be done?
   If no, then find a way to accept or cope or adjust.
   If yes, then go to # 3.
3. Is it urgent--must it be done now or can it wait?

If no, then schedule a time to address the stressor.
If yes, then go to # 4.
4. Have I the resources?
If no, then decide what I need or who can help.
If yes, then work towards resolving the stressor.

(H) And Makes a Judgment

If the person has a good combination of traits, history, current status, and available resources, then s/he is likely to conclude, "I can handle this." In this situation, the stressor is seen as a challenge, an excitement, an adrenalin high, and then the individual goes on to manage the stressor. On the other hand, if s/he has a poor combination of traits, history, current status, and available resources, then the person is likely to conclude, "I can't handle this"; what follows is a state of distress.

(I) Then Has a Fairly Automatic Physiological Response

The human nervous system is divided into the central and peripheral nervous systems, the peripheral into the autonomic and somatic nervous systems, and the autonomic into the sympathetic and parasympathetic nervous systems. When a stressor comes along, all systems play a role in a response that generally is pretty automatic and very rapid (from 0.5 seconds to a couple of seconds). For example, some of the possible changes are noted below.

1. The sympathetic nervous system gets fired up and/or the parasympathetic nervous systems take the brakes off, which leads very rapidly to changes such as an increase in heart and respiration rate, increase in airway passages, larger pupil size, increase in blood pressure, and changes in the blood flow. All of these serve to ready the internal organs to deal effectively with the stressor.
2. The neuro-endocrine system is activated, starting from the central nervous system to the body's glands, which release various hormones. Thus, more glucose becomes available to deal with the energy demands, the immune system is put on alert, and natural analgesics are increased so that the person will not be hindered by minor pain.
3. The HPA axis (hypothalamic-pituitary-adrenal) goes into action to conserve energy (for example by suppressing digestion), suppress sleep, and increase metabolism. This activation leads to many other changes in the brain, as discussed previously.
4. The somatic nervous system is activated so that muscle tension increases in preparation for whatever action is needed--fighting, fleeing, or freezing.

The purpose of these changes within our body is to prepare us to handle the stressor. The systems evolved over thousands of years and allowed our species to survive real threats to our survival. However, our biological system does not distinguish a physical from a psychological threat and so psychological stressors serve also as triggers that put many biological systems on alert.

(J) Then Associated Cognitive Changes

With the total body on alert and often without our conscious perception, cognitive changes can be noted. These include an increase in overall alertness, a decrease in noticing or attending to irrelevant distractions, better vision and hearing, quicker reaction time, and wakefulness rather than being able to sleep.

(K) And Associated Emotional Changes
Depending on a number of factors, we may notice emotional changes as a result of the rapid activation of biological systems. These emotional changes can include feeling pumped and energized, feeling angry, and feeling scared.

(L) And Associated Behavioural Changes
We become ready to act in some way, depending on the circumstances. Generally, our behavioural changes can be described as fighting, fleeing, or freezing.

If the Stressor Is Resolved
Once the stressor has been managed successfully, then gradually all biological systems re-set and we return to our normal state. However, we need to realize that the return to normal biology is much slower than the rapid reaction to a stressor. Depending on the stressor and other factors, it can take from hours to days to return to normal levels. Within the autonomic nervous system, the parasympathetic nervous system must go into operation and/or the sympathetic nervous system needs to deactivate enough so that heart rate and respiration will decrease, blood vessels will dilate again, and the pupils will return to a normal diameter. However, if the person faces continuing stressors, then the levels do not re-set to previous baselines; each new stressor/trauma raises the resting level further, and ultimately it takes less and less potency in a stressor to trigger both physical and psychological reactions.

If the Stressor Is Not Resolved
For a variety of reasons including (a) aspects of the stressor (for example, it is too intense or it continues for too long) and (b) characteristics of the person's physiological processes (such as hyper-reactivity of the sympathetic nervous system), the biological systems of some folks remain on alert for a long time. Since these systems are not intended to remain in an aroused state for too long, a number of things can happen.

1. Functioning deteriorates progressively and sometimes the person experiences burnout and exhaustion.
2. The individual takes steps to try to feel better. However, often the things tried make things worse, such as when the person increases alcohol or drug use, goes on a spending spree, gets away from everything and everybody, engages in impulsive risk-taking, or reacts against others with frustration and anger.
3. Left unchecked with stressors unresolved, s/he is then well on the way to developing symptoms involving combinations of the following:
   a) *physical reactions* in any or a number of systems including gastro-intestinal (upset tummy, vomiting, diarrhea), cardio-vascular (chest pain, migraine headache, rapid heart rate, high blood pressure, and coldness in hands

or feet), pulmonary (hyperventilation, asthmatic symptoms), neurological (dizziness, pain, disorientation, hyper-vigilance, and sleep problems), motor (tremors, muscle soreness, tics/twitches, jumpiness, agitation, and tension headache), sexual (loss of interest, inability to perform), and energy (low energy and being easily fatigued). Moreover, if the person has a pre-existing medical condition (such as essential hypertension, coronary artery disease, or irritable bowel syndrome), unresolved stressors can worsen the condition significantly and even result in death (for example by myocardial infarction in a person with coronary artery disease);

b) *cognitive reactions* including problems in attention (inability to concentrate or focus, being easily distracted, experiencing intrusive thoughts), memory (forgetting, having a slowed retrieval time), thinking (slowed or fuzzy thought processes), decisions (avoidance of decisions or changing decisions after making them), and increased errors in performance;

c) *emotional reactions* such as increased anxiety (feeling fearful, agitated, overwhelmed, tense, or on edge, experiencing excessive worry, having panic symptoms), depression (feeling sadness, loss of interest, withdrawn, helpless, hopeless, worthless, teary, numb, or dead inside), or anger/irritability; and

d) *behavioural reactions* including more alcohol/drug abuse, extreme withdrawal, inability to get any where near work without a major anxiety attack, marital problems, negative interactions with one's children, and suicidal thoughts.

Fortunately, there is a good chance of recovery with competent psychological intervention. And as always, the earlier the interventionm the less distress to the person and the greater the ease of recovery.

So based on the above, below are steps a person can take to manage stressors well.

Recommendation-1. Use Helpful Words
How we label things can have a great influence over how we think about them, feel about them, and deal with them. Some words have a very negative tone that can almost paralyse us; *stress, tragedy, problem, life-threatening, chronic, trauma, terminal, forever damaged,* and *hopeless* are some words that can be paralysing to some folks. Thus, a first step is to find a different word that is more neutral in tone. For many, words like *challenge, issue, situation, task,* and *opportunity* are preferable. Of course, after finding the right word, it is important to use that word consistently from then on and to not slip back to using negative words. So, from here on I will use the word *challenge*.

<u>Recommendation-2. Maintain a Healthy Lifestyle</u>
Looking after yourself is critical to your ability to face and deal with life challenges. This has been considered already in the section on stabilization strategies. They include making the regaining and maintaining of health one of your top three priorities, cutting out the use of illegal drugs, cutting down on unhelpful legal substances, eating sensibly, taking the steps to achieve restful sleep, doing aerobic exercise daily, hanging out with positive people, sharing feelings, letting others know how they can help, making room for recreation, finding opportunities for laughter, and using effective methods of relaxation regularly. Those who maintain a healthy lifestyle by making such critical aspects a consistent part of their daily lives are much more able than otherwise to deal with life's challenges.

<u>Recommendation-3. Keep the Challenge in Perspective</u>
When faced with a challenge, we can become so pre-occupied with it that we can see nothing else. This means that, automatically, we overlook all of the positive things that are going on. Consequently, our lives and thoughts become out of balance and we lose perspective. Ultimately, this makes things a whole lot worse for us than they need to be. The antidote to this is to make a written list of all the things that are positive or are going well in your life; this is what my grandparents meant by "count your blessings." Such blessings can include each positive relationship you have, your health and the health of others you love, the good times you have had, the resources available to you, your work, the abilities you have, your accomplishments, the beauty of the world around you, the warmth of the sun, your motivation to meet the challenge, medical insurance, and the sense of meaning/purpose you have in life. So, make such a list and review it often.

<u>Recommendation-4. Use Self-Talk That Is Positive and Increases Focus</u>
Most of us talk to ourselves; we say things to ourselves about events (e.g., "That's good" or "How terrible"), we worry (e.g., "This isn't going to work out"), we evaluate ourselves (e.g., "I'm doing okay" or "I'm messing this up"), and we give ourselves directions (e.g., "Okay, what I need to do next is to check the switch"). It is normal to do this. Moreover, if done correctly, self-talk can be a positive help to us in dealing with a challenge by keeping us motivated and focussed on the essentials. On the other hand, we can think and talk to ourselves in ways that bring us down and draw our attention away from the challenge itself. The trick is to get into the habit of using self-talk that is positive and that helps to maintain focus.

The first step is to become aware of your self-talk. You can do this in a number of ways, for example, by monitoring your thoughts and noticing what you are saying to yourself or by writing down the kinds of self-talk that are usual for you.

The second step is to have an idea of the types of self-talk that are positive and that also help to maintain focus. Below are some examples that many have found to be helpful.

> What is it I have to do?
> Focus on actions not feelings.
> The situation is not impossible; I can handle this.

Stop worrying; worry isn't going to help anyway.
I have a great many resources I can use.
I can deal with this if I take one step at a time.
I have got through situations like this before and I can do it again.
I will not let this overwhelm me.
I can see this as a challenge rather than a tragedy.
These are the things I need to do to get a handle on this.
Relax. I'm in control. I will do some tactical breathing.
Keep the focus on what is in my control, what I can do.
I can make a good effort. No one can expect more.
I can learn how to deal with this challenge.

The third step is to write on an index card examples of helpful self-talk such as the ones above that you want to become more automatic for you. Then, carry the card with you at all times. Bring the card out 5 to 10 times a day and say those words to yourself as a way of getting them better imprinted in your mind.

The final step is to substitute the positive self-talk each and every time your thoughts are negative or are distracting you from the task at hand. For example, if a negative or distracting thought creeps in, then do the following:

  a) say to yourself "Stop that unhealthy kind of thinking";
  b) get out the index card; and
  c) say out loud or to yourself a number of times the positive thought that fits best.

Recommendation-5. Use Helpful Coping Methods As Needed

There may be times when you will fear that the challenge is becoming overwhelming. That is normal and natural for us humans. The objective here is not to become overwhelmed and instead to find a method that helps you cope at this time. One method that is often effective is to stop everything immediately and do the tactical breathing you have learned until you feel back in control of yourself. Another is to pick from section # 2 above those aspects of a healthy lifestyle that work best for you; for example, a burst of aerobic exercise or a chat with a close friend. A third option is to get away to a positive, relaxing, and engaging environment for whatever period of time is necessary; a long walk in nature, an overnighter with a friend in a different city, or a longer vacation are three ways of taking a break from it all.

Recommendation-6. Know Yourself and Your Limits

Being aware of yourself, your thoughts, and your limits can be very helpful in dealing with stressors. Firstly, knowing how many demands/challenges you can handle at any one time is critical--none of us can do everything all the time. It depends on other things happening in our lives at that moment and all the other factors discussed above. Secondly, it is important to not get caught up in the common myths: for example, in the "shoulds" of this life and the major North American myth that there is nothing you can't do if you put your mind to it. Thirdly, it is helpful to set reasonable time frames and not to expect to do all things immediately. Finally, it is critical to recognize what is happening to you (biologically, cognitively, affectively, and behaviourally) and to take whatever steps are necessary to remain healthy in all realms.

Recommendation-7. Identify the Challenge(s)
Once the suggestions from sections #1 to #6 are in place and are working well for you, it is time to turn your attention to the challenge or challenges you face. Clearly, one critical step is to know exactly what you are facing. So, you need to take the time to very carefully and very precisely identify each challenge you are facing. The best way is to write each one down on a separate piece of paper and to make revisions so that you end up with a statement of the challenge that is in as sharp a focus as you can make it.

Recommendation-8. Rank Order the Challenges
There is a limit to the number of challenges we can tackle at any time. So, if you are facing more than one challenge, it is critical that you put the challenges in some order of priority. Then, follow the steps in the sections below for the top priority. When that challenge is under control, select the next one and apply the sections below to it. And so on. In this way, one at a time, you will be able to face what you must, but you will do it in a manner that increases your chance of success.

Recommendation-9. Determine Why This Is a Challenge
In dealing with any challenge, is it important to know why this is a challenge for you, why you feel so strongly about it, and what is the source of these feelings. By so doing, you may find that it is not a challenge at all or that it really is not that important. If neither of these two insights develop, the process of asking "Why?" may give you some direction in dealing with the challenge.

Recommendation-10. Obtain Relevant, Reliable, and Reality-Based Information
In order to develop a successful plan for managing challenges, it is critical that you make sure that the information you have is relevant to your situation, is reliable, and is based in reality. Many people do not check their facts, consult with people who do not know any more than they do, or are faced with a bewildering array of information without the tools for separating fact from fancy. And if your information is in error, then your solution is very unlikely to help in dealing with the challenge as effectively as possible. Too many times I have seen people who acted on information that was dead wrong, with the result that the challenge was not dealt with as effectively as it could have been. I remember the lady who spent over $100,000 on naturopathic remedies for cancer over a two-year period instead of following the treatment recommended by her oncologist and got so much worse that the methods that could have saved her could no longer be used. Then there was the man I saw following a serious suicide attempt after believing for years what he had been told--that his depression was in his head and he just had to bring himself out of it. While I would be the first to admit that there is much we do not know yet, it is also true that we know a lot. So, in both my personal and professional life, I tend to use one or more of three methods to maximize the chances that the information is relevant, reliable, and reality-based.

One method is to find out who is the appropriately trained expert in the field and go right to the top. Not following the prevailing sentiment in the media and on the street, I take the position that some people (experts) know more than others; that good will and a loving attitude are not enough; that experience, while valuable, is insufficient if unaccompanied by education; and that the status of *expert* requires significant training in a specific field based on data obtained from application of the scientific method and the credentials that result from passing examinations. You are free to consult others, but you do at your peril if the person is not an expert in the restricted and meaningful sense describedabove. The lady with cancer noted above is a good example of picking the wrong helper.

A second approach is to have or acquire the skills so that you can evaluate scientifically any information you find that is related to your challenge. Testimonials ("I tried it and it worked"), case studies, and sales pitches are so unscientific as to be laughable if the consequences of following such advice were not either useless or potentially devastating. The man with depression is an example of someone who followed advice that is unscientific. Basically, the skills to assess information in my field of competence require training in the scientific method, which includes knowing how a proper study ought to be designed. In drug studies, for example, this involves the random assignment of people to one of at least two groups, use of a reliable and valid method of assessing the outcome of interest, utilization of the double-blind method in which neither the subjects nor experimenters know which group is the experimental and which the control, use of a placebo-control in which the control group gets an inert substance while the experimental group gets the medication being tested, and formal but appropriate statistical analysis to determine results. Such an approach then needs to be duplicated by researchers in other centres before results can be accepted as a valid finding.

If both of the above methods prove to be unsuccessful, then a third way is to know how to conduct your own experiment. However, you would need to do this very carefully and follow scientific methods. In situations where the required information is not available or using the known treatment methods have not been effective, I am available to assist clients in conducting experiments to determine if a reasonable idea might prove useful. Generally, suggestions such as those contained in the sections that follow are among the critical elements required.

Recommendation-11. Decide What Is in Your Control and What Is Not
In the challenges we face, sometimes an action plan can be developed because the events, causes, situation, etc., are very much in our control--there is definitely something we can do to change things. For example, we can get our depression under control, we can re-arrange our environment and routine so that sleep becomes more restful, we can learn methods of relaxation, we can prepare ourselves for a career that suits us better than the current stressful one, we can use methods that will lessen our symptoms of post-traumatic stress disorder, and so on.

On the other hand, it is also true that sometimes we cannot control or change the basic challenge. For example, some people develop illnesses that are terminal, in some marriages one partner is not willing to work towards a mutually satisfying solution, our child can have a life-long specific learning disability, we may be informed that a person we love has died, and so on.

Thus, for each of the challenges we face, it is important to conduct a reliable and reality-based appraisal so that we know which parts are in our control and which aspects are not. The results will give effective guidance to the proper mix of methods we can use to face the challenge. For example, we may not be able to control the final outcome if we have a terminal illness but we do have some control over the quality of our lives in the time remaining and in how we face the reality of death. In this regard, many have found the *Serenity Prayer* a helpful way to think about things:

> God grant me the serenity to accept what I cannot change,
> the courage to change what I can, and the wisdom to know
> the difference.

Recommendation-12. For Those Aspects in Your Control, Use a Problem-Solving Format

If the challenge or parts of the challenge are truly in our control, then a problem-solving approach is recommended. If the issue involves conflict with another person who also wants to resolve it, then the same format can be used, although some writers use the term *conflict-resolution* rather than *problem-solving*. However, basically the steps one uses are identical.

The *first step* is to identify one problem at a time and then to define it as precisely as your can. Describing it in ways that are observable and measurable is important. For example, the number of minutes spent each day walking on the treadmill at 3 miles an hour is observable and measurable, while simply increasing exercise is not.

The *second step* is to generate as many solutions as you can. This is the creative stage. However it is very important during this stage that you not stop to evaluate any of the creative ideas or discard any of them. Evaluating as you go along will have the effect of dampening the creative process. So, write down each possible solution, however weird or humorous. If an interpersonal conflict (such as "What shall we do tonight?") is the challenge, remember that this is an exercise in cooperative group creativity. Therefore, whose idea it is does not matter--write it down without evaluating it.

The *third second* step is to look at each proposed solution and list its advantages and disadvantages.

The *fourth step* is to choose one solution. One system for selecting a possible solution is to use the *mini-max principle*. That is, since it is rare to find a solution that is perfect, choose the solution that has the fewest disadvantages (mini) and the most advantages (max). If the original issue was a conflict between people, then choose a solution based on the *win-win principle*; the solution must be one that both parties consider to be fair to both, that both can live with, and so that neither party leaves the discussion thinking that s/he lost again.

The *fifth step* is to decide on the details of how the solution will be put into effect. This means taking the time to consider carefully the answers to such practical questions as who, where, what, when, and how.

A *sixth step* is to decide how you will know if the solution is working or not and when to evaluate results.

A *seventh step* is to put the plan into effect for a reasonable period of time.

The *eighth step* is to evaluate results of the plan. If results indicate that the intervention was effective, congratulations are in order using self-statements such as the following:

> I/we did it;
> Keeping focussed on the issue worked;
> One step at a time worked;
> Following carefully a problem-solving strategy was successful; and
> Feelings motivated me/us to resolve this challenge.

On the other hand, if the solution was not as successful as you had hoped, go back to section # 6 and look at the matter again. Be patient--it took Edison 300 tries before he found the right combination to make the electric light bulb.

Once the issue is resolved, then select the next challenge to address and follow the sequence above again.

Recommendation-13. For Those Aspects Not in Your Control, Use a Coping Format

For those challenges or portions of a challenge that are not in your power to change, the best format to adopt is one that focusses on successful methods of cognitive and emotional coping. Effective methods include the following:

a. dispel stressful myths (e.g.," It's not fair" or " I should have been able to . . . ");

b. decrease cognitive distortions (e.g., by saying, "I did the best I could");

c. use methods of thought control and distraction;

d. truly understand what is happening;

e. find meaning in the situation by asking questions such as:

> "What can I learn from this?";
> "Does this mean that a change in priorities is in order?";
> "How can this make me a better person?";
> "How can I prevent this from happening again in the future?"; and
> "Should I join with others in changing this situation?";

f. use the methods learned to date to regulate negative emotions;

g. share feelings with trusted friends and relatives;

h. maintain a lifestyle of healthy living and healthy thinking;

i. focus on what is controllable (for example, reactions to the situation);

j. make sure that nothing is left undone that should be completed; and

k. engage in pleasure-producing activities.

L 04: Level 2 Stabilization Strategies

As needed and possible, a number of strategies additional to the 28 presented in Step E can help to reduce relapse.

These additional strategies include the following:

Strategy 1:	As Possible, Finding Meaning in the Trauma;
Strategy 2:	Re-engaging;
Strategy 3:	As Needed, Developing Skills in Building/Maintaining Healthy Relationships; and
Strategy 4.	Considering a Cause Greater than Yourself.

As with the stabilization strategies in Step E, these additional strategies are given to clients as required.

Strategy 1: As Possible, Finding Meaning in the Trauma

Trauma seems to have less negative impact on us when we get to the point of realizing that something good came of it.   It may be personal growth in awareness, in empathy for others, in recognizing your strengths, in appreciating the things around you, in realizing the response of others, in bringing you closer to people who care about you, and so on.

Or, for a few, it may be something with far-reaching consequences; as a result of the death of a loved one, Mothers Against Drunk Drivers (MADD) was formed. MADD has had a significant positive impact on our society.   As a result of the senseless murder of their teen-aged son, FACT was formed to press for changes in the justice system as well improvements to how survivors of such trauma are treated.

Victor Frankl, a Jewish psychiatrist imprisoned by the Nazis in World War II, experienced and observed years of brutal and inhumane experiences in the concentration camps.   Closely observing who survived and who did not, he found that survival was not based on youth or physical strength but on the strength derived from purpose and the discovery of meaning in one's life and experience. As a result, he went on to develop a treatment strategy known as logotherapy.

As I think back on some of my own experiences that were absolutely dreadful at the time, sometimes years afterwards I noted that some good had come of them. Being in close touch with my father-in-law during his terminal illness removed the fear of death I once had.   Trying to cope with a nasty supervisor allowed me to appreciate my creativity in coping as well as realizing that I had to accept some responsibility for our relationship. Having my own crises of faith enabled me to search for and find a set of beliefs that give meaning to my life and allow me to be of some help to those who believe there is no ground under them. Knee and back problems forcing me to limit the scads of walking I did for pleasure and stress management, taught me to develop patience, to slow down, to find other methods for managing stress, and increased my empathy for those unable to get around easily.   Going through many years when my daughter had a chronic and potentially life-threatening illness gave me an appreciation of compassion of physicians we saw, made clear where our supportive relationships were, and led to a gain in appreciation of the human spirit of my daughter that has allowed her to achieve in spite of adversity.   Years of struggle with my appetite/eating disorder has allowed me to understand the struggles of those with different kinds of addictions.

So, write down below as many difficult/traumatic times from your past as you can remember. Then beside each one, make a note of what helpful things you learned from them.

Stretegy 2: Re-engaging
By the time we get to this strategy for stabilization, you will be ready to re-engage in a full life style if you have not done so already. As before, there are a number of considerations to review. But remember, you need to set aside time on a daily basis to maintain the interventions that have been successful in your recovery; if you stop them, the odds are very high that you will have a relapse.

One aspect of re-engaging is to take back gradually any family responsibilities that you have had put on hold so as to better recover from the traumas(s). Household chores and parental duties are among these. Again, make a list in the space below, priorize them in consultation with your partner, and then, starting with the most important one, begin to put them in place one at a time.

_____

_____

_____

A *second* aspect of re-engaging, for those who are not retired, is returning to work. An initial step is to consider whether or not it is healthy for you to return to your pre-treatment employment. If it is, then you need to consider if any changes in assignment are necessary in order to minimize a relapse. Once this is settled, I recommend a graduated return to work beginning with a few hours a week and slowly increasing this over a few months until you are back full-time. After being off work and attending to treatment requirements, most clients I know do not realize how tiring it will be to return to work.

If the decision is made not to return to the same work you were doing pre-treatment, then many things need to be done; for some, finances will be manageable by taking a medical discharge, for others funds, from a medical discharge will need to be supplemented by part- or full-time hours at another job, and for others a complete change of career will be required.

Depending on the person, my clinical experience is that each of these solutions can be helpful for clients. Space is provided below for planning.

_____

_____

_____

A *third* aspect is determining the role you will play in the activities and other relationships in which you were involved pre-treatment. Which supportive people do you want to re-connect with? Do you want to return to the bowling group? Do you want to try out a different recreational activity? What provides meaning and happiness in your life? Whatever your choices, I consider it important that they not have a negative effect on others or a destructive impact on the planet that sustains us all.

_____

_____

_____

<u>Strategy 3: As Needed, Developing Skills in Building/Maintaining Healthy Relationships</u>
Sometimes, for any of a variety of reasons, relationship skills are not optimal before a person endures traumatic experience as an adult. For others, the traumatic experience(s) resulted in perceiving others or acting towards them in ways that are not healthy for any relationship. And as noted early on, the healthy functioning of any adult depends in part on having supportive relationships with others.

Consequently, some clients need to assess their inventory of relationship skills, many of which are listed below.

_____01. Am I able to listen to what others say with attention and without interrupting?

_____02. Am I able to calmly consider the perspective of others?

_____03. Am I able to accept and respect differences between myself and others?

_____04. Am I able to recognize another's needs and be supportive?

_____05. Am I able to communicate my thoughts and feelings to others calmly?

_____06. Am I able to give lots of positives to others?

_____07. Am I able to receive graciously the positives delivered by others?

_____08. Am I able to give appropriate affection to others?

_____09. Am I able to receive the affection from others graciously?

_____10. Am I able to be appropriately assertive in regards my own needs?

_____11. Am I able to compromise, negotiate, and problem-solve on important issues?

_____12. Am I able to do things such that others will see me as dependable and reliable?

_____13. Am I able to act such that others would see me as trustworthy?

To use the terms of AA, once you have completed a "fearless inventory" of the skills above, then it is time to correct any deficits. How to do this is something we can discuss and I have found a number of books to be helpful.

Strategy 4: Considering a Cause Greater than Yourself

Once the responsibilities of maintaining intervention plans for PTSD, family needs, and employment are being managed well, you might want to think about getting involved in a cause greater than yourself. The options are endless including:

_____01. Helping out with your child's soccer team;

_____02. Taking part in the outreach activities of your church;

_____03. Joining a community service club;

_____04. Improving skills so you can be the best you can in your occupation or profession;

_____05. Joining with others devoted to enhancing world peace;

_____06. Working with others in trying to save planet earth;

_____07. Contributing to Foster Parents Plan to feed the children of the world;

_____08. Joining MADD to further reduce drunk driving;

_____09. Joining FACT to improve services when children are murdered;

_____10. Working with others to prevent the extinction of wildlife; and

_____11. Reaching out to colleagues you know who are suffering as a result of their trauma.

I like the words of singer/political activist Pete Seeger in the prelude to his song "We Shall Overcome," which he wrote during his involvement with the civil rights movement in the U.S.A. He said something like, "If you would like to get out of a pessimistic mood yourself, go help those people in Alabama or Mississippi. All kinds of things that need to be done. Takes hands, heads, and hearts to do them. Human beings to do them. Then maybe we will see this song come true. 'We shall overcome.'"

Another slogan I like is, "Think globally. Act locally."

APPENDIX I: Informed Consent for Assessment and Treatment

Dr John A. Carmichael R.Psych
Practice in Clinical, Police, and Military Psychology

Introduction

My private practice is restricted to adults with post-traumatic stress disorder, a psychological condition that can result from exposure to a traumatic event. As stated in the *DSM-IV* diagnostic guide, a traumatic event is one in which "the person experienced, witnessed, or was confronted with an event or events that involved actual or threatened death or serious injury, or a threat to the physical integrity of self or others." Thus, traumatic stress would include military combat, deployment to peace-keeping duties, the threat of and actual danger experienced by police officers and the citizens they serve, serious motor vehicle accidents, the reality of a life-threatening illness, the death of a loved one, and physical or sexual assault.

I have been involved in this area of service-provision since the 1980s. To date, I have seen well over 500 clients, most of whom have been from police or military backgrounds. Also, since 1980 I have expended a great deal of time and money in achieving competency in the evidence-informed assessment and intervention methods for PTSD and for the commonly associated conditions, as noted below. Moreover, I have presented full-day workshops on PTSD at international meetings of psychologists, I moderate an annual update for professionals, and I continue to provide professional consultation to colleagues. I have contributed to professional publications and my treatment manual was published in 2010.

PTSD is the most common of the trauma spectrum conditions. However, it co-occurs with a variety of other diagnosable psychological conditions. These include depression, panic attacks, sleep disorders, and psycho-physiological reactions. The latter can include such things as tension and migraine headache, high blood pressure, non-cardiac chest pain, fibromyalgia syndrome, and irritable bowel syndrome. When these co-occur with PTSD, I assess and provide effective interventions for them as well as for PTSD itself. To do so, I call upon my additional advanced training in such speciality areas as clinical psychology, psycho-physiology, and applied neuroscience.

Some of my clients with PTSD have received physical injuries related to their traumatic events and continue to experience chronic pain. Not only is the pain a nasty condition in its own right, but ultimately pain can make any psychological condition worse. And the psychological condition can increase the experience of pain. Thus, a vicious cycle is activated. Fortunately, I am familiar with psychological, psycho-physiological, and applied neuroscience interventions that can lead at least to a significant reduction in the experience of pain.

Sometimes, clients with PTSD experience problems in their cognitive functioning as a result of the trauma. Difficulties with attention/concentration, short-term memory, and information processing are common concerns among those with PTSD. Such cognitive issues can result from processes associated directly with

PTSD which sometimes resolve once the trauma condition is treated successfully. Alternatively, the cognitive problems can be due to functional brain injury from the incident itself or related to a previous assault on the brain. In addition to interim cognitive-management strategies, I can use a specialized biofeedback approach that often leads to positive changes in cognitive functioning for both groups of causes since I have the training, equipment, and professional consultation to do so.

Also, it is known that PTSD can aggravate a variety of medical conditions including diabetes, heart problems, arthritis, and cancer. Moreover, there is some evidence that PTSD may in fact be part of the cause of these conditions.

Court Status
I have been qualified in th Supreme Court of British Columbia as an expert witness in regards to both post-traumatic stress disorder and major depressive disorder. Also, I have testified as an expert in regards to administrative reviews.

Post-Secondary Education
1964: B.A. (double major in philosophy and religious studies), Mount Allison, NB
1966: Undergraduate courses in psychology, UBC
1967: More undergraduate courses in psychology, UBC
1970: M.A. (clinical psychology), University of Victoria, BC
1975: Ph.D. (clinical psychology), University of Victoria, BC
1975: Certificate of Clinical Competence, University of Victoria, BC

Professional Registration
College of Psychologists of British Columbia (0048)
Canadian Health Service Providers in Psychology (2690)

Professional Memberships
Canadian Psychological Association
American Psychological Association
Psychopharmacology Division of APA
Psychological Trauma Division of APA
Military/Emergency Trauma SIG
Military Psychology Division of APA
International Society for Traumatic Stress
Association for Applied Psychophysiology and Biofeedback
International Society for NeuroFeedback and Research
Society for Applied Neuroscience
EEG and Clinical Neurosciences Society

Continuing Professional Education
I am committed to providing competent services that are up to date with the continuing developments in such professional disciplines as psychology,

psychiatry, psycho-physiology, and affective neuroscience in relation to PTSD and associated conditions. To do so, I spend an average of 500 hours each year reading relevant material published in professional journals/books, attending post-doctoral institutes/workshops, attending meetings of the professional associations to which I belong, and in supervision/consultation arrangements with recognized experts whenever I begin a new area of intervention. Dr. R. W. Thatcher is my current consultant in the area of applied neuroscience. The College of Psychologists of BC requires 32 hours per year of continuing education to maintain registration.

Personal Information

Date of Birth: Regrettably a state secret
Citizenship: Proudly Canadian
Marital Status: Married to Lynne, who teaches piano and music theory; we have two adult children: Kathryn, a public health nurse, and Brian, a senior high school math and PE teacher
Hobbies: Listening to classical music, armchair astronomy, reading (philosophy, world religions, and Canadian military history), efforts towards personal fitness, watching televised international soccer matches and national football league games, family life (especially with my grandchildren Matthew and Sydney), travel, and continuing professional education.

Clinical Approach and Philosophy

Simply put, my overall objective is to help clients reduce their suffering and maximize their well-being. Thus, relief from distressing thoughts, bodily sensations, feelings, and behaviours is emphasized along with their replacement with clear and controlled thinking, comfortable and balanced physical functioning, emotional calmness, and successful behavioural strategies including effective stress management.

The method I use during the phases known as assessment, diagnosis, treatment, and follow-up is referred to in the professional literature as the *biopsychosocial* approach, since I take into account the multiple determinants of human thoughts, sensations, feelings, and actions. Such "causes" can be biological, cognitive, spiritual, historical, environmental (physical and social/cultural), psycho-physiological, and related to brain functioning as well as combinations of these factors. In this regard, each client is a unique human being who has an active part to play in the healing process that is facilitated by the therapeutic relationship we are able to establish. One of the client's roles is to implement consistently interventions we agree to try with the objective of taking advantage of the client's own natural healing power.

In regards to treatment, I remain committed to implementing only those methods with clinical or research evidence of effectiveness. This means that my practice is *informed by evidence*. Given my extensive continuing education activity, I remain up-to-date in the research, clinical base, and implementation details of common approaches to the individualized treatment of trauma conditions. For the same reasons, I am aware early on of new developments with good clinical, research, or theoretical support, thereby staying on the "cutting edge." However, whether traditional or newer approaches are used, frequent re-assessments and direct client feedback are obtained in order to determine what is and what is not working.

Typical Clinical Procedures

A careful assessment, including a structured clinical interview, an evaluation of psycho-physiological (mind-body) interactions, administration of standardized psychological tests, and assessment of cognitive status, is the foundation upon which I build a formulation-driven, empirically based, and properly sequenced treatment plan. For some clients, direct measurement of brain processes is completed also. Furthermore, I work closely with physicians, since some of my clients with PTSD require medical management of various symptoms. Sometimes crisis management is required before doing an assessment as, for example, with clients who wait until the last minute before calling me and thus when they attend (a) their psychological condition is usually in the severe range of intensity and (b) other aspects of their lives are out of control.

Once the assessment is completed, I provide clients with a written report. Also, the report is available to the client's physician via the client and often contains suggestions about conditions s/he might want to assess and consideration of medication; psychologists in Canada are not authorized to prescribe medication, although I stay up-to-date in this field in so far as PTSD and related conditions is concerned. In addition, referring agencies such as Veterans Affairs Canada often require a copy of my assessment.

Thereafter, the psychological treatment plan is implemented one step at a time. Clinical observations, repeat testing, information from the client's significant others, as well as client feedback are the basis for moving to the next step or for modifying the treatment plan.

Components of the Treatment Plan I Use Most Commonly
01. Psycho-education so that the client learns about the nature, causes, and treatment of her/his psychological condition(s). This is based in part on the belief that if you know what you have and why you got it, then the chances of following the right path increase;

02. Providing a safe environment both physically and psychologically so that nothing needs to remain hidden, thereby allowing us to work together collaboratively and effectively;

03. Psychological recommendations based on research in such disciplines as health psychology to reduce debilitating symptoms as quickly as possible by such things as attention to diet and exercise. I refer to these recommendations as stabilization strategies;

04. Biofeedback-Assiisted Training in Relaxation / Self-Regulationom (BFT), which forms a substantial part of my treatment approach since I have found it to be so effective in symptom reduction. Procedurally, BFT consists of (a) applying electronic sensors to various parts of the body without pain or embarrassment, (b) processing the information through electronics and computer software, (c) displaying the results on a computer monitor in real time (feedback), and finally (d) providing clients the opportunity and means to learn how to make required changes. Such modifications critical in the successful treatment of trauma-caused conditions include learning how to lower muscle tension (BFT-M) and how to decrease dysregulated autonomic nervous system functioning (BFT-A) characterized by such aspects as respiration anomalies, elevated heart rate, lowered heart rate variation, elevated blood pressure, and restricted blood flow. Similar learning is critical for those who also experience psycho-physiological conditions. For more information about biofeedback, see **www.aapb.org**.

05. Aggressive methods for normalizing sleep and making it restful if the methods above have not been successful enough. This includes sleep hygiene, relaxation training at bedtime, and sometimes the use of an electronic device that allows the person to generate the kind of brain waves associated with restful sleep.

06. As needed, strategies consistent with cognitive-behavioural therapy to examine beliefs and thoughts, to challenge ineffective ones and then substitute effective thinking, to learn workable problem-solving skills, and to discover ways to calm and train the mind. Generally I have found this to be most successful after BFT.

07. Identifying other stressors, implementing testable strategies to change what is within one's control, and then learning specific methods for coping with stressors outside of one's direct control. In a recent review of the records of my clients who had PTSD as a result of duty-related accumulated trauma, all but one of my WWII veterans had achieved remission, 70% of my police clients achieved remission, and all of the peacekeepers I have seen experienced a significant decrease in symptom intensity with about 40% of them achieving remission.

08. Additional methods of treatment from applied affective neuroscience (such as BFT-C) if any symptoms remain after completing the above (see also www.isnr.org) . For example, with BFT-C, the one remaining WWII veteran achieved complete remission, a further 22% of my police clients reached remission, and all the peacekeepers who completed treatment no longer have symptoms of PTSD and none dropped out of treatment.

Session Format
For my part, the client's spouse/partner is welcome to attend meetings. This can serve many purposes including (a) keeping her/him aware of what is happening, (b) obtaining his/her critical input as to what is happening at home, (c) helping the primary client to remember the specific between-session recommendations I make, and (d) making available a forum for the spouse/partner both to discuss

how the condition is impacting family members and to resolve minor stress points.

During treatment I will have an agenda for each session to which both the client and spouse/partner can add. Typically, during each session we re-assess the level of the condition that brought us together, review the data and impressions of the client on how things have gone regarding the specific recommendations from the previous meetings, make modifications to these suggestions as required, insure that the next aspect in the treatment plan is understood clearly and practised in the session, determine how it will be implemented before the next session, discuss matters on the client's agenda for the session, and record items for attention in the next session.

Once treatment goals have been achieved, follow-up sessions are encouraged since(a) they lessen the chances of a relapse, (b) they provide us with an opportunity to head off the negative effect of new stressors as early as possible, and (c) refresher/booster training can be provided as needed.

Accessing My Services
The simplest way to begin the process of therapeutic recovery is to call me so I can provide you with the information you need.

Fees
My fees are $160 per hour and are assessed for any time expended in relation to the client including relevant telephone calls, file reviews, and report-writing. Also, you need to know that I invoice one hour if a client fails to provide me with 24 hours advance notice of cancellation. Some clients have coverage through their employers, via extended health care plans, through WorkSafe BC, or through VA Canada. A call to me is the easiest way to learn about the various coverages. Psychologists are not covered through the BC Medical Plan.

Confidentiality
With the exceptions noted below, the matters we discuss during our sessions are confidential. Still, I do take notes (to aid my own memory) but these notes are not transcribed. In addition, I assume your permission to share relevant information with your physician, partly in case medical issues are involved or medications are required. However, like all of us in the helping professions, I have some exceptions to confidentiality. Thus, I will break confidentiality:
01. if I believe that there are safety issues in regards to yourself or others:
02. if I believe that it is unsafe for you to continue active duty or to operate a motor vehicle;
03. if I suspect that you are abusing another;
04. if I receive a proper court order;
05. in respect to reports required by such agencies as Veterans Affairs Canada when they fund treatment. A copy of any report I send is always provided to clients;

06. if a client initiates legal action against me (this has not happened to date); and

07. if the client instructs me to release information.

Safety

With the abuse that people have reported even from professionals, it is important that I take steps to assure your safety. So firstly, you should know that no action has ever been taken against me in civil, criminal, or professional matters; you can check this if you wish by contacting the College of Psychologists of British Columbia (800-665-0979). Secondly, if the plan we have made is not working out as expected or if you are feeling uncomfortable, please raise the issue with me; if it turns out that we are unable to resolve the matter to our mutual satisfaction or if you wish to terminate with me for any reason, then I will be pleased to give you the names of other professionals and help you to transfer to them. Thirdly, you should be aware of my policy that my wife or another adult female is always here during any appointments I have with ladies. If you wish to reassure yourself that such is the case, just ask. Finally, you are welcome to have a friend or family member come with you.

Personal Emergencies

If there is a personal emergency, you have a number of options. For example, you can call me and leave a message on my secure voice mail; I pick my messages up a number of times each day between 08:30 hours and 1700 hours Monday to Friday so you can expect a delay of up to a couple of hours before I try to return any calls. When I am home on weekends, I check voice mail a couple of times a day. If we are having regular appointments, I will let you know beforehand when I will be away. Also, while I am away, a person I hire will check my telephone messages at least once a day and return your call to let you know that I am away; we will have discussed beforehand the options you have. If the emergency involves *increasing or persistent suicidal or homicidal thoughts* or if you need to talk with or see someone before I am able to contact you, I recommend that you call one of the people on your emergency list, get in touch with your family doctor, visit a drop-in medical clinic, or go to the emergency unit of your hospital.

Negative Reactions

Sometimes during the course of psychological treatment, clients can experience reactions they would prefer not to experience. For example, images of their traumatic experience may arise during and after our brief discussions of them. Tiredness may increase. The beginnings of the state of calmness we are trying to achieve may be followed by a return to previous levels of tension and agitation. If these or any other symptoms develop during our time together, I encourage you to bring this to my attention as soon as you can so, as necessary, we can do something about them.

Informed Consent

I have read and discussed the above to my satisfaction and give my consent for Dr. Carmichael to complete an assessment and continue with the resulting treatment plan once it is explained to me.

Kindly initial the bottom of pages 1 through 6. Then sign below.

Signed:_____ Date: _____

APPENDIX II: CARMICHAEL STRUCTURED CLINICAL INTERVIEW

IDENTIFYING INFORMATION

Date:_____

Name:

_____

DOB: _____ Gender: _____

Home Address: Street:

_____

City / State / Zip code:

_____

Home Tel: _____Cell phone: _____

Work Tel: _____

Family Doctor: _____ Aware/Not | Inform/Not tell

Ethnic Origin: _____

First Language: _____

MARITAL STATUS

Spouse: _____DOB: _____ Years Together: _____

Education: _____

Occupation: _____

any previous
separation:_____

ever consider a separation or
divorce:_____

how do you get
along:_____

any problems in
relationship:_____

what quarrel
about:_____

can you talk to other easily:_____

can you talk about what worries you:_____

can you confide in other:_____

does other talk over issues with you:_____

are you affectionate:_____

is other affectionate:_____

how is sexual side of things:_____

Interests in common:

_____

What appreciate about other:_____

JC SUM of Marital Relationship:

_____

Previous Marriages (with dates, separation/divorce date, reason for break-up)

_____

_____

_____

Children (ages, and stressors; pregnancy/delivery, health/illness, dev./educ., IPX, behaviour.)

_____

_____

_____

Miscarriages:_____

SUPPORT SYSTEM (family, friends; time spent with them, activities together; stability)

_____

_____

_____

RELIGION/SPIRITUALITY/PHILOSOPHY OF LIFE/TURNING POINTS

rel/denomination raised in:_____

current spiritual practices: attend | read | pray | meditate | fast | confess | offering

Other:_____

belief in God/higher power:

_____

most important/meaningful in life/purpose:

_____

code of conduct/values:

_____

how keep sense of hope:

_____

life after death:

_____

can you forgive u:

_____

any guilt | heroes/models | spiritual experiences | turning points/NDE | unresolved / conflict areas:

_____

_____

_____

LEGAL/CRIMINAL/INTERNAL/PCC

Pending:

_____

_____

Past:

_____

EMPLOYMENT (from high school to present with dates + reason for change)
Past:

_____

_____

_____

Current:

_____

Preparation/training/competency:

_____

Satisfaction:

_____

Supervision: supportive, clear reporting relationships, conflict:

_____

Company: how change is managed, supportive:

_____

Physical Environment: space/density, privacy, air, temp, vibration, odour, noise:

_____

Physical Safety:

_____

Task demands: hi demand level | too routine/unchanging | degree control/latitude | skill utilization | adequate resources| amount of change | new technology with no training:

_____

_____

_____

Role demands: ambiguous expectations | value conflict

_____

_____

_____

IP Demands: team pressure | co-worker conflict | unproductive co-worker

_____

_____

_____

Recent changes:

_____

_____

_____

Future Plans + Preparation:

_____

_____

_____

MEDICAL HEALTH STATUS

Current Medical Conditions and Prescribed Medications (including inhalers)

_____

_____

_____

Non-Prescribed/Over-the-counter Medications (OTC)

_____

_____

_____

Health Food Remedies

_____

_____

_____

Relevant Past Medical History

arthritis/FMS_____

cancer_____

diabetes_____

heart/BP

_____

lungs_____

liver_____

digestion_____

prostate_____

**female
reproductive**_____

Pain_____

Childhood illnesses:

_____

_____

_____

Exposure to parasites/tropical travel/other infections

_____

_____

_____

Head injury/head trauma/knocked out/head smack with stars or dazed/hitting head
MVA/other impact to head:

_____

Non-impact acceleration/deceleration/whiplash:_____

Sport accident:

_____

Child (e.g., fall off swing/down stairs):

_____

Adult (e.g., fall off equipment, tree, building):_____

Child (punched in head /shaken):

_____

Adult (fight, beaten, boxing, assault):_____

Pressure wave/blast injury/explosion:

_____

Symptoms:

_____

Loss of consciousness:_____

_____

Concussion (headache, dizziness, tinnitis, nausea/vomiting, slurred speech, phonophobia, photophobia, sleep disturbances):

_____

_____

_____

Post-conconcussive: memory loss for events immediately before or after the event, headache (to 90%), dizziness/light-headedness (to 53%), blurred vision (to 14%), anosmia (5 %), photophobia (7%) phonophobia (15%),

fatigue, tinnitis, irritability/patience/outbursts/mood swings, insomnia, attention span/concentration, working memory, anxiety, nausea/vomiting, speech/communication

_____

_____

Interference of oxygen to brain
Drowning:_____

Suicide attempt
_____

Significant/Chronic
A_____+D_____

CVA/aneurysm:
_____

Toxins/poisons_____

Electrocution_____

Acute respiratory distress:
_____

TIA: weakness/numbness/tingling one side of body, difficulty speaking or understanding words,
vision loss, severe + unexplained dizziness/unsteadiness/lack coordination/sudden fall,
severe + unexplained headaches
_____

Other:
_____

Surgery/Hospitalizations:

_____

_____

Pre-, peri- and post-natal history
mother significant use drugs/alcohol/nicotine in pregnancy:
_____

mother injury during pregnancy:

_____

birth complications:

_____

Infant prolonged high fevers:

_____

Seizures:

_____

Learning disability:

_____

Other:

_____

<u>**Current General Symptoms**</u>
**temperature: feeling cold, feeling warm, excessive sweating, heat
intolerance:** _____

Breath odour, sore throat, flu symptoms, decreased immunity to virus:

_____

Fatigue:

_____

(If significant unexplained chronic fatigue consider interview for CFS)

<u>**Vision**</u>
**Blurred, double vision, spots, movement, increased/reduced blinking, puffy
     red eyes, yellow white of eye, nystagmus, anything unusual about
     the way things have looked:** _____

See things other do not:

_____

_____

<u>**Hearing**</u>
Loss, sounds (water, buzz, ringing), fading:

_____

Anything usual about the way things sound:

_____

Hearing voices or other things that weren't there:

_____

_____

Smell
Sense of smell:

_____

Particular odours, anything unusual about how things smell:

_____

Smell things others could not:

_____

Touch
Sense of touch:

_____

Loss of feeling:

_____

_____

Cardio-Vascular and Cardio-Pulmonary
Heart: fast or slow rate, extra or skipped beats, chest pain, beats that are forceful/pounding: _____

_____

Blood Pressure:

_____

Respiration: shortness of breath wheezing choking cough hyperventilation asthma:

_____

Upper GI
Nausea, vomiting:

_____

Food cravings/intolerance:

_____

Eating disorder: anorexia, bulimia:

_____

Fluid intake:

_____

Weight_____ Height_____ wt/ht index_____

Men: Collar size _____

<u>Lower GI</u>
Diarrhea, constipation, light stool colour, stool smell, flatulence, feel
incomplete emptying abdominal pain distension, bloating, training/urgency
to BM, passage of mucus, rectal bleeding, weight loss, cramping:

_____

_____

<u>IBS:</u> 12+ weeks in past year of abdominal discomfort or pain relieved by
defecation associated with altered stool frequency, and/or associated with
altered stool form:

_____

_____

<u>Functional abdominal bloating:</u> 12+ weeks in the past year of abdominal
discomfort, bloating or visible distension:

_____

_____

<u>Functional diarrhea:</u> 12+ weeks in the past year of loose or watery stools
75% of the time without abdominal pain:

_____

_____

<u>Functional constipation:</u> 12+ weeks in the past year including 2 or more of
the following:

            _____straining > 25% of defecations
            _____lumpy or hard stools >25% of defecations
            _____sensation of incomplete evacuation >25% of
the time
            _____sensation of anorectal blockage >25% of the
time
            _____manual manoeuvres to facilitate >25% of
defecations
            _____<3 defecations / week

<u>Genito-Urinary</u>
urination: increased, retention, dark colour, odour, painful,
incontinence/leakage (e.g., laugh) decreased testicular:

Size/lumps:

Libido:

Arousal / attain erection/lubrication:

Men: Maintain erection:

---

orgasm latency:

Sexual pain:

Women: Menstruation/PMS:

Women: Perimenopause:

Women: Postmenopause:

<u>Skin and Hair</u>
Skin: colour, itching, jaundice, reddening, oily, dry, bleeding, bruising, ulceration, rash, waxy, excessive sweating, inability to tan, poor wound healing, facial obesity, premature mouth/eye wrinkle, swelling, psoriasis:

Hair: texture, loss, oily, dry, itchy, flakiness, itchiness:

<u>Speech and Language (observation)</u>

Articulation:

Voice change:

Expressive aphasia:

Speed:

_____

Understanding/comprehension:

_____

Monotone/modulation:

_____

Affect:

_____

Memory:

_____

Eye-contact:

_____

<u>**Cognitive (S = self-report, C = collateral, O = observation)**</u>

Attention/concentration:

_____

Memory (STM):

_____

Long-term memory:

_____

Multitasking:

_____

Decision-making:

_____

<u>**Thinking**</u>
A. Has your imagination been playing tricks on you in any way?

_____

_____

B. Do you have any ideas that other people might not understand?

_____

_____

C. Have you ever had trouble with your thoughts?

_____

D. Have your thoughts ever been so confused that you lost track of your ideas?

E. Are your thoughts ever frightening or disturbing?

F. Do you ever feel as if you lose control of your thoughts?

G. Do you ever feel that people are watching you, following you, trying to hurt you?

<u>Neurological</u>

Fainting:

Seizures:

Dizziness:

<u>Musculo-Skeletal:</u>
Joints: pain, swelling, rigidity, stiffness:

Muscles: pain, stiffness, tenderness, wasting, spasm/jerks, tingling, numbness:

GM: tremor, coordination, gait, clumsiness, paralysis, balance:

FM: tremor, coordination, dexterity, rigidity:

_____

Activation level:

_____

Posture:

_____

Headache
Location:

_____

How does it occur? sudden/slow

_____

What time of day does it usually occur?

_____

Age when first started: _____ (alarm if onset > age 50, sudden onset, accelerating pattern)_____

Triggers/precursors: yawning, polyuria, mood change, visual disturbances, coughing, exertion, sexual activity, period time, sneezing, stress:

_____

_____

Type of pain: pressing, tightening, non-pulsating, constricting, band/cap, throbbing, pounding:

_____

_____

Severity: 1 = low and 10 =excruciating:

Episode _____ Duration_____frequency_____

Associated symptoms: nausea, vomiting, photophobia, phonophobia, diarrhea, droopy eye lid, dizziness, double vision, loss of muscle coordination, worsened by movement:

_____

_____

What makes it worse: bend, straining, lying down:

_____

_____

What makes it better:

_____

_____

How do you cope with the pain

_____

_____

Treatment to date:

_____

_____

Family history re headache:

_____

_____

Other Pain (_if chronic, consider relevant additional questions_)

Location:

_____

_____

Etiology:

_____

_____

Type of pain:

_____

_____

severity (0 = low and 10 = excruciating:

_____

Treatment to date:

_____

_____

Restrictions because of pain:

_____

Impact: finances | work | marriage | children | friendships | vegetative | recreation | suicide | self-concept

What makes it worse?
_____

What makes it better?
_____

FMS_____

Chronic Illness (*if present, consider interview for CI*)

Disease:
_____

_____

Etiology:
_____

_____

Treatment to date:
_____

_____

Life-threatening:
_____

_____

Predictability/expected course:
_____

_____

Restrictions:
_____

_____

Impact: finances family recreation work mobility self-concept stress/anxiety: _____

What worsens symptoms?
_____

What makes it manageable?
_____

Sleep

I. General Information and Insomnia *(if sketchy/memory problems, do 7-night sleep study)*:

Daytime nap (when and amount)?
_____

Time start relaxing in the evening_____ What activities?

_____

What else do you do in the bedroom?
_____

What time are lights out_____ sleep onset_____

(problem if =/>30)_____

Number of times of night waking and when:
_____

Time to fall back asleep: _____

(problem if =/>30) _____

What do you do to get back to sleep?
_____

What triggers sleep problems?
_____

pain ____ depression ____ PTSD ____ GAD____ PD ____ other
_____

Medications: ___Theophyllline (e.g., bronchodilators) ___stimulants (e.g., Ritalin) ___beta

antagonists (e.g., Propranolol) ___SSRI ____ other
_____

Nightmares:
_____

_____

What time do you usually wake up?_____ What time do you usually get up? _____

Actual number of hours of nighttime sleep_____(prob=/<6.5)

Actual number of hours in bed _____%efficiency_____

Usual/desired number of hours sleep_____
actual/desired = _____%

Do you feel rested when you get up?_____

Do you have trouble with daytime
fatigue/sleepiness?_____

Onset and duration of sleep difficulties:

_____

_____

Epworth Sleepiness Scale:

How likely are you to fall asleep in the following situations?
0 = never/1 = slight chance/2 = moderate chance/3 = high chance
_____sitting and reading
_____watching TV
_____sitting inactive in a public place (e.g., theatre, meeting)
_____as a passenger in a car for an hour without a break
_____lying down to rest in the afternoon
_____sitting and talking to someone
_____siting quietly after lunch without previous alcohol
_____in a car while stopped for a few minutes in traffic

2. Sleep anxiety:

___trouble sleeping in your own bed but okay when
elsewhere
___worry, think, plan while in bed
___watch the clock throughout the night
___feel nervous/tense while in bed
___light sleeper and hear every noise
___panic attacks in the night
___obsessive thoughts

**3. Obstructive Sleep Apnea (1% - 2% of population; M/F = 8/1; > if 40 yrs
and over, >if OW):**

_____micro-arousals (>15 sec but 10-60+/hr))
_____usually preceded by cessation/gasp/snorts

_____snoring (heavy enough to disturb others)
_____a.m. headache
_____ a.m. dry/sore/raspy throat
_____high night peeing
_____high night sweats
_____esophageal reflux at night
_____day forgetfuness
_____ day concentration problems
_____ problems with sex drive
_____ BMI<35
_____ collar size >16.6"

Sleep characteristics:
___sleep onset quick
___ normal total sleep
___ gen okay sleep efficiency
___ continued resperatory effort
___ REM sleep very suppressed
___slow wave sleep decreased

medical causes:
___congestive heart failure
___COPD
___ chronic renal failure
___diabetes
___arthritis
___FMS
___increase in menopause
___alcohol HS
___sedatives HS
___mild= AHI of 5-15 per hour ___moderate = 15-30 ___severe = >30

4. Periodic Limb Movement (PLMB) (21%-36% pop < as age; no gender diff; diag = 5/hr):

___been told that twitch, kick, jerk repeatedly in sleep (more than 5 per hour is abnormal.)
___wake up with the bed all messed up
___from low to frequent micro-arousals (EEG) from sleep (usually not aware)
___sleep onset okay, not wake for long periods, normal total sleep time and efficiency, normal architecture

Treatment:
0.5-2.0 clonazepam___ dopamine receptor agonist (bromocriptine 0.1-0.6/pramipexole 0.125-1.0)___

5. Restless Leg Syndrome (RLS): PLMD also present 80% of time; co-occur with MDD 10-19% of time

5-10% in general population: 10% to 20% in over 65: usually idiopathic
___trouble getting legs comfortable (esp thigh and calf), trouble keeping them still, have to move them
___creepy, crawly, achy, electric, coiled spring sensations in legs when sitting/lying for long periods
___therefore an urge to stretch/knead legs or get up; doing so relieves things for a while before discomfort restarts
___feelings occur mostly when resting
___generally worse in nighttime than morning: between 2000 and 0400 hrs
___these sensations make it hard to fall/stay asleep

Treatment:
        a. no caffeine or alcohol within 12 hours bedtime
        b. reversible causes: iron deficiency, antidepressants,
        c. long-acting dopaminergic (no if psychosis): ropinirole (most
            freq: 2mg HS) pramipexole, pergolide
            augment prn with oxycodone, codeine, gabapentin

6. Narcolepsy (0.03% to 0.16%; gender neutral; familiar pattern;

___struggling to stay awake daytime
___cataplexy (sudden loss of muscle tone–head/jaw drop, buckling knees)
___vivid hypnagogic hallucinations with overwhelming sense of dread or impending death
___sleep paralysis (awake but unable to move/speak)
___often associated: double vision memory problems balance disturbances
___onset awakenings, problem sustaining consolidated sleep, short REM latency, Hi REM in

Treatment:
modafinil 100-400 - if that doesn't work, stimulant meds; doing stuff to stay awake, re-scheduling, naps

7. Circadian Rhythm Sleep Disorders:

Delayed Sleep Phase Type
      ___difficulty falling asleep before 0200 to 0600 and difficulty waking before 1000 to 1300
Treatment:
        a. chronotherapy: progressively delay bedtimes
        b. bright light therapy for 1 to 2 hours in morning
        c. melatonin, 5-10 mg, 5 hours before bedtime

Advanced Sleep Phase Type
      ___sleep onset 1800 - 2100 and wake 0200 to 0500
Treatment:
        a. chronotherapy: progressively advance bed times

b. bright light therapy: evening, 1-2 hours

Shift Work Type
Treatment:
a. bright light (continual or 20 min per hour) during night shift, stopping 2 hours before end of shift then avoiding/minimizing bright light on way home or at home
b. daytime (dark room, quiet, etc.)
c. Modafinil at start of night shift

DRUG AND ALCOHOL

A: WHO Alcohol Test (Score =)

1. How often do you drink?
    0 = never 2 = weekly 3 = 2-3 per week 4 = 4 or more times per week

2. How many drinks do you consume each time?
    0 = None 1 = 1/2 2 = 3/4 3 = 5/6 4 = 7-9 5 = more than 9

3. How often have you had more than 3 (female) or more than 5 (male) on one occasion?
    0 = never 2 = ess than monthly 2 = monthly 3 = weekly 4 = daily

4. How often during tha last year were you unable to stop drinking once you had started?
    0-never I = less than monthly 2 = monthly 3 = weekly 4 = daily

5. How often during the last year have you failed to do what was expected of you because of drinking?
    0 = never 1 = less than monthly 2 = monthly 3 = weekly 4 = daily

6. How often during the last year have you needed a first drink in the am to get yourself going after a heavy drinking session?
    0 = never 1 = less than monthly 2 = monthly 3 = weekly 4 = weekly

7. How often during the past year have you had a feeling of guilt or remorse after drinking?
    0 = never 1 = less than monthly 2 = monthly 3 = weekly 4 = daily

8. How often during the past year have you been unable to remember what happened the night before because of drinking?
    0 = never 1 = less than monthly 2 = monthly 3 = weekly 4 = daily

9. Have you or someone else been injured as a result of your drinking?
    0 = never 2 = yes but not in the last year 4 = yes during the last year

10. Has a relative, doctor, or other health worker been concerned about your drinking?
    0 = never 2 = yes but not in the last year 4 = yes during the last year.

B. Drug use
Past: _____ Current:
_____

C. Caffeine (coffee, tea, cola, chocolate, cocoa)
Past: _____ Current: _____

D. Tobacco (>10/day may indicate dependence)
Past: _____ Current: _____

Lifestyle Variables
    a. **Exercise/sports:**

    _____

    b. **Hobbies/leisure time:**

    _____

    c. **Activities with others:**

    _____

    d. **Diet:**

    _____

Recommend see dietitian?
_____

Mental Health History

**Previous
therapists**_____

**Past
Medications:**_____

**Symptom
Onset:**_____

Diagnosis_____
Symptom Course:
_____

History in Family:
Father:_____

_____

_____

Mother:_____

_____

_____

Siblings:_____

_____

_____

Paternal grandfather:

_____

Paternal grandmother:

_____

Maternal grandfather:

_____

Maternal grandmother:

_____

**Any Current Suggestive Symptoms (*if suspicious, complete* DSM-IV *format)*
   1. MDD/MDE:**

_____

   2. **GAD:**

_____

   3. **PD:**

_____

   4. **Phobia:**

_____

   5. **OCD:**

_____

**Suicide Potential (*if suspicious, complete suicide checklist)*
1. Feel at all that life is not worth living?**

_____

2. Ever thought of harming self?

_____

3. Are suicidal thoughts present?

_____

4. Has there been a previous suicide attempt?

_____

Homicidal Potential
1. Outside of duty, ever kill another?

_____

2. Do you have anger such that you ever think of killing?

_____

3. Do you have any current homicidal thoughts?

_____

4. Do you ever blank out in a rage state?

_____

5. What stops you from murder?

_____

FAMILY HISTORY
Mother:
Current status:

_____

Concern/Stressor:

_____

Child relationship (available, supportive, affection, discipline, etc.)

_____

Father:
Current status:

_____

Concern/Stressor:

_____

Child relationship (available, supportive, affection, discipline, etc.)

_____

Parent's marital relationship (:affection, abuse, issue resolution, separation):

_____

Losses (who, what, when, where, impact then, impact now)

_____

_____

_____

_____

Childhood (evaluation, positives, negatives, problems, perceived impact

_____

_____

Adolescence (evaluation, positives, negatives, problems, perceived impact)

_____

_____

<u>EDUCATIONAL HISTORY</u>
Last Grade Completed:

_____

Special classes, tutoring, held back, modified class, learning disability, accelerated:

_____

Best subjects:

_____

Post high school education/ training/courses:

_____

Educational goals:

_____

_____

CURRENT METHODS OF STRESS MANAGEMENT
Talking/sharing:

_____

Healthy living:

_____

Calming/relaxation:_____

Problem solving strategies:

_____

Temporary distraction:

_____

Escape/don't think about it:

_____

Hope for best:

_____

Exercise:_____

Trauma History - Have you ever experienced any of the following)?

Types of Trauma

__A. child physical abuse: punch/kick/stab/burn/bite/hit/threat
_____

__B. childhood sexual abuse: touch/make touch, insert hand/tongue/penis in mouth/vagina/anus

_____

__C. Childhood emotional abuse: criticizing, belittling:
_____

__D. Were you aware of any of these happening to siblings or parent (including VIR)? _____

__E. Any current physical/sexual/emotional abuse including spousal?
_____

__F. Criminal victimization: armed robbery, assault, kidnap, rape, date rape, threatened?

__G. Forced to evacuate from natural disaster: earthquake/flood/avalanche/tornado/landslide?

_____

__H. Forced to evacuate from human/occupational disaster: fire/chem spill/industrial/explosion

_____

__I. Military ops: combat, POW, concentration camp, UN peacekeeping
_____

__J. Life-threatening/serious injury in transportation accident: car/train/plane (self or loved one)?

_____

__K. Life-threatening illness (MI, CAD, ICU, HIV, CA)–(self or loved one)?
_____

_____

_____

__L. Surgery: awareness under anesethesia?
_____

__M. Traumatic childbirth/invasive procedures, death neonate, miscarriage/abortion? _____

_____

__N. See/handle dead/mutilated bodies?
_____

__O. Refugee (torture, displacement)?
_____

__P. Shock because one of events on this list happened to someone close to you? _____

__Q. Impact because of your work with victims of trauma/offenders:
_____

_____

R. Loss of a loved one through accident, homicide, or suicide?
_____

S. Adult abusive experiences?

_____

_____

Trauma details related to above (duration, frequency, injury, consequence to perpetrator, thoughts, actions, parental/support response, how coped, impact, any continuing effects):

_____

_____

_____

Guilt/shame/anger_____

Trauma that occurred to parents/grandparents

_____

_____

_____

Stressors, Perception and Management:

Proximal Stressors Noted to This Point

Any Other Proximal Stressors
Family health:

_____

Family change (e.g., seperation, empty nest):
_____

Finances:_____

Work/promotion/retirement:_____

Health limitations:

_____

Housing/neighbourhood:

_____

Major Disappointments:

_____

Other:_____

Appraisal of Proximal Stressors (See stress management file--Peacock + Wong, 1990)

___A. Controllable-by self (___have ability to do well, ___have what it takes,
      ___will overcome problem, ___have skills necessary)

___B. Centrality (___important consequences, ___will be affected, ___serious implications,
      ___long-term consequences)

___C. Controllable-by-others (___someone I can turn to, ___help is available,
      ___ resources are available, ___is someone who will help)

___D. Threat (___threatening situation, ___feel anxious, ___outcome negative,
      ___impact negative

___E. Challenge (___positive impact, ___eager to tackle, ___can become stronger,
      ___excited about outcome)

___F. Uncontrollable (___totally hopeless, ___outcome uncontrollable,
      ___beyond anyone's power, ___problem unresolvable)

APPENDIX III: CONSENT FOR BFT-C

Dr. John A. Carmichael R.Psych (0048)
Practice in Clinical, Police, and Military Psychology

BFT-C FOR SELF-REGULATION OF THE BRAIN: INFORMED CONSENT

About Consent
Please note that the information contained in both the Informed Consent for Assessment and the Informed Consent for Treatment remain in effect. This paper is additional to those and presents information about BFT-C as well as for purposes of informed consent.

Previous BFT
In the biofeedback training (BFT) you have completed already (sometimes referred to as peripheral biofeedback), you will remember that sensors were attached to such surfaces as your fingers, hand, muscles, or chest to gather psycho-physiological information. The biological information that was collected was processed through electronics to a software program so you could see on a computer screen what was happening within you in real time. Then you were able to practice such self-regulation techniques such as changing breathing patterns, re-positioning muscles, gentle stretching exercises (like yoga), and passive awareness, so that you were able to make desirable changes in your autonomic nervous systems (BFT-A) and/or your level of muscle tension (BFT-M). Both of these methods of BFT are well-established methods for relaxation and self-regulation and are used as one component in my treatment plan for PTSD where there is clear evidence of autonomic nervous system hyper-arousal and/or hyper-reactivity and/or evidence of excessive muscle tension. Other components we have completed to this point include seeing your physician, learning about the condition you have, putting into place stabilization strategies based on research in health psychology, establishing restful sleep, and elements of cognitive behavioural therapy (CBT). Taken together, I refer to these interventions as my Basic Approach.

Results of My Basic Approach
As of November 2007, an analysis of the military veterans I had seen with PTSD (with and without other psychological conditions), indicated that 70% to 80% achieved complete remission after completing my basic approach; that is, they no longer met the diagnostic criteria for PTSD. To date I have seen veterans of WWII, some who saw combat action in Korea, some who were in a combat unit in the first Gulf War, a number who had been on peace-keeping deployments (Gaza, Cyprus, the former Yugoslavia, and Rwanda), and some who were injured during training operations. Military veterans I have seen since that time have achieved the same results

Also in July 2007, I conducted a chart examination of 28 police officers with PTSD (and often with associated conditions) as a result of duty-related accumulated trauma and on whom I had complete psychological test data. Some were still serving and some had retired. With implementation of my basic approach, about 70% of them became symptom free; they no longer met the diagnostic criteria for PTSD. Officers I have seen since that time have experienced the same outcome.

Furthermore, between 70% and 80% of the civilians I have seen with PTSD and related conditions were in remission after the basic approach. Their trauma stemmed from such experiences as motor vehicle accidents, assault, and life-threatening illness.

In addition, I have noticed increased positive results with BFT-C among those whose depression, pain, or cognitive impairment was less than that desired after completion of my basic approach.

All groups of clients experienced some relief of symptoms with the basic approach. However, from 20% to 30% still had enough symptoms to conclude that they had not yet reached remission. These clients too have always been a concern to me. Consequently, for many years I searched for additional treatment methods that might lead to remission for these clients.

BFT-C: The Basics

In 2000 or so, I became aware of clinical approaches stemming from a sub-speciality area known as applied neuroscience. So I delved into the published professional literature, attended many professional conferences, and consulted with experts in the field. Applied neuroscience takes the EEG signal as the starting point in understanding and modifying conditions such as PTSD, depression, pain, and cognitive impairment. The EEG is one aspect in the operation of the central nervous system (the brain).

The EEG measures the brain's electrical activity as it is recorded on the surface of the scalp. I and some of my psychology and neurology colleagues use the qEEG which refers to the mathematical processing of the digitally-recorded EEG. Studies have shown that measurement by qEEG has high reliability and significantly higher than visual inspection of the EEG. A 2004 ISNR (International Society for Neurofeedback and Research) committee of psychologists and neurologists chaired by Dr. Cory Hammond had this to say about qEEG:

> "The committee is of the opinion that such comprehensive evaluations may yield additional scientifically objective information, including a quantitative description of the relationships between different brain regions. We believe that this may prove extremely valuable in many cases in guiding and individualizing subsequent treatment, especially in treatment-resistant cases. . . . In comparison with costly and less available neuroimaging modalities, some of which require exposure to radioactive material, the qEEG provides a relatively inexpensive, culture-free, non-invasive assessment of brain function."

They went on to write

> "In the last ten years several hundred well designed EEG and qEEG papers, with sizeable samples and normal controls, have documented that there are electro-physiological abnormalities in a high proportion of psychiatric/ psychological conditions."

Such studies now number in the thousands.

Moreover, many studies as well as the experience of professionals in the field indicated that biofeedback training based on the EEG signal (BFT-C) was very successful in altering a range of symptoms characteristic of depression, seizures, attentional difficulties, sleep problems, chronic pain, and cognitive impairment (including issues resulting in mild traumatic brain injury (mTBI). See also the 2009 definition of BFT-C (www. isnr.org) for which I was instrumental in researching and writing. More information about the relation of qEEG and PTSD is found in this book.

Also, I learned that as far back as 1995, the Board of Professional Affairs of the American Psychological Association (of which I am a member) determined that psychologists using qEEG and BFT-C

"have training and experience in areas of tests, measurement, research and statistics. qEEG is one of many techniques that have been investigated and utilized by psychologists. . . APA concludes that it is appropriate for psychologists who are trained and practicing within the scope of their competence to use qEEG and biofeedback."

So, the use of qEEG is well established in the research literature with thousands of publications to date. Moreover, the link between qEEG results and both three-dimensional neuro-imaging and neuro-psychological functioning is well established. Furthermore, the qEEG has been accepted in most courts in the US even when challenged. Finally, the use of the qEEG is supported by many professional organizations such as the American Neuropsychiatric Association, The EEG and Clinical Neuroscience Society, the Association of Applied Psychophysiology and Biofeedback, and the International Society for Neurofeedback and Research. I am a member of the latter three organizations and thus follow their code of ethics, receive their regular publications, and attend at least one of their annual conferences each year.

The research, theory, and practice of BFT-C is as follows. Our brain receives messages from the various systems within our bodies, processes input about our environment from our senses, stores and accesses information, directs muscle activity, communicates within various centres in our brain, regulates emotion, determines the sleep-wake cycle, determines heart rate, and so on. These functions are accomplished by the movement of electro-physiological messages from one brain cell (neuron) to another, which is facilitated both by the activation of neuro-transmitters and by the physical growth of dendrites on neurons. This results in the increase in the activity of some sets of brain cells and the simultaneous inhibition of other sets of cells. A percentage of this activity is associated with the generation of different frequencies and amplitudes of brain waves (EEG) as well as other characteristics such as coherence and phase over specific sections of the brain or within networks. For example, fifteen specific brain areas are involved in various aspects of attention, different aspects of memory involve the hippocampus and areas of the temporal cortex, the brain stem pays a critical role in heart rate and blood pressure, and such brain areas as the amygdala and anterior cingulate are involved in the symptoms of PTSD.

Our brain operates without our conscious control just as do other organs such as heart and immune system. As long as all parts of our brain are working well, we are able to do what we have to do; we are able to learn, we are able to change thoughts and actions, we are able to regulate our emotions, we are able to act or hold back depending on the circumstances, and so on. On the other hand, when brain functioning is not optimal, then we can suffer from symptoms which we experience as depression, PTSD, worry, memory problems, and so on. Moreover, even very small changes in the amplitude of any brain wave frequency at a relevant site can be the difference between having symptoms and being free of them. Such small differences can not be detected by a human eye no matter how well trained it is. So we use the power of computers, electronics, and extremely complicated software programs.

Also, the 2004 position paper by ISNR, to which I belong, suggests that you be informed as follows:

> "It is important for you to understand that what I do is not the same as a "clinical EEG" which is used in medical diagnosis to evaluate epilepsy or to determine if there is serious brain pathology such as tumor or dementia."

In the qEEG, I am concerned with the brain's electro-physiological functioning. Thus, what I do is (1) measure the amplitude of various brain wave frequencies at specific scalp locations as well as the connectivity between one part of the brain and another, (2) compare this with what one would normally expect, and then (3) use this along with additional information to devise a BFT-C plan so the client can make the required changes in brain wave activity. This can lead to positive changes to the thoughts, feelings, and behaviour associated with the psychological condition of concern. I am happy to be a psychologist and I am not nor do I present myself as a neurologist.

BFT-C: The Assessment Process

Thus, with a lot of technical training, the purchase of expensive computerized equipment, and consultation with an expert for each client I see, in 2003 I began to use BFT-C with the hopes that it would address the symptoms that remained in spite of completion of my basic therapeutic approach. Initially, I consulted on each case with Dr. J. Horvat, a licensed psychologist in Texas with BCIA credentials and who is both well-known and well-respected in the field. Regrettably, he passed away in January 2008. Currently, I am fortunate to obtain consultation on every client I see for BFT-C with Dr. R. W. Thatcher, an acknowledged expert in the field who has published widely (6 books and over 200 articles to date), lectured internationally, and who developed the NeuroGuide software used by many of us in the field to analyze the qEEG.

At this time, I discuss the possibility of trying BFT-C with clients only if my basic approach has not been optimal and there is reason to believe that BFT-C might be helpful.

Procedurally, I complete a qEEG (quantitative EEG) which is neither painful nor embarrassing. This involves placing a cap on the client's head under which are sensors that measure and record brain wave activity between 1 and 30 Hz simultaneously, at 128 samples a second, and at the 19 scalp locations of the accepted 10/20 placement system. That's about 73,000 bits of information per second in an assessment that goes on for 28 minutes. The resulting information is digitized and stored on my computer. Then this qEEG along with other important information (such as diagnosis, remaining symptoms, medications, and any history of head injury) is web-mailed to Dr. Thatcher who examines it and also removes any artifacts (such as muscle movement) electronically. The results are then processed through his established NeuroGuide scoring system which includes comparing the client's data with that of a normative sample.

Also, Dr. Thatcher processes my data through a second software scoring system known as LORETA (Low Resolution Electromagnetic Tomography) which further localizes any electro-physiological abnormalities in three-dimensional space. LORETA has a sub-millisecond time resolution with spatial resolution similar to the fMRI. As Dr. Thatcher wrote on 18 June 2009 on his internet forum

> "There is a vast qEEG literature showing good correspondence between the surface EEG and the location of strokes, tumors and lesions as well as abnormal activity in "modules" associated with clinical problems. . .There are over 500 qEEG LORETA studies and the vast majority of these studies show good correspondence with fMRI and PET and MEG...related to localization of function, or actually "functional systems" or modules."

The resulting "brain maps" from NeuroGuide and LORETA are web-mailed back to me. In part, the information highlights any deregulation with respect to such aspects as amplitude, connectivity, and phase. In addition, Dr. Thatcher makes BFT-C recommendations for consideration.

BFT-C: The Treatment

I review the brain maps and resulting suggestions and develop an BFT-C plan designed to alter deregulated patterns based on (1) the symptoms needing attention which have already been assessed, (2) the current state of knowledge in regards the relationship between specific symptoms and the function of localized brain regions, (3) results of the individualized NeuroGuide in relation to the normative group, (4) results of the LORETA analysis, and (5) approaches with scientific merit which have been reported to be successful in the treatment of PTSD and/or related conditions.

The resulting BFT-C plan is next explained to the client and then implemented using the hardware and software of the DeyMed diagnostic system. Once attached to the equipment as was done for the baseline qEEG, the client receives real-time feedback on the brain wave metric of interest such as amplitude or connectivity and at the frequency and site of interest. Given this information, the client is able to gradually make the desired changes in brain wave activity. Also, although we train at the limited number of predetermined sites as indicated above, experience to date indicates that changes in one brain location often results in positive changes in other sites due in part to the significant inter-connections within our brains.

Typically between 20 and 40 weekly sessions are required before the brain waves associated with PTSD change in the desired direction. After we have completed a few sessions, these changes seem to continue between training sessions and once training is completed. Additionally, clients are encouraged to continue with those prior recommendations which had a positive effect including stabilization strategies, BFT-A, BFT-M, attention to sleep, and effective stress management. These too can exert a positive influence on brain functioning.

Checks on symptoms and any possible side effects are made after each session and before the next one. While side effects are rare with the manner in which I do BFT-C, they can occur. Since adding BFT-C to my practice in 2003, it has been necessary to terminate this procedure only three times; twice because of complaints of headache and one after the client complained of increased problems in short-term memory. Headaches disappeared after termination of BFT-C and the client complaining of memory problems never returned. Also, drop-out rates are very low; I think maybe 6 to 8 over a 7 year period. One reason for drop-outs was client perception that BFT-C was not helping.

After each group of 15 to 20 sessions and more often as required, standardized psychological tests are re-administered. This provides us with an objective way to evaluate BFT-C and make changes as needed. At one time, my re-assessment included a repeat qEEG. Since results every client indicated that the desirable changes in the electro-physiology at specific locations had been made, I abandoned doing repeat qEEGs both because of these consistent findings and because of the associated expense of around US$400 for each qEEG. Now, I continue BFT-C sessions until remission is achieved, or the client is satisfied with results to date, or there is no further progress after a sufficient number of sessions. With the procedures I use, treatment planning based on the qEEG rather than attempting "canned" protocols, the obtaining of expert consultation for each client, and client feedback on results, no lasting harm has been reported by any of my client.

BFT-C: The Results

In 2007 and again in 2009 I reviewed results on all clients who had received BFT-C. The data indicated that about 95% of those with PTSD achieved remission of symptoms when they had not done so with my basic approach. This high percentage was true also for the co-existing symptoms that some with PTSD can experience including depression, sleep problems, and panic. Additionally those with co-existing symptoms of pain reported at the very least as being less bothered by pain than before BFT-C. Similarly, there was convincing evidence that those with cognitive impairment (including as associated with mTBI) experienced significant improvement in cognitive functioning with BFT-C.

BFT-C: Further Information

Further information will be provided to you as we proceed, and additional information may be obtained from any of a number of Web sites maintained by professional groups such as the following

www.isnr.org: the official Web site of the International Society for Neurofeedback and Research which includes 750+ licensed professionals such as psychologists and neurologists from all over the world. I am a member and former director of this professional organization;

www.aapb.org: the official Web site of the Association of Applied Psychophysiology and Biofeedback and includes licenced practitioners from all over the world. I am a member of this group which includes those practicing various types of biofeedback including for muscles, the autonomic nervous system, and brains;

www.appliedneuroscience.com: Dr. Thatcher's Web site which includes a list of some of his very many publications. Without specialized knowledge, you may have some difficulty in understanding many of the articles he has placed there; and

http://psyphz.psych.wisc.edu/web/index.html: a Web site by a prolific affective neuroscience researcher, Dr. R. Davidson and colleagues, who are now at the University of Wisconsin. Using the EEG and neuro-imaging techniques such as fMRI, he was one of the first to discover one of the brain patterns associated with major depression.

Also, a member of one of my professional associations, Dr. Glenn Weiner, made an invited presentation to a medical college in the US. The presentation includes Glenn talking as well as some very good slides you can see at the same time. You need to get to the website below and click in the box at the upper right of the screen. It takes a little while to load. The Webs ite is

http://www.biofeedbackforthe brain.com/neurofeedback-research/

Although it deals mostly with attention deficit, the basics are there and presented as clearly as I have seen.

BFT-C: Why Other Professionals Do Not Use It

I acknowledge that BFT-C is not yet widely used in the treatment of PTSD and associated conditions. The reasons can be many including (1) insufficient knowledge about applied neuroscience in general and BFT-C in particular, (2) inability or unwillingness to spend the money for the specific training for weeks at a time while no income is being generated but costs of travel, accommodation, and course fees are incurred, (3) inability or unwillingness to spend at least $12,000 for a good basic system, and (4) inability or unwillingness to pay consultant fees for each assessment (about US$400 each) are among the reasons few psychologists in BC use BFT-C. Also, in spite of the evidence, some continue to believe either that (1) it is an unproven method or (2) it should at best be considered as an experimental method.

BFT-C: Follow-up

Even after successful completion of BFT-C, under conditions of significant stress a percentage of those who had reached remission may experience a return of some symptoms. With my clients, this is more likely among those with prior co-existing symptoms such as depression and chronic pain than those who presented with only symptoms of PTSD. For example, analysis of my data indicated that from 20% to 25% of those with a co-existing depression suffered a relapse within 12 months following the termination of our sessions.

Consequently, now I encourage clients to continue with follow-up sessions once acute treatment is concluded. This is consistent with a continuing care model and involves a gradual decrease in the frequency of sessions as gains are maintained; we begin with twice a week, then decrease to monthly, then every other month, then quarterly, and finally to twice yearly check-up sessions. By following this strategy there have been no major relapses.

Follow-up sessions are devoted to (1) assessing symptoms, (2) intervening aggressively as soon as signs of relapse are present, (3) identifying and managing new stressors effectively as soon as they arise, (4) adding some new stabilization strategies, and (5) providing regular booster sessions of BFT-C. Additionally, for those who choose to do so, instruction in mindfulness meditation is available; research suggests that by doing so there is a decrease in relapse rates among those who have suffered from anxiety and/or and depression.

BFT-C: For Whom Is BFT-C Available

With all of the above in mind, at this time the option of BFT-C is offered only to selected clients as follows:

1. The client does not have a history of seizures, psychosis, or multiple personality and

refrains from any use of illegal drugs as well as excessive alcohol;

2. We have tried my basic approach (1) without achieving a complete remission of psychological symptoms characteristic of such conditions as PTSD and depressions, and/or (2) without a sufficient decrease in the experience of chronic pain, and/or (3) without achieving a reasonable level of cognitive functioning;

3. Based on information to date, there is some reason to believe that BFT-C may be helpful;

4. The client accepts that BFT-C is not yet a common approach in the treatment of his/her condition;

5. The client agrees to assume responsibility for informing me immediately if s/he is experiencing any distress during training (at which point the training will stop and be re-assessed) and informs me in a timely fashion of any distressing after-effects;

6. The client follows instructions to maximize obtaining a good qEEG recording; and

7. The client signs below giving informed consent to a trial of BFT-C

BFT-C: Informed Consent

Since BFT-C may be helpful for you, I need your consent to proceed. By your initials at the bottom of each page and by your signature below, you certify (1) that you have read the document, (2) that details of BFT-C have been explained to your complete satisfaction, and (3) that you agree to proceed with BFT-C.

Signed:_____ Date:_____

APPENDIX IV: DIAGNOSTIC CONUNDRUMS

That there is controversy in the set of criteria for the diagnosis of PTSD is not surprising.

First, new information from research studies is being published at a very high rate.

Second, the *DSM-IV* used in North America is severely out of date with this research since the guide book was published in 1994. The fifth revision is not expected before 2013.

Third, in addition to considering available data, the criteria in 1994 were determined by democratic processes within the committee responsible for *DSM-IV*. Like any other document produced by a group, accommodations to interest groups were made.

So, there is nothing holy about the *DSM-IV*. But it does give clinicians a frame of reference.

However, other than being an academic issue, the consequences of the conundrums below are significant in that they have an impact on who receives the diagnosis and who does not as well as having an effect on specifying the intensity of the conditions. This is critical where the diagnosis relates to court proceedings, administrative decisions, and insurance matters. For example, if a client does not meet all of the required criteria for the *DSM-IV* diagnosis of PTSD, then s/he may be denied pension or insurance payments. Relatedly, economic interests of agencies are served by non-payment or reduced payment of claims.

Unfortunately, although similar symptoms to *DSM-IV* are noted, due to the absence of clear decision rules, diagnosis by the alternate ICD-10 diagnostic system has problems too.

> Typical symptoms include episodes of repeated relieving of the trauma in intrusive memories . . . occurring against a backdrop of a sense of "numbness" and emotional blunting, detachment from other people, unresponsiveness to surroundings, anhedonia, and avoidance of activities and situations reminiscent of the trauma. Commonly there is fear and avoidance of cues the remind the sufferer of the original trauma. . . . There is usually a state of autonomic hyperarousal with hypervigilance, an enhanced startle reaction, and insomnia. Anxiety and depression are commonly associated with the above symptoms and signs, and suicidal ideation is not infrequent. (World Health Organization, 2007)

ICD-10 is in the process of revision also.

Conundrum-1
The formal diagnosis of PTSD requires the presence of specific combinations of symptoms. At minimum, a person would need to have any symptoms of 5

symptoms from Cluster B, any 3 of 7 symptoms from Cluster C, and any 2 of 5 symptoms from Cluster D. Via mathematic calculations, this means that there are 1750 different symptom combinations all still being called PTSD. To my mind, it is difficult to believe that the same biological processes are operative for all of these possible combinations. It is much more likely that some combinations of symptoms involve some factors while other combinations are related to different brain factors. This may explain in part different findings that have emerged in respect to the causes of PTSD, why some qEEG findings occur with one client but not with another, and why various interventions are effective with some clients but not with others.

Conundrum-2

Symptoms can be distressing or disruptive and worthy of therapeutic attention even if their number or combination do not meet the diagnostic criteria for PTSD. Thus, in the clinical literature increasing attention is being paid to what now is being called "*Partial or Sub-Threshold PTSD* ." Since the *DSM-IV* does not recognize this, those who do not meet the minimum criteria for PTSD must be classified as having either "*anxiety disorder NOS*" or "*adjustment disorder.*" Three to four times the number of those who meet the full criteria for PTSD will have partial symptoms that are still significant clinically such that treatment is recommended. That being said, studies have indicated that those who meet the current full diagnostic criteria for PTSD usually are suffering more than those with partial symptoms. As a result, some have proposed that PTSD be considered as a disorder with symptoms distributed along an intensity continuum of mild to severe rather than using an arbitrary cut-off point to decide that some have PTSD and others do not. Broman-Fulks and his group (2009) (a) reviewed the research literature and referenced many studies among adults such as combat veterans that used taxonomic mathematic methods and (b) determined that these studies supported a dimensional conceptualization of PTSD.

Conundrum-3

There is controversy among clinicians and researchers as to whether or not some of the numbing/dissociative symptoms (C3-C7 in the *DSM-IV*) should be required for the diagnosis of PTSD.

First, most of those with partial PTSD meet the A, B, and D criteria, and exhibit some avoidance but do not have a sufficient number of the numbing/dissociative symptoms on criterion C. For example, the 2008 study by Norris and Aroian of 453 Arab-Muslim immigrant women to the U.S. who reported experiencing multiple traumatic events (such as military combat, serious accident, imprisonment, ethnic persecution, life-threatening illness, and torture) indicated that 82% met the criteria for re-experiencing, 62% had arousal symptoms, but only 43% had 3 avoidance/numbing symptoms. Moreover, all but the avoidance symptom of trying to not think or talk about the trauma occurred less frequently than re-experiencing or arousal symptoms. This is in line with other studies and the suggestion is that the avoidance/numbing symptoms are very influenced by culture. Relatedly, there is reason to believe that there are clear neurobiological mechanisms to account for re-experiencing and arousal symptoms but not for avoidance/numbing symptoms. Research in emotional processing theory holds that some individuals use avoidance/numbing approaches as a way to cope with the other two clusters of PTSD symptoms.

Second, mathematic processing by factor analysis of symptoms typically show that symptoms of avoidance and symptoms of numbing/dissociation represent separate clusters along with a dimension of intrusive re-experiencing and one of anxious arousal. Most studies to date (such as McDonald and associates, 2008) have found that PTSD symptoms cluster around these four separate factors.

Third, a series of well-designed studies by Frewen and Lanius summarized in their 2006 paper point to significant differences between those with PTSD who do and those who do not report numbing/dissociative symptoms such as derealization, depersonalization, and feelings of emotional detachment. When provided with script-driven traumatic cue imagery, those without dissociation showed *decreased* activity in a number of brain areas such as the medial prefrontal cortex, the anterior cingulate cortex, and the thalamus consistent with less regulatory control over affect arousal. Those who experienced dissociative symptoms displayed *increased* activation of the inferior frontal gyrus, the medial prefrontal cortex, and the anterior cingulate. Thus, some have argued that we ought to be considering two forms of PTSD: one with and one without dissociative symptoms.

This is not to conclude that numbing symptoms are unimportant. The 2008 study by Malta and colleagues followed disaster workers who had moderate PTSD. The reserchers found that only the severity of numbing symptoms predicted PTSD at the 2-year follow-up point.

Conundrum-4
Similarly, given the high frequency (averaging 50% across various studies) with which depression accompanies PTSD, the fact that there are a number of symptoms in common, and the additional challenges to treatment presented by a co-occurring depression, many experts believe that the diagnosis of PTSD should be divided into two other groupings: PTSD with and PTSD without depression. Interestingly, one 2007 research study concluded that nearly 80% of those with depression showed symptoms of PTSD and about one-third of them were unable to name a single traumatic event that could have caused the PTSD symptoms. Also, some factor analytic studies (see McDonald and colleagues 2008) have labelled the fourth cluster as dysphoria that also included numbing-related symptoms as well as sleep problems, irritability/anger, and difficulty concentrating. The last three, along with dysphoria and some of the numbing symptoms, are part of the *DSM-IV* diagnostic criteria for depression. Clearly, continuing research is required to sort out such matters. In the interim, the diagnostic alternative is to continue the practice of providing multiple diagnoses, one for each condition additional to PTSD.

Relatedly, evidence to date suggests that most of the time other psychological conditions that co-occur with PTSD develop after the onset of PTSD or around the same time. This is true for such conditions as major depressive disorder, generalized anxiety disorder, substance abuse, and panic disorder. Thus, if such conditions develop along with PTSD as a consequence of trauma, it is a rare occurrence that they develop before the onset of PTSD. However, as a direct result of traumatic experiences, this does not mean that such conditions cannot develop in the absence to PTSD.

Conundrum-5
Some with PTSD have a history of childhood abuse before additional traumatic experiences as an adult. Usually, the abuse, both physical and sexual, continues for many years of the child's life. In this situation, researchers and clinicians have proposed the terms Complex *PTSD* or *Disorders of Extreme Distress*. Moreover, they argue that such a concept ought to be included in diagnostic manuals as one type of PTSD.

Conundrum-6
Southwick and associates (1997) reported an interesting study suggestive of two different types of PTSD depending on which biological system was activated. One is an adrenergic type and the other is a serontonergic subtype. They report that PTSD symptoms among those with the adrenergic type worsened after an infusion of the agonist yohimbine but not after MCPP. On the other hand, those with the serononergic type had increased PTSD symptoms following MCPP but not after receiving yohimbine. The results suggest that PTSD may be an end-stage syndrome that can occur via any of a number of biological processes.

Conundrum-7
Studies such as the 2008 one by Adler and her colleagues call into question the validity of the *DSM-IV* A-2 criterion (that the trauma response involve intense fear, hopelessness, or horror) when the client has been trained to deal with life-threatening events. This would include police and military personnel. Their study of 202 soldiers returning from a year in Iraq indicated a sizeable percentage who did not meet the A-2 criterion. Nevertheless, this sub-group of soldiers had the same significant symptoms of PTSD without meeting the A-2 criterion compared to those meeting A-2 criterion. Those not meeting the A-2 criterion credited their training and were likely to experience anger as a response to trauma. The authors' position is that insisting that the A-2 criterion be met leads to an underestimation of the number of those who might benefit from treatment. Also, the authors cite studies of police and combat veterans wherein 90% of police in one study and 70% of older military veterans in another experienced A-2 symptoms after traumatic experiences. However, it is true also that two other studies found that those with combat experiences were less likely to meet the A-2 criterion compared to people exposed to other types of trauma. The latter two studies as well as a number of others reported other possible criteria that could be included as A-2 criteria. For example, police officers meeting all the criteria for PTSD except A-2 identified reactions included senselessness, isolation, shock, numbness, guilt, and hostility.

In 2009, Resnick and Miller reviewed studies providing evidence (a) that fear and horror were not significantly related to PTSD severity but ratings of helplessness and numbing were, (b) that the response of anger shame, and guilt were independent predictors of PTSD, (c) that fear accounts for about 11% of the predictiveness while other responses including anger account for 24%, (d) that guilt is strongly associated with PTSD among combat veterans, battered women, and rape victims, (e) that PTSD among prisoners of war was associated with shame rather than guilt, (f) that anger was associated with PTSD, especially among military personnel (as found in 39 studies), (g) that anger among female rape or assault victims was positively related to the development of PTSD, and (h) that anger after criminal victimization or a motor vehicle accident predicted PTSD severity. These and other findings lead to the inescapable conclusion that many emotional reactions can be associated with PTSD and that the current criterion A-2 triad of fear, horror, and helplessness often occur less frequently than anger and other reactions.

The 2009 review by Kilpatrick and co-authors supports the conclusion above. Moreover, evidence is cited in line with the view that information about criterion A-2 does not greatly increase predictive utility beyond criterion A-1. Additionally, the study by the authors of adults with PTSD after a hurricane supported this conclusion. About 84% of known PTSD cases were included if both A-1 and A-2 were required for the diagnosis while about 97% of cases were captured if A-1 was met regardless of A-2 response.

Conundrum-8

The *DSM-V* is in planning stages and is expected to be available in 2013. There is a proposal circulating for a new category to be developed called *stress-related fear circuitry disorders*. This diagnosis would include what we now know as PTSD, panic disorder, and social phobia. The basis for this proposal is some commonality in neuro-circuitry, cognitive alterations, and neuro-hormonal changes. Others are arguing that emphasizing the fear circuitry is ill-advised in that depression co-occurs very commonly with PTSD and may not be related to the fear circuitry and that depression itself can be the result of trauma. My own concern with the label is that it will serve to decrease further the number of males (particularly police and military personnel) willing to come forward with PTSD symptoms.

In 2009, Resnick and Miller provided a good rationale for placing PTSD along with acute stress disorder, adjustment disorder, and perhaps complex PTSD in a spectrum class of *traumatic stress_disorders*. Their reasons are based on evidence that PTSD is not really an anxiety disorder, the grouping in which it is placed currently. Specifics include (a) that fear is just one of many common emotions in PTSD, (b) that guilt, shame, and anger are associated strongly with PTSD, (c) that reactions to trauma-related cues may not involve fear or anxiety, and (d) that anxiety conditions such as generalized anxiety disorder and panic disorder either develop after PTSD or at about the same time as PTSD. Interestingly, they cite studies suggesting that (a) generalized anxiety disorder may have more in common with depressive conditions than other anxiety disorders, and (b) obsessive-compulsive disorder shares essential features with a variety of non-anxiety disorders.

APPENDIX V: INTERVENTION METHODS I DO NOT USE

Some professionals use methods that I do not include in my psychological treatment of PTSD. These include the following.

V-01.   Exposure Therapy/Prolonged Exposure Therapy
V-02.   EMDR
V-03.   Virtual Reality Exposure Therapy (VRET)
V-04.   Cognitive Reprocessing Therapy
V-05.   Interpersonal Therapy
V-06.   Hypnosis
V-07.   Health Food Remedies
V-08.   Verbal Psychotherapies

V-01. Exposure Therapy

I do not use the method known as *exposure therapy* or *prolonged exposure therapy* although it is considered as an appropriate method within the general framework of cognitive behaviour therapy (CBT). Some psychologists use it routinely, sometimes to the exclusion of other interventions.

This technique involves graduated, safe, and prolonged exposure (usually via imagining) to reminders of the trauma and is usually provided in combination with additional components of treatment. Many well-controlled published studies among civilians indicate treatment combinations that included exposure therapy are effective in decreasing both the arousal and avoidance cluster of PTSD symptoms. A review of exposure therapy by Bradley and colleagues published in 2005 indicated that 68% of those who had compled exposure therapy no longer met the diagnostic criteria for PTSD. However, my inspection of the reported research indicates that many of the clients were victims of a single-episode trauma. One specific example of a research study with such civilians is the 2008 one reported by Bryant and his research team. Large symptom decreases were obtained with the combination of imaginal exposure, *in vivo* exposure, and cognitive restructuring. The study in 2008 by Rauch and associates reported that prolonged exposure among female victims resulted in positive changes in both health perceptions and social functioning.

There are a few studies that included exposure therapy as part of a multi-component approach with military veterans. For example, a 2009 clinical case series by Rauch and co-investigators reported symptom reduction in 80% of their 10 combat veterans after methods including prolonged exposure such that 50% of the group no longer met the criteria for PTSD. Also, Rademaker and colleagues (2009) reported on a multimodal group therapy approach including exposure therapy with Dutch peacekeepers in which PTSD symptoms were decreased. Also, that group summarized other such studies with military personnel and concluded that generally exposure therapy in combination with additional components was successful in reducing PTSD symptom levels. However, their own study and the others cited did not lead to remission.

For a number of reasons, I do not use exposure therapy with my clients.

First, as far as I can determine, symptom reduction among military personnel, while noticeable, is still quite minimal. For example, in one study, about 40% of those with PTSD had a reduction of 10 or more points on the CAPS (and 60% did not). Many studies show reductions with end points well above the CAPS criteria for PTSD; symptom intensity was lessened somewhat but not to the point of remission. Also, Dr. Bryant reported at the 2005 ISTSS conference that his usual method of cognitive behaviour therapy (CBT), which included prolonged exposure and was found to be effective with civilians, was ineffective, made symptoms worse, or resulted in patients dropping out of therapy when implemented with personnel from the Australian Defence Forces who had PTSD. A subsequent revision of procedures in an open trial giving 12 service personnel skill training before prolonged exposure led to positive changes for many, although 37% still met the criteria for PTSD afterwards. The reasons for negative findings for some clients may relate to reduced NAA levels in the ACC, as reported by Schuff and colleagues (2008) in the above section on the pre-frontal cortex.

Second, I am bothered with reported drop-out rates as high as 30% in studies using exposure therapy. In rebuttal, major developers and proponents of exposure such as Dr. Foa suggest that drop-out rates are not any worse than with other forms of treatment. In contradiction, my drop-out rate is well below 5% and with remission rates better than those reported in research studies.

Third, my police and military clients have not wanted to try exposure therapy in any formal, structured way (a) in part because of their experience that prolonged focus on the trauma that brought them to my attention serves to increase symptoms and (b) in part because of the impracticalities presented by the very large number of traumatic events they have experienced.

Fourth, the few times I tried the technique, I observed what to me was an alarming increase in hyper-reactivity across many aspects of the ANS functioning I was monitoring such that I discontinued the procedure. This is in keeping with both medical and psychological practice that accept the dictum of "first do no harm" as a principle of treatment.

Finally, many of my police and military clients are able to talk spontaneously about their traumatic experiences once they have mastered tactical breathing and have sleep under good control while others do so during or after BFT-C trials. Those who do so are able to talk about the traumatic experiences without clinical signs or monitored psycho-physiological indicators of distress. Additionally, most are able to pass by scenes of previous traumatic events and be confronted with other cues without relapse. For some, once symptoms are in the remission range, we might set up a graduated series of steps towards a return to work.

Interestingly enough, I am not alone in the hesitation to use exposure methods in the
treatment of PTSD. The review by Cook and Niederehe in the 2007 book *Handbook of*

PTSD: Science and Practice led to this conclusion "In fact, there appears to be caution
in the traumatic stress field more broadly about the use of this technique, as evidenced
by reports that frontline clinicians in 'real world' settings rarely use this treatment" (Friedman et al., 2007, p. 264).

Results of a group-based field test with veterans selected for the project was reported by Ready and his team (2008). In groups of 9 to 11, clients attended 3 hours with the group for 16 to 18 weeks. Clients received presentations on symptoms, breath re-training, thought-stopping, grounding techniques, and so on before sharing their war experiences for 2.5 hours or so with the group. Then they spent about 30 hours listening to their own presentations and an additional 27 hours listening to others' experiences. Prior to the interventions, 73% were taking psychotropic medication; during the intervention, that number increased to 100%, and at the conclusion, 99% were taking medication. Meaningful reductions in symptoms were reported and the drop-out rate was 4%. Given the range of intervention components, it is difficult to judge the relative contribution of the exposure component on outcome.

V-02: EMDR

The intervention method known as *EMDR* is used by many well-respected professionals. Generally, it has demonstrated positive results in research studies with civilian populations. The intervention includes talking about the trauma while the client's eyes follow the therapist's finger as it moves from side to side. In this way, it is believed that clients become able to dissociate emotional reactions from traumatic memories. Results with EMDR among combat veterans have been mixed with some studies showing improvement, some studies showing no change, and some studies showing no difference over other approaches. I am not aware of any studies with police officers. Some research comparing EMDR with and without eye movements has shown that the eye movements are unnecessary, thereby making the technique more like a graduated form of exposure therapy.

I do not provide EMDR for a number of reasons. First, I have not yet taken the required training in order to do it as prescribed. Second, both police and military groups are reluctant to engage in exposure therapy anyway. Third, I consider my use of methods based on clinical psycho-physiology and applied affective applied neuroscience to be more grounded in basic science and to make more sense than EMDR.

V-03: Virtual Reality Exposure Therapy (VRET)

VRET is an exposure-based, technology-delivered intervention developed for PTSD among military personnel. It is introduced at that point in the range of interventions when exposure therapy would normally be introduced. Procedurally, while watching a computer-generated view and sounds of battle, a therapist instructs clients to recall traumatic memories of their tours of duty. The therapist also controls the various stimuli such as helicopter sounds, gunfire, voices telling personnel to "move out," and so on. Across a wide range of studies, participants engaging in VRET have experienced emotional responses, which is taken to mean that it is effective in activating the fear circuitry.

The few studies to date report positive results of VRET. In the Rothbaum and co-worker study of VRET (1999) with 10 U.S. veterans, results indicated a reduction of arousal symptoms ranging from 15% to 67% as measured by CAPS 6 months later. Intrusion and avoidance symptoms were not different from baseline at 6-month follow-up. A further 2001 open clinical trial by Rothbaum and colleagues reported a reduction in PTSD symptoms among 16 male Vietnam veterans after an average of 13 sessions over a 5- to 7-week period. For example, overall average CAPS scores decreased from 68.0 before treatment to 57.8 after VRET, to 47.1 6 months later. These are meaningful reductions, but in my CAPS scoring system, a score of 43 or above means that PTSD is not in remission. A case report of a soldier returning from Iraq (Gerardi et al., 2008) indicated that VRET resulted clinically and statistically in decreases in PTSD scores. Powers and Emmelkamp (2009), in a meta-analysis of 13 studies, found that virtual reality exposure was more effective than imaginal exposure and as effective as *in vivo* exposure in treating a range of anxiety disorders including PTSD.

I do not use VRET due to the equipment requirements, cost, and programming required. Also, this falls into the category of prolonged exposure to trauma cues, which I do not include in my approach.

V-04: Cognitive Re-Processing Therapy
Cognitive re-processing therapy is one approach within the general group of interventions known as cognitive behaviour therapy (CBT). It involves walking the client through the traumatic experience with a view towards understanding it in a different and more healthy way. The hope is that as a result of doing so, the PTSD symptoms will decrease. Often this procedure requires that clients retell details of the trauma many times.

There is some research literature indicating that including this element along with other methods of CBT can be effective in reducing the symptoms of PTSD.

While, I deal with distorted beliefs and cognitive processes throughout my basic treatment approach, I do not include cognitive processing therapy. *First*, with the hundreds of traumatic events experienced by my police and military clients, it would not be possible to do this. *Second*, when patients have talked about traumatic experiences, direct measurement has sometimes shown significant psycho-physiological reactivity such that I became concerned about the levels displayed. *Third*, the methods I utilize are at least as successful as cognitive re-processing therapy and I do not run the risk of re-traumatizing clients in respect to what is known about the operation of both the rACC and the amygdala.

However, often when the PTSD is in remission, police/military clients choose to discuss some of their traumatic experiences, perhaps as a way of cognitive re-processing. At this point, clients are very much in control; for example, previous affective elements are absent and clients know exactly what works best for them in managing any reactivity. Moreover, traumatic memories do not replay automatically as before.

V-05: Interpersonal Psychotherapy (IPT)

IPT is a structured psychotherapy that was developed to address interpersonal and social problems. Thus, it was natural to extend IPT to clients with PTSD since such issues are among the reasons that clients seek help. A 2004 pilot study by Robertson and co-workers found that IPT resulted in improved social functioning but with limited effect on specific symptoms of PTSD. IPT is considered by many as one of the reasonable methods of treatment for depression. While brief discussions of interpersonal matters take place from time to time with my clients, generally ITP is not necessary, as such issues resolve themselves with either my basic approach or with the addition of BFT-C. Very infrequently, clients and their spouses are referred to a colleague specializing in marital therapy. Even less frequently, some couples have decided to end their relationships.

V-06: Hypnosis

Hypnotherapy using light or deep trance techniques has been used clinically for decades to treat combat-related stress disorders. Clients are induced into a comfortable physical and mental state and then receive suggestions to distance themselves from the traumatic event. In this way, it is believed that the event and the reaction are separated. In a 2008 study by Abramowitz and co-workers, Israeli combat veterans with chronic PTSD received anti-depressant medication and supportive psychotherapy and were then divided into two groups. One received zolpidem 10 mg each night and the other group was treated with symptom-oriented hypnotherapy. Compared to the zolpidem group, those receiving hypnotherapy had more decreases in intrusive re-experiencing and avoidance as well as improvement in sleep.

V-07: Health Food Remedies

There is no evidence that any of the so-called "health food remedies" are effective in the treatment of PTSD, although well-conducted studies indicate that generally *kavakava* can reduce symptoms of anxiety and *valerian root* may help with sleep. However, caution is urged since there is research indicating that some health food remedies interfere with the effectiveness of medications including antibiotics and those employed in the treatment of cardiovascular conditions. Furthermore, the reality is that health foods or remedies are neither standardized nor controlled so that within and between brands one does not know how much of the alleged active ingredient, if any, is present. Also, information is absent as to what other ingredients may have been added as well as the effects of added ingredients.

V-08: Verbal Psychotherapies

One of the early pioneers of talk therapy was Freud. His approach was expanded further by a number of his followers. Later came other verbal psychotherapies. The basic tenet was that if you explored the meaning of the trauma and linked it to early and other experiences, then insight would develop, after which the symptoms would decrease. For some therapists, the analogy was the steam engine; if the steam builds up inside, then the steam must be released or the boiler will blow up.

While my clients and I do talk about matters of importance, I find no compelling reason to rely on verbal psychotherapy. Although treatment approaches based on this analogy have been around for a long time, generally they have received little support when scrutinized through research studies using proper controls. It is more reasonable to believe that talk therapy will serve only to maintain or worsen PTSD symptoms. Evidence shows that repeated exposure to the trauma may serve to re-engage the amygdala and related fear structures while at the same time putting inhibitory controls off-line,. My clinical experience has been that clients usually initiate conversations about their traumatic experiences and other life-altering events without negative effects after they have learned to regulate their psycho-physiological systems. Under these circumstances, whatever cognitive closure is required can then be achieved.

APPENDIX VI: PREVENTION OF PTSD

Contents

VI-01. Propanalol

As mentioned previously, a number of studies have indicated that enhanced memory of arousing or emotional experiences is mediated by the effects of central beta-adrenergic activation on brain structures such as the amygdala. Moreover, three studies have now been published involving the administration of *propanalol* (a noradrenergic beta blocker) shortly after exposure to acute physical injury. At follow-up one month later, all of those administered propanalol in the 2006 RCT study by Pitman and collaborators showed evidence of decreased physiological reactivity to scripts of the trauma while this was true for 8 of the 14 given a placebo. The 2003 case series reported by Vaiva and co-workers found that those who were administered propanalol had fewer symptoms of PTSD two months later compared to those who declined a trial of this medication. On the other hand, the 2007 study by Stein and associates found no significant differences in depression or PTSD symptoms between those who took propanalol compared to those administered a placebo.   Finally, a recent unpublished study indicated that when propanalol was administered even years after the traumatic event, it led to a noticeable decrease in physiological reactivity when recalling the traumatic incident.

VI-02. Critical Incident Stress Debriefing (CISD)

An intervention known as CISD is sometimes provided within a few days of a single traumatic incident to all personnel involved in the same event or all members of a military unit.

Typically, this single group intervention includes a review of the incident, sharing of thoughts and feelings about it, a presentation on possible effects of the trauma, and some recommendations to facilitate recovery.

Sound research studies on the effectiveness of a single session of CISD have not found this approach superior to no intervention in preventing either later PTSD or long-term negative outcomes. Two 2002 meta-analyses of studies by research teams led by Rose and by van Emmerik found no evidence to support the use of CISD to decrease psychological distress or prevent the onset of PTSD. Adler and her team (2008) reported an RCT study wherein 952 U.S. military peacekeepers were assigned to standard CISD, stress management classes, or survey only. Based on self-reports, (a) CISD did not cause undue distress as compared to the other conditions nd (b) was associated with more alcohol problems. Perhaps since the fear structures can easily be re-triggered when talking about the emotional factors of the trauma so soon afterwards, some studies have found that up to one-third of participants are worse after the group session than before.

Given this information, when as the local police psychologist I was called upon to *do something* following a traumatic incident involving police officers, I adopted a revised format. Wanting to avoid re-triggering fear structures, I used a psycho-education approach consisting of (a) exclusive focus on the objective details of the event so that all participants left with a clear sense of what happened and could thereby achieve cognitive closure, (b) consideration of what went well and what needed change should another similar incident occur, in part so that the event could be seen as having some benefit without blaming anyone, (c) a brief review of possible psychological and physiological reactions so as to normalize these experiences, and (d) provision of specific recommendations on how best to maximize mental health after a trauma. The latter consisted of reviewing some of the stabilization strategies mentioned previously in Step E. Participants seemed to achieve closure in this way but I have no data as to the impact on any subsequent symptoms of PTSD; some participants called for appointments regarding the incident up to two years later and others never made further contact with me.

VI-03. Multi-Session Interventions

Some multi-session post-incident methods have been implemented with one targeted person at a time. Usually this has followed a specific traumatic event when the person had symptoms of Acute Stress Disorder (ASD). This *DSM-IV* condition has many similarities to PTSD and can develop within the first month following a traumatic event (for example, see Bryant and associates, 2003). Intervention elements described as "early trauma focussed cognitive behavioural therapy" have included education about trauma and possible effects, relaxation training, exposure therapy, and identifying and subsequently changing maladaptive interpretations or beliefs about the event. The 2008 publication by Kornor and co-authors reviewed the studies to date and concluded (a) that there is evidence that this approach is effective in preventing chronic PTSD in such clients and (b) that replications are required since all the evidence originates from one research team.

At the 2005 ISTSS conference, Dr. R Bryant, an acknowledged expert in ASD, recommended that treatment for ASD be provided only if an assessment indicated that significant symptoms were present for 2 weeks after the traumatic incident. He noted that 75% of those with ASD develop PTSD subsequently. However, he noted also that 50% develop PTSD without first meeting the diagnostic criteria for ASD, partly because they do not experience the symptoms of dissociation required for the diagnosis of ASD.

Consistent with these thoughts was the 2008 results of a study by Palyo and associates of 345 victims of motor vehicle accidents. Those who had symptoms of numbing (loss of interest, detachment, and restricted range of affect) at least one month after the event were more likely than otherwise to have some of the PTSD symptoms of hyper-arousal (sleep difficulties, irritability, exaggerated startle response, and concentration problems). This co-relational relationship held even after controlling for factors such as depression, re-experiencing, and avoidance.

VI-04. Stress Inoculation Training

The U.S. military is very involved in studies to find strategies so that troops can withstand combat stress. For example, one approach has been to use computers and electronics to create a virtual reality. With this technology, soldiers are exposed to repeated and realistic portrayal of the signs and sounds of battle long before being deployed to an actual combat zone. This is seen as a form of innoculation training in that the presumption is that soldiers will become desensitized to combat situations amd develop cognitive-behavioural ways of coping.

Stress management training (SMT) is another term that is used for this approach. Some studies of SMT were reviewed in 2009 by McKibben and collaborators. Most reports conducted generic coping strategies as well as time management, muscle relaxation, and controlled breathing to reduce overall levels of arousal. The authors cite a 1996 meta-analysis that revealed that such SMT training led to improved performance under stress. Also, they note that a number of studies suggest that SMT showed promise in the treatment of PTSD. In their own study, results were compared between soldiers in an infantry division who received SMT prior to deployment to Iraq and those who did not receive SMT. Although many of the size effects were quite small, results indicated that SMT led to "lower levels of PTSD and physical symptoms, as well as higher levels of morale, retention intentions, leadership, and marital functioning" p. S73) Also, it was not known why some soldiers had not taken SMT.

APPENDIX VII: ABOUT DEPRESSION

V11-01: Introduction

I have had an interest in depression for a very long time and have provided services of assessment and treatment of this psychological condition to adults since the early 1980s. Most recently I have continued to provide these services to police and military veterans when either their depression co-occurred with PTSD (post-traumatic stress disorder) or when it was clear that the depression followed from duty-related traumatic experiences.

The symptoms of depression can follow any of a number of stressors, in which case the term *reactive depression* has been used in the past. For example, the symptoms of depression can result from traumatic stress, such as when one's life has been under serious threat. Relatedly, from 30% to 70% of those with PTSD also meet the diagnostic criteria for depression. At other times, depression stems from chronic stress such as long-term illness and marital relationships fraught with problems over a considerable period of time. Also, depression can follow the loss of a significant person in one's life. Generally, reactive depression is viewed by professionals as the end-product of difficulties with emotional stress, which acts to destabilize brain circuits that are involved in the regulation of mood, thinking, and behaviour. As noted below, various other factors can play an important role in the development of symptoms.

However, sometimes no specific cause can be found and in a percentage of these situations there is a family history of depression. This has been referred to in the past as *endogenous depression* with the assumption that genetics are a factor; the person inherits the capacity to develop depression. As with reactive depression, those with endogenous depression demonstrate dysregulation of nervous system processes.

VII-02. *Clinical depression is a very democratic illness.* It strikes people of both genders, at all ages, from all races and socioeconomic groups, in all religious faiths, at all educational levels, throughout recorded history, and independent of sexual orientation. Some very famous people have struggled with this condition including Biblical persons (King David), political leaders (Churchill and Lincoln), music composers (Handel and Beethoven), authors (Poe, Williams, and Wolfe), artists (van Gogh), comedians (Dick Cavett and John Cleese), creators (Charles Schultz of *Peanuts* fame), business tycoons (Ted Turner), TV reporters (Mike Wallace), and poets (Leonard Cohen).

VII-03. In fact, *from 10% to 25% of women and from 5% to 12% of men* will experience at least one clinical depression in their lifetimes. Thus, at any time, between 2% and 4% of the adult population are experiencing a significant depressive condition such as Major Depressive Disorder. The rate among adolescents is as high as 6%. In addition, between 1.4% and 5.7% suffer from brief or minor depression and two to three times as many folks have depressive symptoms that do not meet formal criteria of the two accepted diagnostic systems, the *DSM-IV* and ICD-10. Unless related to a single trauma or accumulated traumatic events, the first episode of depression often occurs in the late teens and early twenties and is often preceded by a series of "minor" depressions.

VII-04. *The female-male differences* in rate of depression noted above have been the subject of a number of studies. The conclusion has been that women have higher rates of depression than men, certainly (a) in developed countries, (b) between the child-bearing years of adolescence to menopause, and (c) across most countries and most ethnic groups. In addition, research indicates that the duration and relapse/re-occurrence rate between the genders does not differ in adulthood but that women have a higher number of first onsets compared to men. At this time it is not clear why there are gender differences, although a number of possibilities have been explored as noted below.

*The impact of sexual crimes is significantly greater for females than males. In addition, studies show that 7% to 19% of girls are victims of childhood sexual abuse as compared to 3% to 7% for boys. Also, 13% to 15% of women are victims of completed rape. Research studies have been consistent in demonstrating a link between such sexual assaults and depression.

*Women are more likely than men to have incomes below the poverty line and the rates of depression increases as income/ socioeconomic status lowers.

*Compared to males, females may have more problems with their roles in society, such as lack of choice, overload, and competing social demands. All of these are sources of chronic stress and chronic stress is known to increase the likelihood of depression.

*Compared to men, women tend to admit to more symptoms of depression and are more likely to have co-occurring conditions such as panic disorder. Research has demonstrated that the severity of the depression as well as the presence of additional psychological conditions have a negative impact on such factors as complexity of treatment and the length of time required to respond to treatment.

*There is a suggestion of gender differences in neuro-transmitter systems. Estrogen and progesterone both have effects on the central nervous system and changes in the levels of these reproductive hormones may directly affect such neuro-transmitters as serotonin and nor-epinephrine. Both of these neuro-transmitters have been implicated in depression..

*The presence of different hormones may be such that any genetic predisposition to depression may be "switched on" during puberty. However, some studies have not found this to be the case.

*In contrast to males, many females may have increased sensitivity to adverse experiences and feelings of sadness, possibly due to distinctive interpersonal orientation, social circumstances, and a tendency to process such experiences with a ruminative coping style. For example, one study indicated that relative to men, women are more likely to feel strong emotional ties with a wide range of people in their lives, to view their roles with others as central to their self-concepts, to care what others think of them, and to be effected emotionally by events in the lives of other people.

*Some studies have found that gender, menopausal status, and age can affect response to medication used in the treatment of depression while other studies have failed to find such differences.

*Previous explanations that males have a somewhat different pattern of symptoms than women do not appear to be a major factor in gender differences. However, excessive amounts of sleep, psychomotor motor slowing, feelings of worthlessness or guilt, anxiety, over-eating/weight gain, and somatic complaints are reported more by women than men. Also, women attempt suicide three times more often than men although women are less likely to die from suicide than men. Men are more likely than women to have a co-occurring alcohol or substance-abuse disorder.

VII-05. At present, about 20% of depressions are first diagnosed before the age of 25 and 50% before the age of 39. However, current evidence indicates (a) that 50% of folks seeing a physician for the first time for a depression have had a previously untreated episode and (b) that about 80% of those adults with a depression had a previous episode as a child or adolescent.

VII-06. There is evidence that depression in the mild to moderate levels of severity for both genders is *occurring earlier in life* than was once the case. This fits with my experience of seeing a number of children with depression during the years when I directed a day treatment program for children ages 2 years to 5 years with significant psychological problems. Furthermore, studies also suggest that the *rate of depression is increasing*. While many possibilities for these findings have been discussed, at this time there is no definitive answer. The changes do not seem to be related to increases in the knowledge about depression or to improved methods of screening or assessment.

VII-07. Among the most common *factors that seem to be associated statistically with the likelihood of depression* are:
    *previous low mood or depression;
    *experiencing an anxiety disorder as a child;
    *a family history of depression;
    *exposure to traumatic events especially multiple ones, as is true among police and military;
    *childhood neglect and physical and sexual abuse;
    *chronic or repeated daily stressors;
    *loss of important persons in one's life, especially during childhood;
    *chronic marital/partner problems;
    *the presence of other psychological conditions; and

*the presence of any of a number of significant medical conditions that interfere with life.

* Current thinking is that these risk factors can ultimately lead to changes in brain and other biological functioning as noted below, which in turn produce the symptoms of depression.

* The more risk factors there are, the greater the likelihood of developing depression, the greater amount of time required to treat the depression, the greater the chance of a suicide attempt, and the higher the relapse rate. Also, the more risk factors there are, the greater the severity of symptoms.

VII-08.   There are *many sub-types of depression*, so the term "depression spectrum" is often used.   Some of the sub-types described in the diagnostic guides are:
*mood disorder due to a medical condition (such as underactive thyroid);
*premenstrual dysphoric disorder
*substance-induced mood disorder (such as side effects of medications or use of illegal drugs);
*minor depressive disorder;
*recurrent brief depression;
*adjustment disorder with depressed mood (such as after a divorce);
*dysthymic disorder (a low grade depression that continues for years); and
*major depressive episode or disorder (including post-partum onset or seasonal pattern).

VII-09. Regardless of the sub-type, as compared to those without a depression, typically people with a depression will have a combination of such symptoms as:
*abnormality in emotion regulation with persistent sadness/teariness and/or a high frequency of irritability/lowered patience even in the absence of negative events;
*loss of interest and/or pleasure in things once enjoyed, blunted response to things most people find pleasurable;
*increase or decrease in appetite and/or increase or decrease in weight;
*being slowed down or speeded up;
*fatigue/loss of energy;
*loss of motivation/increase in not caring;
*feelings of worthlessness/low self-esteem;
*exaggerated sense of guilt;
*problems with attention/focus/concentration including inability to shift focus as needed;
*short-term memory impairment ;
*difficulties with planning or decision-making;
*impairment with information-processing, particularly if effort is required;
*withdrawal from others/self-isolation, taking no pleasure in associating with others;
*feelings of hopelessness;
*loss of sexual interest;

*constant focus on negative aspects of self and situations, inability to see or focus on positives;

*excessive response to failure, amplification of its significance, difficulty in suppressing thoughts associated with failure, increased self-focus after learning about failure rather than focus on the event, increased depressed mood after an negative event;

* excessive seeking of reassurance;

* setting unrealistic/unattainable goals; and

* recurring thoughts of death/suicide.

VII-10. For a number of reasons, *people often do not recognize that they are becoming depressed.* The symptoms of depression often begin insidiously rather than with a bang, initially symptoms are of low grade in terms of severity or are intermittent, symptoms increase gradually, and symptoms are accompanied by a loss of perspective and critical thinking that is part of the depression itself. Thus, early on, clients often resist suggestions from others that anything is wrong. One study showed that the average time between the onset of symptoms and a professional's identification of the problem was about six months. Chances are this is even more true for men than women, given the male tendency to ignore what is considered to be weakness. This certainly has been my experience with police officers, 911/dispatch operators, and military folk, who tend to wait until things have reached crisis proportions before seeking help. The unfortunate outcome of delay is that treatment is more difficult and longer than when symptoms are acted upon early in the developmental progression of a depression and thus clients suffer much more than is necessary.

VII-11. People, unfortunately including too many in the helping professions, have *many myths about depression.* A myth is a fixed but false idea. Some of the most common myths are noted below. The impact of myths is very serious and includes under- or inadequate treatment, higher than necessary intensity of suffering for the person in the midst of a clinical depression, and an increased likelihood of suicide.

*One myth is that depression is a sign of weakness; if the person really wanted to, s/he could pull out of it. A recent study indicated that an astonishing 71% of those surveyed really believed depression to be the result of emotional weakness, 45% believed it to be the sufferer's fault, and 35% saw it as the end result of sinful behaviour. The fact is that neither strength of character nor social standing will prevent the development of depression.

*Another myth is that depression should not be taken as seriously as "real" health problems such as cancer or heart disease. Knowledge of the consequences of depression does not support such a belief. Unfortunately, the result of this myth is that the person with depression often does not believe that s/he needs or deserves help since the opposite of "real" is either "faked" or "phoney." Moreover, well-intentioned family and friends often avoid talking about the condition with the person who is depressed for fear that talking will make it worse. The fact is that when a person is depressed, a supportive listener can be very helpful.

Otherwise, the sufferer may withdraw further from people, develop a terrible sense of isolation, experience a consequent increase in the intensity of the depression, and be at increased risk for a suicidal act.

*A third myth is that the solution to depression advocated by many sufferers as well as those s/he knows is as simple as "Stop wallowing in self-pity," "Adopt a more positive attitude to life," "Get more absorbed in work or family or activities," and so on. The fact is that depression is not a problem in attitude or outlook. It is a real psychological condition that requires psychological attention. Most people with depression simply cannot enjoy the activities that used to make them happy and because of such symptoms as problems with concentration as well as the desire to be alone, they cannot do things as well as before and they do not get the satisfaction they once did. Moreover, often they are too tired and/or unmotivated to do much of anything. However, when the time is right, thinking and actions will need to be altered; this needs to be done therapeutically instead of within an angry or blaming approach, other things will need to be in place, and changes will need to be made gradually.

*A fourth myth related to those above is that depression is not a real illness because there are no usual signs of illness such as loss of consciousness, major cuts or bruises, body parts in a cast, and so on; the person looks the same as always. But depression is a true psycho-physiological illness, as noted below, with associated chemical and electro-physiological changes in the brain. In fact, it is often helpful to think of depression as similar to diabetes. In diabetes, problems in the pancreas lead to symptoms and in depression, brain functioning is involved. Folks with diabetes often have to take regular shots of insulin to combat the illness, while those with depression often have to take medications or do something to re-adjust brain functioning. In order to live lives that are as full and happy as possible, those with diabetes have to make life changes (such as diet, rest, and exercise) and the same is true for those with depression; changes in lifestyle, thinking patterns, and stress management for those suffering from depression are critical to both getting better and preventing a relapse.

*A fifth myth is that the person has no reason to be depressed because s/he has such a great job, a wonderful spouse, talented children, a nice home, a nifty car, an easier life than many, and so on. The fact is that depression can attack anyone no matter how "great" others see his/her life to be. Also, when folks are depressed, they are unable to notice, remember, appreciate, or be influenced much by positive things around them. Moreover, typically they are unable to see the blessings that surround them, and when they do notice them it makes no difference in their mood or behaviour.

* A sixth myth is that, because it is out of character for her/him, the depressed person would never commit suicide even if s/he makes some reference to death or suicide or seems obsessed with dark themes. The fact is that, when depressed, the person is not like his/her usual self. Many folks with depression have associated suicidal thoughts and far too many of them (from 15% to 17%) die by suicide. Moreover, 70% of suicide completers are depressed and most of these folks saw their physicians within 6 weeks of the attempt.

Many factors increase the risk of suicide including a previous suicide attempt or having family members who have attempted suicide, impulsive-aggressive traits, high levels of subjective depression, an overall sense of hopelessness, perceiving few reasons for living, low levels of the neurotransmitter serotonin in the ventromedial prefrontal cortex of the brain (which is involved in impulse control), a lifetime or current overuse of alcohol and/or drugs, a history of childhood physical or sexual abuse, and a history of head injury or neurological disorder. Also, suicide attempts are more likely during the spring and summer months.

*The most common methods used by males are firearms (63%), hanging (17%), and poison (6%), while for females it is firearms (40%), poison (25%), and hanging (16%).

Being in treatment decreases the chances of a suicide attempt considerably. Consequently, one of the first things to be done is to help the client put things into place in order to reduce the risk of suicide. Regular treatment sessions where suicidal thoughts can be talked about openly and risk factors assessed regularly, removing the methods most likely to be used in a suicide attempt (such as guns), reducing risk factors by doing such things as increasing social supports, informing key people in the client's life that the client is at risk for suicide, having these folks on a telephone call list that the client can use when suicidal thoughts are high, keeping the physician informed, putting in place treatment methods noted below, and knowledge of professional after-hours resources such as the local hospital are among the elements that need to be put in place. Sometimes, but rarely in my practice, it is necessary to have the client admitted to a hospital until risk of suicide is lowered sufficiently.

VII-12. Depression has a number of *significant consequences or impacts.*

The personal consequences of having a depression as compared to those who are not depressed are: less educational attainment, less income earned over a lifetime, and lower overall rating of satisfaction with life. These are in addition to the symptoms themselves and the impact of believing myths about the condition. In fact, many studies have demonstrated that the personal consequences of depression are rated as more severe that those associated with other significant chronic illness including heart problems, diabetes, asthma, high blood pressure, and arthritis. Moreover, since the onset of depression is usually in the late teens and early twenties when folks should be moving towards an adult role, depression can interfere with role transition, as studies have shown. Furthermore, the chances of an earlier death is higher among those with depression as contrasted to those without, in part due to greater rate of smoking, the medical outcome and increased risk-taking associated with greater alcohol and drug use, and suicide completions. Added is the sad reality that many folks with depression wait a full decade before seeking treatment and few of those who do seek professional help receive adequate care. Moreover, under standard care, only 30% respond to treatment, as many as 30% have a depressive episode that lasts more than 2 years, and even after 1 year of enhanced treatment. only 11% of patients achieve remission of symptoms (e.g., Leuchter and associates, 2009).

The health consequences in contrast to those without depression are significant, such as more hospitalizations, double the normal risk of serious heart problems (especially for men), a three-fold increase in dying from a heart attack, double the normal risk of developing diabetes, greater likelihood of blood flow problems to the brain, the possibility of an increase in blood pressure with resulting stroke, suppression of some parts of the immune system with accompanying increased risk for infectious disease, and poorer medical outcome from many types of chronic illness the person may develop. Moreover, in studies of people who have both depression and a illness such as diabetes, inflammatory bowel disease, cancer, chronic obstructive pulmonary disease, or coronary heart disease, the number of related and unrelated symptoms are much higher and the degree of impairment is much higher than among those without depression. This may be due to a number of factors including a high focus on somatic symptoms and the finding that those suffering from depression are three times less likely to comply with their doctors' advice than those without depression.

The family/interpersonal consequences are significant among those with depression as compared to those without depression: withdrawal from the family with resulting negative impact on family life; greater expressed anger/irritability; decreased chances of maintaining and enjoying relationships with friends; less awareness of natural human relations; a contagion effect (increased symptoms of depression in those living with the person); greater chance of having an affair (more for men than women); greater chance of engaging in other destructive patterns of behaviour such as gambling, over-spending, and over use of alcohol; and marital problems leading to signitifant rates of separation and divorce.

The consequences to society of depression include higher medical costs, lost productivity, and missed work days. Depression has been found consistently to be among the most costly health problems in the world: a U.S. government study in 1990 reported that the direct costs such as medical care were estimated at 26 billion. Another study in 1996 estimated that the annual salary-equivalent costs of depression-caused lost productivity exceeded 33 billion U.S. dollars. The 2009 study by Leuchter and colleagues set the cost at in excess of 80 billion U.S. dollars. A 1999 Canadian study set the annual cost at $500 million. The World Health Organization rates depression as the fourth leading cause of years of productivity lost and predicts that it will be the second overall greatest cause of lost years of productivity in adults by the year 2020. Already it is responsible for more disability than any other condition in the middle years of life and is at least as disabling as other general medical conditions. Studies also show that adequate treatment has a significant positive impact on societal consequences; for example, those with severe depression who received treatment reduced impairment days from 79 to 51 days a year with a corresponding decrease from 62 to 18 days a year for those with a moderate depression.

VII-13. *Other psychological conditions can be present* along with the depression and usually they are referred to as co-morbid conditions, although co-occurring might be a better term since the concept of co-morbidity implies that the two stem from different causes or processes, which may not be true. For example, in one recent study of those with depression, the authors found that 74% had one or more co-occurring symptoms and 32% had three or more. Of the latter, 57% had additional symptoms of anxiety and 39% were experiencing problems with substance abuse. Other studies have concluded that anxiety and anxiety disorders such as panic disorder, sleep disorder and disturbances, and post-traumatic stress disorder co-occur most frequently with depression.

*The reasons for the co-occurrence is not entirely clear at this time. Among the possibilities that have some support are (a) high prevalence rates for both conditions so that they would be expected to occur together just by chance, (b) some overlapping of symptoms, (c) the co-occurring symptoms cause such stress or disruption to the individual that depression results, (d) both conditions represent different aspects of the same illness, (e) the two conditions having a common underlying cause or process, and (f) one condition, such as dysthymic disorder, makes the person more vulnerable so that s/he develops a depression when faced with additional stress.

*Regardless of the reason, co-occurrences (a) increase the person's suffering, (b) increase the degree of social and occupational impairment, (c) increase suicide risk, (d) add complications to the treatment plan, (e) increase the duration of treatment, (f) often result in residual symptoms once treatment is completed, and (g) increase the likelihood of a relapse or re-occurrence after treatment is concluded. Fortunately, with appropriate psychological and other treatment of co-occurring symptoms, they can usually at least be decreased in severity and often a full remission can be achieved.

*As I read the professional literature, the relationship between depression and personality or personality disorders is not yet clear even though community samples have shown that more than 20% of folks with a depression also have a diagnosable personality disorder. At the clinical level, the results of personality tests are likely to yield distorted findings due to the mind set and cognitive impairments characteristic of depression. Also, many of the admitted negative personality characteristics seem to diminish with successful treatment of the depression. On the other hand, personality disorder seems to be one risk factor for the later development of depression and for increased rates of relapse for reasons not yet understood but which could include increased stress due to problematic personality styles. Studies on the impact of personality on treatment outcome have been inconsistent. My approach is to first treat the depression and then later determine if attention to personality characteristics is required.

VII-14. Because successful treatment and the prevention of relapse depends so heavily on attending to all relevant symptoms as well as other details such as those noted above, *my initial assessment_includes* the use of number of measures. Generally, I begin with a structured clinical interview reviewing such things as medical symptoms, family history, lifestyle, exposure to trauma, and current stressors as well as asking the client's partner to provide information. Second, the Beck Depression Inventory (II) is administered, which yields a depression score, and is re-administered from time to time throughout treatment as a quick way to determine progress. Third, I ask clients to complete standardized psychological tests including the Personality Assessment Inventory (PAI), The Beck Anxiety Scale, and the Beck Hopelessness Scale. Fourth, a psycho-physiological assessment is completed wherein muscle tension and characteristics of the autonomic nervous system are measured directly, as noted below. Fifth, clients complete the MicroCog, a computer-administered standardized test of attention, memory, and cognitive processing, functions that are often impaired in depression. In many cases these cognitive features improve with decreases in symptoms of depression. Finally, depending on symptoms, sometimes other assessment methods are added such as (a) those required for the assessment of sleep-disordered breathing such as obstructive sleep apnea and (b) a qEEG to determine electro-physiological abnormalities in the central nervous system.

VII-15. Once the assessment is completed, the client and I meet for an *initial information session*. The client learns about the results of the assessment, the diagnosis, and the causes and processes associated with his/her symptoms. This psycho-educational approach often is helpful in dispelling myths and other errors of thinking that continue to be addressed throughout our therapeutic relationship as needed. Then we review the treatment plan and the prognosis. Additionally, during this meeting we pay careful attention to safety issues as required, consider how emergencies could be handled, plan for whatever practical assistance is needed, address work issues, learn what help is available from relevant agencies and how to access this help, talk openly and honestly how best we can work together, and decide what (if anything) to tell friends and relatives.

VII-16. The structured clinical interview provides information that could lead one to consider if there might be a *genetic component* to depression, a tendency to inherit the susceptibility or vulnerability to develop depression. This component is estimated to be a factor in up to 70% of those suffering the most severe types of depression. For example, if a parent suffered from depression, the natural offspring are three times more likely to develop a depression than if family history is negative for depression. Among the best types of studies are those that compare identical (monozygotic) and non-identical (dizygotic) twins who have been adopted early in their lives and thus raised by people other than their natural parents. Results to date indicate that if one of a pair of dixygotic twins has a depression, the likelihood of the other twin developing a depression is 19% to 28%, whereas if one of a pair of monozygotic twins has a depression, the likelihood of the other twin developing a depression is between 44% and 58%. Thus, overall, the best current guess is that roughly 40% of depression is due to genetics and about 60% to environmental factors. However, these statistics apply to groups, so it is not yet possible to make conclusions at the level of the individual about genetic versus environmental factors responsible for depression. Also, exactly what is inherited is not fully known at this time. Fortunately, even if genetics play a significant role, the depression can still be treated successfully. Moreover, knowledge of a possible genetic link is helpful to clients both to decrease their personal sense of responsibility for their depression and to alert them to be watchful for signs of depression in their children.

VII-17. For reasons not yet entirely understood, a history of excessive use of alcohol increases the likelihood of subsequent depression fourfold even in the absence of other drugs. Also, in part since it is a central nervous system depressant, use of alcohol can result in symptoms of depression as well as making recovery from depression more difficult than it needs to be. Additionally, both *excessive alcohol use and even weekend use of drugs* such as marijuana, cocaine, and crystal meth result in irreparable brain damage to the frontal and pre-frontal areas of the brain, which are critical to such aspects as emotion, impulse control, and judgment. For some, alcohol is used as a way to de-stress and help with sleep onset and is accepted as an appropriate strategy by colleagues. Thus, my structured clinical interview includes questions about current and past use of alcohol and other drugs. If assessment indicates that they are a continuing factor, then the client is advised to discontinue their use; studies have determined that some of the largest decreases in depressive symptoms occurred following a period of abstinence. If clients are unable to abstain on their own or with the help I can provide with my limited training and experience in addictions, then we discuss options.

VII-18. The structured clinical interview provides medical information that is helpful since there can be a relationship between depression and *both medical conditions and some medications* used to treat these conditions.

*Symptoms of depression are more common among those with medical conditions than among those without medical problems. Some of the various medical conditions and their associated risks for depression as found in studies are noted below.
    *cancer: 24%,
    *stroke: 25% to 70% depending on site of stroke (e.g., 70% if left front),
    *renal disease: 5% to 22%,
    *diabetes: 9% to 27%,
    *respiratory diseases such as asthma and emphysema: 51% to 74%,
    *chronic pain: 35% to 50%,
    *epilepsy: 20% to 30%,
    *Parkinson's disease: 8% to 50%,
    *multiple sclerosis: 27% to 54%,
    *lupus: 43%,
    *HIV/AIDS: 4% to 22%, and
    *heart conditions: 20% to 30%.

*Disease characteristics including the chronic/life-threatening nature of some conditions (clearly a traumatic event in one's life), variability in the course of the disease with times of hope followed by times of despair, the uncertainties about both the process and the outcome, the intrusiveness of the illness in daily living, and the chronic pain/discomfort of the illness as well as some of the medical procedures are among the factors that can contribute to the development of depression.

*Additionally, there is reason to believe that some of the illnesses involve biological processes similar to those found in depression. For example, dysregulation of the sympathetic nervous system is found in depression and in some forms of heart disease. In fact, the consistent conclusion by the large number of published studies indicates a bi-directional relationship between depression and coronary disease; untreated depression is a risk factor for heart disease and heart disease increases the chances of a depressive episode. Moreover, sometimes the depression precedes signs of the medical illness; I have diagnosed a number of my clients with depression months and years before they developed conditions such as arthritis, diabetes, and multiple sclerosis. However, in these situations, it may be that the relationship between the depression and the medical problem was due simply to the base rates of the two conditions.

*In addition, various conditions of the endocrine system can produce some of the symptoms found in depression:
>   *male hypogonadism with loss of libido, low mood, and low energy;
>   *acromegaly with hypersomnia and loss of libido;
>   *Cushing's and Addison's diseases with dysphoric mood;
>   *hyperprolactemia with low libido;
>   *hyperthyroidism with anxiety, irritability, and psychomotor agitation; and
>   *hypothyroidism with depression and cognitive impairments. Although there do not appear to be conclusive data on the overall role of thyroid function on depression, a subgroup of those with depression do have subtle anomalies of thyroid levels. For such clients, there is reason to believe that adding thyroid medication to the anti-depressant may be helpful.

*Regardless, the combination of a serious illness and depression leads to a significant reduction in the quality of life and social relationships, a decrease of compliance in the treatment of medical conditions, a lowered probability of recognizing and thus treating the depression, added difficulty in treating the depression, and for some, a higher mortality rate. For example, those with both heart disease and depression have a three- to four-fold increase in mortality as compared with cardiac patients without symptoms of depression. For these and other reasons, both the medical condition and the depression need to be treated appropriately. Fortunately, regardless of co-occurring medical conditions, standard treatments for depression are still helpful.

*Also, several types of *medication* such as those below *can produce some* of the symptoms of depression:
>   *analgesics and anti-inflammatories;
>   *antibiotics and antifungals;
>   *blood pressure and heart medications;
>   *drugs used in the treatment of cancer;
>   *neurological and psychiatric medications;
>   *steroids; and

*hormones.

*If there is reason to believe that medical conditions and/or medications are exerting an influence on depression, then the client's physician is informed so that appropriate medical investigations can be done as required and subsequent medical treatment begun as needed. *Clients are encouraged not to discontinue any medications on their own* even if these medications are on the list above. Rather, they need to discuss the matter with their physicians as soon as possible and discuss options that can include switching to different medication or adding another medication to manage the side-effect of depression.

VII-9. A common professional opinion with some research evidence is that those with depression *have dysregulation of the neuro-transmitter system* from which the common term "chemical imbalance in the brain" emerged (although recent formulations looking at the same evidence propose that problems of information processing within the neural networks might underlie depression). Neuro-transmitters allow communication between one brain cell and another. If the level of these is insufficient in areas of the brain responsible for such components of depression as mood, then depression is the result. Current evidence indicates that, depending on the person, neuro-transmitters such as serotonin (and there are 14 distinct sub-types), GABA, nor-epinephrine, and dopamine are implicated in depression. Direct assessment of neuro-transmitters is rarely done due in part to the risk that the required procedures (such as cerebrospinal taps) pose to the client. However, studies (a) where anti-depressant *medications* have been prescribed and (b) where there is some evidence that they act on the neuro-transmitter systems often describe improvements in symptoms of depression over and above any placebo effect. Also, following medication, subsequent neuro-imaging methods have shown positive changes in key areas of the brain such as increases in prefrontal areas, decreases in the hippocampal area, and changes in the subgenual cingulate.

* However, recent studies completed in the U.S. indicate that:
     a. only between 9% and 25% of those with depression contacted their physician;
     b. of those, fewer than 50% were recognized as having a depression, in part because during their appointments patients complained only about chest pain, fatigue, dizziness, headache, back pain, or insomnia rather than the key symptoms of depression;
     c. of those recognized with depression, only 30% were prescribed the right medication as outlined in accepted treatment guidelines;
     d. fewer yet were prescribed a sufficient dosage and/or instructed to continue the medication over sufficient period of time;
     e. a significant number did not take the medication or did not take it as consistently as required;
     f. very few were referred to professionals with specific expertise in the psychological treatment of depression; and
     g. consequently, only about 40% recovered in 4 to 6 months when the criterion was just a 50% reduction in symptoms of depression. Fewer yet achieved remission.

A 2006 study among more than 3000 clients of the U.S. Veterans Administration found similar results. To date, I have not been able to locate comparable studies among Canadian physicians.

*The most common antidepressants approved for use in Canada include the SSRIs (Celexa, Prozac, Zoloft, Paxil, and Cipralex), Effexor, Wellbutrin, and Remeron. Generally, the SSRIs are tried first and typically both women and the elderly require a lesser dosage than men in their thirties to sixties. For a variety of reasons, about 33% of folks with a depression will not respond to the first anti-depressant tried.

*At this time, there are no sure methods for determining beforehand which medication will be effective for which client. Research is continuing on this issue with current hints that increasing serotonin levels reduces anxiety, increasing NE levels improves pleasure and motivation, tricyclics exert a positive effect on melancholic symptoms, duloxetine reduces subjective ratings of pain and other physical symptoms, Remeron increases sleep, and Wellbutrin increases energy, pleasure, and sleep usually in combination with an SSRI. Otherwise, the choice of where to start has been made on the basis of (a) previous experience (which anti-depressant was effective in a previous bout of depression), (b) which medication has been effective in first-degree relatives, and (c) which potential side-effects are desirable and which need to be avoided. However, a 2008 exhaustive review of medication commonly used in the treatment of depression determined that *Cipralex* and *Zoloft* outperformed the others.

*Any medication (including aspirin) can have side-effects and the anti-depressants are no exception. For example, tricyclics can worsen pre-existing heart conditions and SSRIs often result in sexual side-effects such as a loss of libido and performance difficulties that are sometimes permanent. Fortunately, tables are available that estimate the likelihood of various side-effects. Still, about 50% of folks stop taking anti-depressant medications within the first month because of side-effects. However, many of the possible minor side-effects go away within the first four weeks and often those that do not diminish can be managed either by adjusting the dosage, adding another medication, or changing the antidepressant. Except among those with a particular medical history, research to date with most adults has not demonstrated any long-term life-threatening side-effects associated with the anti-depressants noted above. However, due to insufficient studies with humans, generally physicians do not recommend antidepressants for pregnant or lactating women unless the mother's depression is so severe that it is likely to have a negative impact on the child (for example, the mother is suicidal or is seriously malnourished). Also, side-effects can occur depending on such factors as the simultaneous use of other medicines, but these can be minimized by reference to available information.

*Thus, in order to be on an effective medication at the right dosage and to manage side-effects as soon and as effectively as possible, it is recommended that clients keep in weekly contact with the physician for the first few weeks after medications are begun, regularly after that, and with each change in medication or dosage. Re-administration of the Beck Depression Inventory can provide helpful information to the physician.

*Some studies to date suggest that (1) medications be adjusted until the symptoms of depression are within the normal range (called remission), (2) that the same level of medication be continued for the next 9 months or so (called the continuation phase) if the current episode of depression is the first one (one review of 31 research studies concluded that doing so reduced the odds of a relapse by 70%), (3) that the medications be continued even longer if the person has experienced one or more previous episodes, and (4) that careful thought go into whether or not medications should be decreased gradually after the end of the continuation phase or be continued. A common recommendation for folks with three or more episodes of depression is that they remain on a maintenance dosage for life. However, generally such studies have not factored in the results of psychological treatment.

*Many physicians have adopted a step-wise strategy until the maximal effect is achieved although the various treatment strategies have yet to be evaluated thoroughly. A common approach is to:

> (i) start with an SSRI;
> (ii) maximize the anti-depressant effect by increasing the dosage as high as the client can tolerate as long as there are no serious side effects;
> (iii) if there is some positive effect but not maximum effectiveness and if the SSRI dosage cannot be increased, prescribe an augmenter such as lithium, which has proven to be very effective with the TCA type of medication;
> (iv) if the first SSRI is not effective, then switch to another SSRI medication and again as required increase the dosage and then augment as needed;
> (v) is still ineffective, then switch to a different class of medication;
> (vi) if still ineffective, then try combinations of antidepressants;
> (vii) if still ineffective, then try an anti-depressant in combination with an atypical neuroleptic medication (for example, Prozac plus olanzapine); and
> (viii) if still ineffective, then consider electro-convulsive therapy

(ECT).

*Dr. Tomarken and colleagues summarized anti-depressant medications in the 2007 book edited by Rottenberg and Johnson as follows:,

> Across medication classes, 45% to 65% of patients are responders, whereas 20% to 40% of patients respond to placebo. A closer examination of the data, however, reveals that the benefits due to the active pharmacological effects of medications are not large. If such benefits are quantified as the difference in response rates between medication and placebo conditions, it appears that only 20% to 25% of the patients administered a given anti-depressant benefit from its active pharmacological properties. Similarly, when outcome is quantified on a continuous metric, average effect sizes for drug-placebo differences are typically in the small to medium range. (p. 264)

*My own clinical experience with depression since 1985 is really quite mixed. Some of my clients have recovered from depression without medication, others have had a very good and often dramatic positive response to medication, still others have obtained no benefit from multiple trials of medication. Once psychological interventions are well in place, some clients have been able to decrease medication dosages sometimes to complete elimination, others have suffered a relapse when tapering medications and have needed to return to initial dosages, and some have done well on medications initially only to find that they were no longer effective after a period of several months (known as the poop-out effect).

*Given all of the above, my current approach is to discuss the information above with clients. Then it is between them and their physicians as to whether or not to attempt a trial of medication. If asked for my personal view, when the depression is in the mild to moderate range of severity and if clients can cope with the symptoms, then I suggest that clients consider holding off on medications while giving the psychological approach a try. On the other hand, if symptoms place the depression in the severe range of intensity, I suggest that the client consult with a physician for consideration of trial of medication while simultaneously we are putting psychological interventions in place – a practice that has much research support. During my conversations, I remind clients that I am not a psychiatrist whose medical speciality has to do with medications. Moreover, at this time no Canadian psychologist can prescribe medication. I do keep up in the research literature regarding medication management via membership in the American Psychological Association section on psychopharmacology, through reading of the medical literature, and through attending conferences.

VII-20. The structured clinical interview is helpful as issues of health behaviours are considered. This is important since there is evidence that factors related to what I call *stabilization strategies* can make a significant difference. Often, those with a depression do not engage in some of the critical elements of a healthy lifestyle and if these are adopted then depression levels often improve.

* some symptoms of depression are lessened if the person's diet contains sufficient amounts of the B vitamins (especially folic acid) or if these nutrients are obtained through appropriate supplements.

* it has long been known that depression is lessened if the levels of omega-3 polyunsaturated essential fatty acids are adequate and the omega-3 levels are much higher than the intake of omega-6 or omega-9. Omega-3 is selectively concentrated in synaptic neuronal membranes and regulates many functions that affect the central nervous system. A 2006 review by Freeman and her associates of all published studies and presentations at scientific meetings supported the value of omega-3 and specifically the omega-3 components known as eicosapantaenoic acid (EPA) and docosahexaenoic acid (DHA) in the treatment of depression. Omega-3 can be obtained only through diet and the richest sources of omega-3 are some fish (salmon, sardines, mackerel, swordfish, and

bluefish), walnuts, flax (cereal, seeds, and oil), cod liver oil, and canola oil. Omega-3 supplements are available but the consumer must insure they contain a much higher ratio of omega-3 to omega 6 or 9 (some suggest ratios of 2:1); otherwise they have little positive effect on depression. Weekly omega-3 amounts ranging from 7 to 11 grams seem most effective. In addition to its effect on mood, other data suggest that omega-3 plays an important role in the prevention of gastrointestinal conditions, rheumatoid arthritis, respiratory illness, cancer (breast, prostate, and lung), and various cardiovascular conditions (arrhythmias, thrombosis, and high blood pressure); also, it helps to sharpen memory.

*depression is decreased if intake from cane and corn sugar is low.

*the low caloric sweetener *aspartame* (nutrasweet) may increase symptoms of depression, chronic pain, and fibromyalgia since it is a excitatory neurotoxin. *Xylitol* may be preferable.

*depression is decreased if cholesterol levels are neither too high nor too low.

*depression is decreased with adequate daily levels of aerobic exercise, probably because it results in positive changes in brain chemistry including a decrease in "stress" hormones secreted by the pituitary gland. This has been a consistent finding in research studies. Moreover, one recent study indicated that among patients who are more severely depressed, the results of behavioural activation were comparable to the results of taking anti-depressant medication; behavioural activation included increased physical exercise.

*for some, depressive symptoms (often overeating, carbohydrate craving, weight gain, lethargy, and oversleeping) appear or worsen between November and April and sometimes if there is a succession of dark, cloudy days; this is known as SAD (seasonal affective disorder). Over 60 controlled studies have attested to the effectiveness of sitting in front of a SAD light over and above any placebo effect. The SAD light is most effective when used early in the morning for 15 to 30 minutes daily or, for a minority of clients, during both early morning and early evening. The device must filter out UV light and be at 10,000 lux in intensity. Side-effects reported by some include headache, eyestrain, nausea, and jumpiness, which can be minimized usually by strategies such as starting for short periods of time in October and increasing the time of exposure gradually or sitting a bit farther than usual from the light box. I have made arrangements so that SAD lights can be rented from me for a trial period and then purchased at my cost (about $150 versus the regular price of $200); I make no profit on this. Additionally, increased exposure to sunlight including being out in the early morning sun, bright inside light, high reflective surfaces in the home (such as white walls, light carpets, and light furniture), and taking winter vacations in sunny climates is helpful. In addition to those with SAD, I have found that many of my clients with depression also profit from the use of the SAD light during the winter months. A recent study confirmed my experience with folks whose depression occurred after a stroke.

*depression is reduced when the person can take advantage of many opportunities for laughter.

*depression is lessened if the person deliberately smiles more (even though not feeling happy); research shows that our external facial expression sends back signals to the brain that help us to feel the corresponding emotion.

*depression is decreased by the scheduling of pleasant activities or hobbies.

*depression is decreased with increased time spent with supportive family and friends.

*Consequently, issues of healthy living are assessed in my approach and corrective recommendations implemented as required.

VII-21. Once we have completed the above, the next aspect usually is an attempt to reduce current stressors in the person's life, since the presence of *stressors and demands above the person's coping capacity* can worsen a depression and also lead to a relapse. The most commonly understood situation is one in which the individual is burdened by stressors imposed from without such as chronic exposure to trauma, job demands, family problems, illness, and chronic pain. Moreover, studies and clinical experience indicate that for some folks with a depression, smaller and smaller stressors, some of which are simply characteristic of life in the twenty-first century, become sufficient to exert a negative effect far beyond the seriousness of the stressor. Our knowledge of the second situation, known as *stress generation*, has been the result of decades of very fine research by Dr. Constance Hammen at UCLA. Basically, she and her colleagues have demonstrated conclusively that some folks with a depression are more likely than those without the condition to actively but unintentionally generate stressful situations for themselves. For example, they can select partners who are rejecting or abusive, use child-rearing strategies guaranteed to be unsuccessful, pick jobs for which they are totally unsuited, hold beliefs that are not reality-based, be self-critical over their failed attempts at perfection, or not utilize effective skills in problem-solving and conflict-resolution. The net result is increased stress, which then increases the likelihood of recurrent or chronic depressive experiences.

*Regardless of whether is it externally imposed or self-caused, stress is not just something that we experience in our thinking or feeling. Rather, it involves a number of psycho-physiological processes, which themselves can worsen what is happening to us. For example, stress activates the central nervous system, causing the release of CRH from the hypothalamus, which leads to the release of ACTH from the anterior pituitary with the resulting increased secretion of cortisol from the adrenal cortex atop the kidneys. With the release of these stress hormones, other changes occur such that our body is prepared to respond by fighting, freezing, or fleeing. As reviewed in a 2005 paper by Burke and her colleagues, the hormonal response to even minor stressors is different between those with depression and those free of this condition. For example, folks with depression had much higher cortisol levels after the stressor was over (they remained stressed) and this pattern was most noticed in older clients and those with more severe depressions.

*The reactivity from stress hormones extends to other aspects of our biology, which require attention as noted below on the section dealing with biofeedback and the autonomic nervous system.

*Thus, as part of my assessment and intervention, stressors are identified early on in treatment and then effective methods to manage those stressors that can be controlled are implemented. Coping strategies are considered for those stressors that are beyond our control. However, sometimes the depressive condition and associated impairment in cognitive abilities are such that some stressors cannot be managed successfully at this point in the treatment program. If this proves to be the case, usually they can be addressed successfully after other interventions are in place such as biofeedback training for self-regulation of the autonomic nervous system (BFT-A), other aspects of cognitive behaviour therapy, or BFT-C for self-regulation of the central nervous system, all of which are described below.

VII-22. Studies have found that many with depression demonstrate hyper-activation of the HPA Axis, the hypothalamic-pituitary-adrenal feedback system. This is one of the most consistent findings in depression. Normally, in response to acute stress or threat, the neuro-hormone CRH is released from the hypothalamus in synergy with AVP and then transported to the pituitary gland. This results in the release of peptides such as corticotropin and endorphin from the pituitary. Then, corticotropin stimulates the adrenal gland atop the kidneys to release cortisol. In this manner, the autonomic nervous system is activated (see below), attention and other cognitive processes come on line while other processes become disengaged, fuel for energy is released, and large muscles become prepared for action. Thus, the person becomes ready to fight, flee, or freeze. When the stress is over or the threat removed, typically levels return to normal gradually. This is referred to as recovery or homeostasis.

*However, many studies indicate that recovery does not occur for those with depression and other trauma-caused conditions. The long-term net result is *dysregulation of the autonomic nervous system* (ANS) as indicated by such characteristics as elevated heart rate, elevated blood pressure, decreased peripheral blood flow, increased sweat gland activity (skin conductance), and decreased heart rate variation as compared to those without depression. Such characteristics are often present even during resting times (hyper-arousal), especially when experiencing or just thinking about a traumatic event or stressful life circumstances (hyper-reactive), and many times do not return to baseline levels (hypo-recovery) after the stress is over. All of these characteristics occur when there is an imbalance between two branches of the ANS: the sympathetic (arousal) and parasympathetic (relaxation).

*Furthermore, due in part to chronic elevation of the stress hormone cortisol, common long-term results can include gradual loss of minerals from bone, abdominal obesity, increased risk of heart disease, problems with memory (likely in part as a result of atrophy of the hippocampus brain structure), deficits in other cognitive processes, and hyper-activation of the amygdala (a brain structure associated with fear/anxiety).

*Hence, my assessment of depression includes the use of relevant psycho-physiological methods that I learned during post-graduate training and my two years of supervised practice, and for which I have the necessary (and expensive) high-tech equipment. The assessment involves placing electronic sensors on the surface of the skin without pain or embarrassment, recording activity levels of the ANS (see below) under a number of conditions, and displaying ongoing activity in real time on a computer screen.

*If the psycho-physiological assessment indicates ANS dysregulation, then the subsequent use of biofeedback-assisted training in relaxation/self-regulation (BFT-A) for identified sub-systems becomes part of the overall treatment plan. Well conducted studies have indicated clearly that the BFT-A techniques I use including *Tactical Breathing-1, 2, 3, and 4* are helpful in decreasing the symptoms of depression and their reactivity to stressors as well as having other positive effects.

In *Tactical Breathing-1*, while connected to the biofeedback equipment and observing their own breathing characteristics, clients learn to alter breathing paying particular attention to such features as type of breathing, rate of respiration, amplitude of each breath, level of oxygen saturation, and the form of each breath. Breathing the right way affects the vagus nerve, which in turn allows better activation of the parasympathetic nervous system. The end result is decreased hyper-arousal and hyper-reactivity as well as recovery after stress.

Tactical Breathing-2 is also known as RSA or RF breathing. In this training, clients are hooked up with sensors that measure respiration, other aspects of the autonomic nervous system, and a three-channel electrocardiogram so that the relationship between respiration and normal heart rate variation can be determined. Based on a lot of good research over the past 5 to 10 years, the objective is bring heart rate variation, blood pressure activity, and respiration into synchrony, which is associated with improved parasympathetic controls on the sympathetic branch of the autonomic nervous system. Studies have shown consistently that by doing so, reductions are achieved in symptoms of depression as well as of panic, PTSD, emotional arousal in response to stressors, chronic pain including fibromyalgia, blood pressure, irritable bowel syndrome, inflammation, asthma, and functional heart problems. In addition, studies have found increases in gas exchange efficiency in the lungs, increased overall oxygen saturation levels in the blood, increased cerebral perfusion of oxygen, and increased heart rate variability, vaso control, and autonomic flexibility.

Tactical Breathing-3 involves breathing in such a way that CO_2 levels during exhalation (ETCO$_2$) are within the range that results in regulation of the ANS.

*Usually in the initial training sessions, clients experience a noticeable sense of relaxation that follows from using these integrated breathing techniques. Moreover, they are able to see on the monitor for themselves that the changes in breathing result in positive changes in other measures of the autonomic nervous system.

*In-session training and then home practice of the techniques multiple times a day allow clients to live in a calmer state than before. With practice, the new breathing methods become more and more automatic. Moreover, they become easily extended to additional times of the day such as while standing in line in a bank, during television commercials, while on the telephone, while 911 runs the licence plate, during meetings, and so on. As one instructor of mine noted, "You have to breathe anyway, John. So why not do it right all the time?"

Tactical Breathing-4 is the use of the combination of 1, 2, and 3 above when stressors mount or the person begins to experience depression, anxiety, hyper-arousal, or hyper-reactivity. By the consistent use of this breathing during stressful time, a sense of calm returns and the connection between the stressor and the response is weakened.

*Extensive information on this topic is available in earlier sections of my book on PTSD.

VII-23. In contrast with those who do not suffer from depression, folks with depression usually experience one or more *sleep problems*. In fact, 80% of those with depression complain about sleep. Sleep problems at the very least increase the symptoms of depression and, for some and for a variety of possible reasons, pre-existing sleep problems seem to increase the risk for later depression. There is an increasing trend among professionals knowledgeable about sleep disorders to consider them to be a separate but co-occurring condition requiring specific treatment along with the depression rather than hoping that treating the depression will automatically lead to positive changes in sleep. Regardless, the evidence is clear that sleep must be normalized because evidence indicates clearly that doing so alleviates suffering and improves outcome. Research to date has determined that those with depression differ from those without as follows:

*more problems with insomnia such as taking a long time to fall asleep, more frequent night waking, more difficulty returning to sleep after night waking, early morning waking without being able to return to sleep at all, decreased sleep efficiency (total sleep time divided by total time in bed), waking feeling tired and unrefreshed, and daytime sleepiness or hypersomnia (sleeping for unusually large amounts of time during each 24 hour period);

*more problems with disturbing dreams as well as panic attacks during the night;

*decreased amount of delta brain waves (0.5 to 3.5 Hz) during sleep, especially during the first NREM period. During the night, delta waves are characteristic of being in a deep, refreshing sleep state;

*at the same time, increased incidence and/or amplitude of alpha and beta EEG activity during sleep, sometimes referred to as sleep intrusions. Such brain wave activity is characteristic of an awake state;

* beginning REM sleep (rapid eye movement sleep during which dreaming takes place) much earlier than the typical 90 minutes after sleep onset, as early as during the first few minutes of each sleep cycle and thus known as sleep-onset REM. Also, increased REM density (number of eye movements per minute of REM sleep) has been reported;

*decreased coherence in various EEG frequencies between left and right hemispheres of the brain (interhemispheric coherence) and in both fast and slow wave EEG activity recorded from the same electrode sites (intrahemispheric coherence). Women with depression are more likely to show low temporal coherence while men are more likely to experience reduced slow-wave activity or delta activity especially in the first non-REM period of sleep;

*greater EEG amplitude in the right hemisphere than the left during REM sleep; and

*EEG sleep differences increasing as a function of age.

Interestingly, most of the EEG differences are true also for clients with other psychological conditions. However, the facets most characteristic of depression as compared with other conditions are the increased amount and percentage of REM sleep and decreased phasic REM activity.

Moreover, like other folks, those with a depression may experience any of a number of specific sleep disorders. *One is sleep disordered breathing* such as obstructive sleep apnea wherein clients stop breathing for significant amounts of time throughout the night due to collapsing of upper airway tissues. After a period of time, corrective processes are initiated automatically and include a brief period of awakening. Clients can have hundreds of micro-awakenings in the night without being aware of them. In this way, sleep is disrupted. *Another sleep disorder is periodic limb movement* where jerky movements continue throughout the night, usually accompanied by micro-arousals. *A third is restless leg syndrome* wherein clients make frequent adjustments in order to get their legs into a comfortable position. As many as 53% of those attending sleep clinics have high rates of depressive symptoms. The net result of any sleep disorder is a decrease in restful sleep with at least the increased potential for worsening of depression.

*Currently, I assess sleep characteristics initially by analysis of information provided by clients and their bed partners. Then, depending on findings, a

number of options become available in my aggressive approach to resolving sleep problems as quickly as possible:

*specific suggestions for restful sleep (sometimes referred to as sleep hygiene);

*decreasing autonomic nervous system dysregulation with tactical breathing;

*use of a device (AVE) that produces restful brain waves;

*consideration of appropriate medication; and

*referral to a clinic for a complete investigation of sleep known as polysomnography (PSG).

*More information about sleep is available in another section of my PTSD book.

VII-24. Many people with depression also have beliefs (about the world, others, and self), *thinking styles,* and *other aspects of cognitive functioning* that are not working for them. For example, they may have a very low opinion of themselves, may be constantly self-critical, may fail to see any positives in life, may see the world as always frightening, may have no consistent effective strategy for solving problems, may have no goals consistent with their core values, may process information on the basis of assumptions that are not reality-based, may not know how to develop personal contentment and serenity, may be unaware of their strength of character and talents, may avoid contributing to positive institutions like the family, or may not be involved in other activities that provide meaning to life.

*It is still not known if these factors are the cause of or the result of the depression. However, research suggests that modifying the unhealthy thinking (another component of CBT) can have a positive effect on depression. Moreover, there is evidence of positive EEG changes with CBT.

*Thus, on a session-by-session basis I am alert to signs of unhealthy thinking. Then, as required, CBT is included in the treatment regime within our sessions as issues arise. While some psychologists put most of their treatment emphasis on CBT, I have not found this to be effective, perhaps because of the high intensity of symptoms my clients have in relation to the generally lower levels of depression found among clients in the CBT research literature. This view was supported by a recent article in a professional journal that evaluated all of the studies using CBT in the treatment of depression; the authors' conclusion was that CBT works well for those whose depression is of lower severity but is not effective as a stand-alone method of intervention with those whose depression is more severe. Given that many of my clients have depression scores in the higher ranges as well as PTSD, CBT is but one of the components of my treatment plan.

*Also, as needed and when they are ready to profit from them, my clients have found that several read-and-do self-improvement CBT books I have recommended are helpful, depending on characteristics of their depression. This approach is sometimes referred to as *bibliotherapy* and at least 17 good research studies have demonstrated that it is a reasonable way to proceed. The books I recommend most commonly are listed below. Generally, clients work through the first four in the order below and others are added as required. All but the first one come in paperback.

The origin of everyday moods: managing energy, tension, and stress. (1996). Dr. Robert Thayer. Oxford University Press, NY. Hardback ISBN 0-19-508791-7. Reviews mood origins you have some control over and how best to achieve optimal mood.

Don't believe it for a minute: forty toxic ideas that are driving you crazy. (1993). Drs. A. A. Lazarus, C. N. Lazarus, and A. Fay. Impact Publishers; San Luis Obispo, CA. ISBN 0-915166-80-2. Describes unproductive or counterproductive thoughts and then reviews strategies to change them.

When perfect isn't good enough: strategies for coping with perfectionism. (1998). Drs. M. Antony and R. Swinson. New Harbinger, Oakland, CA. ISBN1-57224-124-1. Looks at the trait of perfectionism, provides a way to assess it in your life, and considers ways to change it when it is unhealthy.

Don't take it personally; the art of dealing with rejection. (1997). Dr. Elayne Savage. New Harbinger, Oakland, CA. ISBN1-57224-077-6. Considers sources of the feeling of being rejected, looks at why they have a negative impact, and then provides strategies for making changes and moving on.

Your perfect right: assertiveness and equality in your life and relationships, 8th edition. (2001). Drs. R. Alberti and M. Emmons. Impact Publishers. ISBN 886230-28-5. Excellent for those who have difficulties with appropriate assertion. Considers why people have difficulty, looks at why assertion (not aggression) is okay, provides strategies to practice, and then help others deal with the new, assertive you.

Six keys to creating the life you desire: stop pursuing the unattainable and find the fulfilment you truly need. (1999). M. Meyerson and L. Ashner. New Harbinger, Oakland, CA. ISBN 1-57224-125-X. The title says it all.

VII-25. Research indicates that *with appropriate traditional treatment*, (which usually is a combination of medication and CBT or interpersonal therapy), about 40% of those with a depression recover within 3 months, 60% within 6 months, and 80% within one year. Also, at this time depression is now commonly considered to be a condition that comes and goes throughout the life span; between 40% and 85% of those experiencing a first depression will have an average of four additional episodes. Studies have demonstrated that 60% to 68% will have a second depression within 5 years, 75% to 80% will have as second bout within 10 years, and 87% will have a second one within 15 years.

*Researchers have discovered that those whose onset of depression occurred at an early age, or whose depression lasted for a long time, or whose symptoms were more severe, are more likely than their counterparts to have a more chronic course of the condition as well as less likelihood of a complete recovery.

*With each episode, the chances of another one increases and the time between episodes decreases; an episode may occur while the individual is still on the medication that once was effective. This is known as a relapse or a *break through depression,* which usually happens very slowly so that often the client does not notice it until it reaches a certain level of severity.

*Also, there is some evidence that each successive episode of depression is more severe than the previous one as well as being more difficult to reverse.

*However, the *chances of a relapse/breakthrough* are lessened to the extent that (a) all of the symptoms of depression are managed successfully (the risk of relapse/re-occurrence increases to the extent that some symptoms remain untreated), (b) all treatment methods successful in managing the previous episode of depression are maintained consistently throughout life (dropping elements of the original treatment plan increases the chances of a relapse/re-occurrence), and (c) regular follow-up sessions are instituted that serve to help maintain intervention plans to date, to manage successfully any new stressors that arise, to detect signs of relapse and facilitate aggressive early intervention, and to consider new methods of treatment developed since the end of the client's intervention phase.

However, these statistics were obtained before the use of biofeedback training noted above, before the approaches to sleep and nightmares I now use, before MBCT, which was devised to prevent relapse (discussed later), and before other interventions described below.

VII-26. *Most clients begin to experience signs of recovery in 4 to 10 weeks.* However, progress often seems painfully gradual and there are ups and downs throughout the recovery period; for example, a short period of feeling good is followed by a longer one of feeling bad again and some time later a good day is followed by a few not-so-good days. Thankfully, over the long run, on average the good days begin to outnumber the bad ones, the bad days are not as bad as they once were, and the time between bad periods gets longer and longer. This is the normal pattern of recovery. Unfortunately, recovery can be very slow, depending on complicating factors such as whether the person experiences a double depression, whether symptom onset was slow rather than rapid, whether the symptoms are in the severe range, whether symptoms were present for a long time before being recognized, whether the first episode began during adolescence, whether there are co-occurring medical or psychological conditions, whether the client does not respond to a trial of a number of medications, and whether the person experiences more stressors that s/he can tolerate during the treatment period .

VII-27. However, after implementing the above, which I refer to as my basic approach, some of my clients do not achieve remission of their depression. In 2007, I collected data on all referrals between 2006 and 2007 who had a diagnosis of major depressive disorder without clear signs of any accompanying diagnosis of such as PTSD. Clients included police officers, 911/dispatch operators, and civilians. Some had taken medications as prescribed and some had not. For all clients, symptom severity was reduced, and for 77%, complete remission was achieved. This means that the basic methods were not successful for about 23% of the folks I saw.

VII-28. Thus, beginning about 2000, I began searching for evidence that yet another component might be helpful for those who had yet to achieve a remission of their depression with my basic approach. As a natural outgrowth of my interest in both psycho-physiology and neuropsychology, I became more aware of the new professional speciality of affective neuroscience. This field is concerned with the relationship between brain functioning and emotion.

VII-29. Research in affective neuroscience has demonstrated that *anatomical differences in the brain*, usually determined postmortem, were found between those with and those without depression. However, as yet it is not clear why such differences are found. For example, they could be due to the cause of the condition, the result of the condition, the consequences of repeated depressive episodes, the result of medications, or matters incidental or irrelevant to depression.

VII-30. Also, there is evidence that many of the areas of the brain with such anatomic abnormalities are the same areas noted in studies of functional brain abnormalities noted in #31 and following sections below. Among the most consistent anatomical findings among those with depression to date are:
>*reductions in size and density of neurons in the dorsolateral and orbital prefrontal areas of the brain as well as the anterior cingulate;
>*increased neuronal density, decreased neuronal cell body size, and increased packing density in the hippocampus;
>*other abnormalities in neurons related to monoaminergic, glutamatergic, and GABAergic neurotransmitter systems;
>*reductions in the density and number of glial cells in the anterior cingulate, dorsolateral pre-frontal cortex, and orbitfrontal cortex; and
>*impaired neuroplasticity and resilience.

VII-31. In addition, there exists considerable professional literature indicating *functional brain abnormalities* among those living with depression that are not present in those without the condition. Some of the differences were revealed in studies using *functional neuro-imaging* such as PET, SPECT, and fMRI brain scans. For example, there have been many research reports as well as presentations at professional conferences demonstrating that folks with depression show decreased cerebral blood flow, decreased use of glucose and oxygen in specific brain processes, and reduced cortical volume.

VII-32. Also, quantitative EEG (*qEEG*) methods have revealed differences in brain wave characteristics such as amplitude and connectivity between those with and those without depression. The differences are found in particular parts of the brain that one or more studies have indicated are related to depression (such as the orbital pre-frontal cortex, the ventro-medial pre-frontal cortex, the dorso-lateral pre-frontal cortex, the hippocampus, the amygdala and related parts of the striatum and thalamus, and the anterior cingulate cortex) as well as other locations not formerly thought to be associated with depression (such as central and parietal areas of the brain). None of the functional brain abnormalities noted above should be confused with major neurological conditions such as brain tumours, aneurisms, and neurological diseases that are investigated by neurologists, nor are they life-threatening.

VII-33. Many studies have been published in respect to a particular type of functional abnormality. With more than 25 years of experience in the field, Davidson and various colleagues reported that an aspect variously described as positive emotions, approach motivation, or goal-directed action planning is associated with more amplitude of beta to alpha waves in the left frontal area of the brain. In contrast, negative emotions and avoidance motivation were related to more amplitude of beta to alpha waves in right frontal area of the brain. The importance of this research is the consistent finding that, compared to those without, people with depression have little in the way of positive emotions and approach behaviour while at the same time experiencing high levels of negative emotion and high rates of behavioural avoidance.

Moreover, several subsequent studies indicate that those with depression have excess alpha power on the left side of the brain and insufficient alpha power on the right side (a combination known as alpha asymmetry). Furthermore, a number of publications have demonstrated that BFT-C (biofeedback training based on the EEG signal discussed later) aimed at altering this EEG pattern resulted in a decrease in symptoms of depression. Moreover, a follow-up study by Baehr in which depression had been reduced by such alpha asymmetry BFT-C indicated that the positive results maintained over a three-year period. Davidson's current view is that the location of EEG sensors in the alpha asymmetry approach most likely is measuring activity in the dorsolateral sectors of the pre-frontal cortex.

Relatedly, a number of studies from Davidson's affective neuroscience lab have demonstrated that individuals with more activity in the left prefrontal cortex have lower levels of the stress hormone cortisol, lower levels of CRF, and higher levels of natural killer activity in the immune system, they recover more quickly following a negative event, and they report higher levels of psychological well-being than those with less activity in that brain region.

In the same vein, Dr. Rybak and colleagues have found that relatively greater left frontal brain activity (reduced alpha waves) than right frontal brain activity was associated with higher levels of behavioural activation but led to increases in aggression. This was confirmed in a recent review article by Harmon-Jones. Davidson noted that the form of anger demonstrated might well facilitate the rapid removal of obstacles that are thwarting goals but that further research is needed to clarify the findings to date.

In the 2007 book edited by Rottenberg and Johnson, Santerre and Allen summarized the asymmetry research as follows:

> Findings from more than 40 studies suggest that resting frontal EEG activity may serve as an indicator of a trait-like diathesis to respond to emotional situations with a characteristic pattern of emotional negativity and behavioral withdrawal. Studies assessing resting EEG activity reveal that relatively less left than right frontal brain activity characterizes individuals with depression both when symptomatic and when well...Such findings suggest that frontal EEG asymmetry may be more than simply a correlate of a depressive episode, potentially serving as a liability marker for the development of depression or other emotion-laden psychopathology . . . . (p. 67)

Tenkeab and colleagues studied grandchildren in families with and without a history of depression and found that children in the high-risk group showed greater alpha over the right parietal areas than the left as compared with grandchildren with low risk for depression. There were no significant differences between the groups over frontal and central sites.

VII-34. Other studies of folks with depression have not found evidence of alpha asymmetry. Some have found other functional abnormalities as described below. In my own work with members of the police community whose depression appeared due primarily to duty-related accumulated trauma, I have yet to find a single instance of alpha asymmetry but have found a variety of other functional abnormalities based on analysis of the client's qEEG by the NeuroGuide and LORETA spftware. Also, during one seminar at a professional conference, it became clear that those suffering from depression can have one or more of at least 6 different abnormal EEG patterns. Moreover, these colleagues reported that the symptoms of depression decreased only if the treatment plan altered the individual's specific EEG pattern.

Holmes and his associates demonstrated that the sensitivity to negative feedback and errors typical of folks with depression are related to reduced gamma activity (36.5 to 44 Hz) in a brain area known as the anterior cingulate cortex. Also, the response of various areas of the brain have been studied after sad stimuli are presented. Those without depression show increases in the ventral and subgenual areas of the cingulate while those with acute depression or a history of depression demonstrate increases in the dorsal cingulate as well as in medial and orbital areas of the frontal cortex.

An intriguing 2004 study by Graff-Guerrero and colleagues measured cerebral blood flow with SPECT among those with depression in relation to scores they had obtained on the Hamilton Rating Scale for Depression. They found that the dorsolateral frontal cortex was associated with depressed mood, late insomnia, somatic anxiety, and gastrointestinal symptoms. The insula was involved in symptoms of agitation and weight loss. Guilt and anxiety were correlated with the anterior and medial cingulate cortex. Middle insomnia and the left amygdala were related.

The 2008 literature review by Drevets and co-authors of an area of the brain known as the subgenual ACC confirmed (a) reduced grey matter volume associated with a decrease in glia cells at post-mortem among those with depression, (b) increased metabolic brain activity as determined by PET scans when comparing before and after scores of those whose depression had improved, (c) reduced activity in this brain area after effective anti-depressant treatment, and (d) decreased levels of deep brain stimulation.

At professional conferences, colleagues have reported that BFT-C at other brain locations and/or frequencies were successful in decreasing the symptoms of depression. Increasing amplitude in the 13 to 16 Hz range or in the 16 to 20 Hz range in the left frontal areas was reported to be successful. Still others reported success by decreasing amplitude in all frequencies between 1 Hz and 30 Hz at brain locations T3 and T4 (temporal lobe) while simultaneously increasing the amplitude of that frequency the client found to be most pleasant. Another group indicated success by increasing coherence (theta or alpha or beta) at one of three pairs of sites (C3-C4, F3-F4, or F7-F8).

At the Annual ISNR Conference in 2008, a large, well-controlled study by Dr. Beauregard's group in Montreal reported significant decreases in depression when treated by BFT-C protocols based on NeuroGuide evaluation of the quantitative EEG. Patients varied from one to the other in terms of both the location of differences from the norm as well as the frequency range of these differences.
In another well-controlled study, this research group identified electromagnetic abnormalities via the LORETA software among those with depression. They reported remission of depression but only among those where the BFT-C was successful in normalizing EEG activity in cortico-limbic-paralinbic regions of the brain. Also, they reported a co-relation between the percentage of change in depression symptoms and the percentage of reduction in high beta activity.

As of my last check, there are now hundreds of research articles relating EEG to depression.

VII-35. Given the emerging evidence that BFT-C might be helpful in the treatment of depression, early in the twenty-first century I provided BFT-C to my police clients who had not achieved remission of their depression after completing my basic approach. This involved using the hardware and accompanying software of the DeyMed Diagnostic System upon which I had trained.

The first step is to explain things to the client's satisfaction. Information above is mentioned as well as the fact that BFT-C has been judged to be "probably efficacious" in the treatment of depression and is well established as an effective treatment for such conditions as attention deficit, seizures, and some aspects of brain injury. In addition, I note my results. To the end of 2009, my clinical experience to date is that all but two who have completed about 25 sessions of BFT-C are in remission; it was not effective for one person and a second had depression symptoms return six months after ending successful treatment. Very few have dropped out of treatment. As well, I indicate that studies to date have not shown any negative side-effects of BFT-C as long as the client is not prone to seizures or psychosis, s/he is not taking any illegal drugs, the treating professional has received adequate training/consultation, and any effects are properly monitored on a session-to-session basis by that professional.

Upon receiving client consent to proceed, the second step is to complete a 19-channel qEEG (quantitative EEG). The client simply sits for about 15 minutes with eyes open and then another 15 with eyes closed while the recording is being from each of 19 sensors placed on the scalp. The system records EEG activity 128 times a second over 30 frequency ranges from each of the 19 sensors.

In the third step, I send the resulting digitized data to an acknowledged expert for analysis and recommendation. Until his untimely death, I worked with Dr. J. Horvat for about 18 months, which included a 1-hour telephone consultation for each client. Currently, I web-mail the digitized information to Dr. R. Thatcher, a world expert in EEG with hundreds of publications in the area. He does a visual inspection of the data and removes electronically any artifacts (such as muscle movement) from the data before processing it through two software programme. One is the normative NeuroGuide which he developed which he developed. It specifies aspects and brain areas of EEG dysfunction in comparison to the normative group. The other is LORETA which assists further in localizing the source of the dysfunction in three-dimensional space.

The fourth step is the resulting "brain maps" that Dr. Thatcher e-mails to me with analysis and recommendations. Subsequently, a BFT-C plan is developed based on the combination of factors such as the brain maps, the specific symptoms of concern, the current knowledge of brain functioning in regards to these symptoms, and with reference as needed to approaches others have found to be effective.

VII-36. Once BFT-C is underway, ongoing evaluation is part of my approach including client verbal reports, comments by significant others, my clinical observations, changes in brain wave amplitudes and/or connectivity observed during BFT-C sessions, standardized psychological testing, and periodic repeats of the qEEG.

VII-37. Yet another indicator that depression is the result of psycho-biological processes beyond the conscious control of clients is the evidence from *ERP research*. When there is a stimulus such as a novel sound, there are measurable normal changes in the brain's initial EEG response known as an evoked response potential (ERP). This is related in part to activation of attentional processes such as awareness of the event, orienting to it, and devoting brain resources to it. An early part of this sequence known as P3 or P300 is a positive deflection in EEG that occurs about 300 milliseconds (about one-third of a second) after the stimulus. As contrasted to those without depression, one frequently reported finding is that those suffering from depression have a delay of more than 300 msec before signs of the P3 wave. A 2006 research study concluded that the demonstrated reduced fronto-central P3 response likely reflected a deficit in orienting and alerting among those with depression, a finding consistent with clinical experience. I am unsure of the practical value of this finding other than at some time I would like to be able to determine ERPs both before and after treatment. This might provide an objective measure of a fundamental change in characteristics of depression.

VII-38. The availability of *supportive relationships* is one key to good mental and physical health. Sometimes, clients do not experience such positive relationships for reasons that can include interpersonal role disputes, role transitions, or lack/loss of key interpersonal skills required to develop and sustain effective relationships. Generally, I refer clients with such issues to another psychologist in the community who specializes in interpersonal therapy since this not an area of expertise for me. Research to date indicates clearly that improvements in interpersonal areas requiring attention are effective in decreasing the symptoms of depression. When the issues are of major importance, the difficult decision must be made as to timing of interpersonal therapy; for example, after completion of my approach or at the same time.

VII-39. In this section, information is presented regarding other specific interventions that may or may not prove to have a role to play in the treatment of depression. They are placed here because my view is that there is insufficient data to evaluate their potential contribution. For some, there exist encouraging *preliminary data* and for others I am not aware of any data having been presented.

*In studies the date, two health food remedies have been found to be effective in the treatment of depression. A meta-analysis of all studies using *Sam-e* (S-adenosyl-methionine) found that effects were comparable to those of the tricyclic anti-depressants. Also, meta-analysis of clinical trials as well as several double-blind studies indicate that *St. John's wort* was effective among those with mild to moderate levels of depression and had fewer side effects than medications. However, a difficulty with health food remedies is that the products are not standardized. Thus, from one batch or brand to another you do not know how much of the allegedly effective ingredient is present nor what other ingredients have been added. In addition, there is good reason to believe that St. John's wort has a negative impact on both cardiovascular medications and those used to fight bacterial infections. None of my clients report having tried Sam-e . The few times my clients with moderate to severe depression have chosen to try a course of St. John's wort, no positive effects on depression have been noted.

*There is good evidence that estrogen influences brain functions in depression. However, there is insufficient evidence that depression levels are changed by prescribing this hormone, including to women in the postnatal or perimenopausal stage.

*There is preliminary research data that *mindfulness-based cognitive therapy (MBCT) for depression* can decrease both the symptoms of depression as well as the chances of a relapse. As an extension to the demonstrated success of the approach in the treatment of anxiety and chronic pain initiated by Kabat-Zinn while he was professor of medicine at the University of Massachusetts Medical School, MBCT involves Eastern meditation strategies directed at keeping one's mind calm and focussed on what is happening in the here-and-now (as contrasted with a mind that is all over the place, known in the Buddhist literature as "monkey-mind"). As applied to depression, the strategies include staying in touch with one's own bodily sensations, feelings, and thoughts as a non-judgmental observer. One aspect involves noting the depression-producing negative self-judgments that flit through consciousness at amazing speed. The next step is not getting caught up in the content of such thoughts, neither grasping at them nor avoiding them. The final step involves gently returning attention fully to the act of breathing. As a long-time practitioner of Eastern meditation and philosophy, I have found great personal benefit in this discipline, which is consistent with the considerable published research. Thus, I am able to guide interested persons in this approach. For those deciding to pursue this approach, as a start I recommend the 2007 publication *The Mindful Way through Depression: Freeing Yourself from Chronic Unhappiness* by Williams, Teasdale, Segal, and Kabat-Zinn (Guildford: paperback ISBN 13-978-1-59385-128-6), which includes a CD containing examples of practices. I can provide additional written resources as requested on the understanding that mindfulness meditation is something that you do rather than something that you read about.

Vagus nerve stimulation (VHS) involves surgically implanting a power generator the size of a pocket watch into the left chest wall, which delivers low-frequency chronic intermittent pulsed electrical signals to the left cervical vagus nerve (cranial nerve X). The vagus nerve travels to the larynx, esophagus, trachea, gastrointestinal organs, heart, and aorta where as part of the parasympathetic nervous system it allows for regulation of the autonomic nervous system. Afferent fibres carry sensory information to the dorsal medullary complex in the brain stem and especially the NTS portion. From there information is relayed to other structures in the brain including the parabrachial nucleus (PBN), cerebellum, raphae, periaquaductal gray, locus coeruleus, limbic and paralimbic systems, and cortical regions. The projections to the locus coeruleus and raphae nuclei involve the availability of norepinephrine and serotonin. The PBN relays information to the hypothalamus, thalamus, amygdala, and nucleus of the stria terminalis. Projections from the NTS reach structures that are associated with the regulation of mood and emotion, seizure activity, anxiety, intestinal activity, and pain sensitivity. Since 1994 in Europe, VHS has been used successfully in the reduction of seizure frequency in patients with refractory, partial-onset seizures. Among the related effects of VHS has been de-synchrony of EEG over a wide area of the scalp, increased cerebral blood flow to some areas (to hypothalamus, insular cortex) and reduced blood flow to other areas (hippocampus, amygdala, and posterior cingulate). The idea of using VHS in depression began with observations of mood improvement in patients with epilepsy who had received this intervention. In my view, the results of VHS are still preliminary and have been mixed. In one uncontrolled pilot study, 30% of clients with treatment-resistant depression showed a decrease of 50% or more on the Hamilton Scale for Depression while 15.3% achieved remission. One year later, the response rate had increased from 30.5% to 44.1%. A second study, completed with treatment-resistant clients who were maintained on anti-depressant medication, found Hamilton depression score decreases of 50% or more in 15.2% of the VNS group and in 10% in those receiving a sham in the control group. Given that starting Hamilton scores were 29, the low rate of 15.2% score change, and the relatively minor outcome differences between VNS and sham control procedures, one reviewer concluded that short-term efficacy of VNS was not demonstrated. Data from the second study at 12 months indicated that differences between the VNS group, which now included other treatment methods, and the control group who received usual treatment methods without VNS were not significant: response rates were 19.6% versus 12.2%. Nevertheless, although the rationale for which remains a mystery to me, regulatory agencies in Europe, Canada, and the U.S. successively have approved VNS in the treatment of depression resistant to other forms of intervention.

*In *repetitive transcranial magnetic stimulation (rTMS)*, a powerful magnetic field is created by pulsing high-intensity current through an electromagnetic coil placed on the scalp. The non-painful current does not cause a seizure and is able to penetrate the skull and induce secondary currents in the brain area under the placement of the device. The result is a depolarizing of neurons, thereby increasing or decreasing brain excitability at that location for a period of time afterwards. The head location used most often is the left prefrontal cortex, generally using currents at or greater than 10 Hz. Many studies to date have not supported the efficacy of rTMSs. Furthermore, where changes have been noted, the clinical benefits have been relatively minor.

*A 2006 report by Kosel and colleagues outlined their use of *magnetic seizure therapy (MST)* for clients with depression who had not responded to either anti-depressant medication or evidence-based psychotherapy. The procedure uses rTMS with stimulation intensity and frequency much higher than in rTMS so as to produce a seizure intentionally but in a more focal form than ECT (electroconvulsive therapy), with the hope of having fewer cognitive side effects than ETC. Their well designed study with the first 10 clients indicated that MST was superior to ECT in that it had fewer side effects and clients experienced faster recovery of orientation, better attention, and less memory loss after the procedure. Unfortunately, the efficacy of the technique in regards depression was not presented. However, ECT has been found to be effective for some patients with treatment-resistance depression.

*In a method known as *deep brain stimulation (DBS)*, a thin electrode is inserted directly into the brain and different currents are applied at various depths. This evolved from the treatment of Parkinson's disease since mood changes were noted. Accordingly, the hope emerged that DBS might assist those with treatment-resistant depression. To date, only a few pilot studies have evaluated the benefit of DBS. Using the criteria of 50% or more reduction in Hamilton depression scores as a criterion for success, both Greenberg and colleagues in 2003 with a case series of 5 clients and Mayberg and associates in 2005 with a case series of 6 reported success. Greenberg and associates reported an average drop of scores from 31.4 to 15.8 while the Mayberg group indicated that 5 of the 6 met the success criteria. The downside is that the required neurosurgery poses considerable risk for intracranial hemorraging, infection, and death.

*A variety of *other medicines* have been administered in the treatment of depression. For example, a search of the experimental literature completed in 2006 by Dr. Young suggested encouraging improvements in symptoms of depression and some areas of neuro-cognitive function with a class of medications known as anti-glucocorticoids, although this group of medications has yet to be approved by federal authorities for the treatment of depression. Other not-yet-approved medications that have some empirical support include bromociptine, beta blockers such as propanalol, and mifepristone.

In summary, it is clear that new methods are evolving regularly in the hopes of finding improved methods of treating depression (especially when it is resistant to more traditional strategies) and that a number of these newer methods are consistent with or extensions of existing methods. For example, *MBCT* is an extension of cognitive behavioural therapy, *VNS* is consistent with *Tactical Breathing-2*, and *rTMS* is consistent with the early research on *BFT-C*.

VII-40. Sleep and Depression

Studies show that up to 90% of folks with a depression will have a sleep disorder that can be verified by EEG tracings. Research regarding the interplay between sleep and depression is still in its infancy, particularly in regards to the nature of the cause-effect relationship. For example, some of the strong findings to date include the following:

a) sleep disturbances affect mood in both depressed and non-depressed persons;

b) depression is one of the two most common causes of insomnia;

c) sleep disturbances may precede the development of depression;

d) some sleep disturbances are resolved once the person recovers from the depression;

e) some sleep disorders do not resolve even after the depression is treated; and

f) depending on the type of antidepressant used, there is a wide range of effects on sleep (some positive and some negative).

As a result, the growing consensus among experts is that sleep disturbances/disorders definitely require specific treatment in addition to effective intervention for the other symptoms of depression. For example, adding an appropriate sleep medication for insomnia may prove helpful if the combination of (a) the implementation of proper sleep hygiene, (b) biofeedback for improved autonomic nervous system regulation (BFT-M), and (c) BFT-M for reducing muscle tension is not effective. Furthermore, a medication such as *trazodone* is often effective when the objective is to alter certain aspects of sleep architecture or to manage sleep disturbances that sometimes can be side-effects of some anti-depressant medications. The use of combinations of medications in other medical conditions (such as cancer and some heart conditions) is not unusual and so should not be unexpected in the treatment of a complex condition such as depression.

Below are listed some of the findings to date. They indicate how complex this issue remains. Some of the contradictory findings may be due to differences in the data based on client's verbal reports about their sleep versus the results obtained when objective measures are utilized, as in a sleep clinic.

A. Many studies have noted that adults with a certain combination of depression symptoms and anxiety have sleep patterns that are very different from the typical pattern, with some studies noting EEG abnormalities in up to 90%. For example, Drs. Ware and Morin provided this chart in the book *Understanding Sleep*:

Sleep Parameter	Normal	Anxiety	Depression
Sleep Onset Latency	5-15 Minutes	more	More
Wake after sleep onset	0-15 minutes	more	More
Total sleep time	6.5-8 hours	less	Less
Sleep efficiency (more fragmented sleep)	90% +	less	Less
REM latency	70-90 minutes	same	less (shorter latency)
% Stage 1 sleep	5-10%	more	Same
% Stage 2 sleep	50-60%	more	Same
% Stages 3 & 4	10-15%	less	Less
% REM sleep	20-25%	same	More

B. Some studies have found REM abnormalities to normalize once the depression is treated successfully and other researchers have not found this to be the case. Results from many investigations support the conclusion that having REM abnormalities may indicate an increased vulnerability to depression.

C. Reduced REM latency and a selective decrease of slow-wave activity during the first NREM period has been associated with higher than normal rates of relapse/re-occurrence after recovery from a depression. Reduced REM latency in relatives of individuals with depression has been reported to triple the risk of developing a depressive illness compared with normal REM latency.

D. In comparison with controls, some depressed patients have increased global cerebral metabolism during the first NREM sleep as well as relatively greater right hemispheric metabolism during REM sleep.

E. While both men and women with a depression show reduced slow-wave sleep compared to controls, depressed men have less slow-wave sleep and lower delta wave counts than depressed women. Both genders with major depressive disorder show lower amplitude of slow-wave activity in the first NREM sleep period as well as abnormal accumulation and dissipation over all NREM sleep. Unfortunately, there are few studies and conflicting results in regards to the value of currently approved anti-depressants to alter slow-wave sleep. A few studies indicate that ritanserin, a 5HT (2a/2c) antagonist, increases slow-wave sleep in folks with a depression.

F. In regards other gender differences, men with a depression show more stage 1 sleep and more awake time while women have more beta and delta activity during sleep.

G. Folks with a depression have been shown to have phase-shifted or blunted rhythmicity in a number of neuro-endocrine systems including cortisol, growth hormone, melatonin, and testosterone during sleep. For example, in contrast to those free of the condition, people with depression show increased flow of cortisol and a surge in growth hormone at the beginning of sleep. The release of cortisol normalizes after recovery from depression but the activity of growth hormones does not. Effects of the latter on EEG during sleep have been noted.

H. Anti-depressant medication is often effective in the short run in changing various aspects of sleep problems in folks with a depression. However, because of extensive individual differences as well as the complex interplay between sleep and depression, it is often necessary to try a number of medications until the desired results are achieved. However, there are not enough studies to indicate if positive changes to sleep are maintained over the longer term. To my knowledge, results of research support the following conclusions:

a) Prozac (fluoxetine), an SSRI anti-depressant with the most research studies, is associated with a 3% to 5% decrease in total REM time, a doubling of the latency to REM, an increase in stage 1 sleep, an increase in the total number of awakenings, a decrease in sleep efficiency, a decrease in slow-wave sleep, an increase in eye movements and other motor anomalies during REM sleep, and sometimes an increase in bruxism (teeth grinding);

b) Paxil (paroxetine) is an SSRI anti-depressant. The results of the 2 studies I have seen indicated a 9% decrease in REM sleep, a doubling in REM latency, an increase in awakenings, a reduction in sleep time, a reduction in sleep efficiency, and an increase in sleep stage shifts;

c) Zoloft (sertraline) is an SSRI anti-depressant. The results of the one published study I saw revealed prolonged sleep latency, decreased total REM time, and decreased total sleep time;

d) Luvox (fluvoxamine) is an SSRI anti-depressant. The results of the only published study I found indicated longer sleep latency, more awake time, triple REM latency, more than 10% reduced REM sleep time (which decreased 4 weeks later), and increased stage 1 sleep;

e) Celexa (citalopram), an SSRI anti-depressant, was evaluated in one uncontrolled study that reported a decrease in REM sleep, a decrease in REM latency, no changes in sleep continuity, and increased stage 2 sleep;

f) Desyrel (trazodone) is an SARI anti-depressant. Studies using a sufficient dosage have demonstrated that it decreases sleep latency, increases sleep efficiency, decreases night wakening, suppresses REM sleep, lengthens REM latency, increases delta sleep, but may lead to daytime sedation. Because of its significant positive effect on many sleep variables and lesser significant effects on other aspects of depression, this medication is commonly used effectively in combination with another anti-depressant rather than used alone;

g) Wellbutrin (bupropion) is an SDRI antidepressant. The results of the only study I have seen indicated shortened REM latency and increased REM time by 8%;

h). Remeron (mirtazepine), an NaSSA antidepressant, was shown to decrease sleep latency, improve sleep efficiency, and had no effect on REM sleep in a single clinical study with 5 patients; and

i). Effexor (venlafaxine), an SNRI anti-depressant, showed decreased sleep efficiency, an increase in wake time, a decrease in total REM sleep, and an increase in REM latency in the one study I have seen.

I. In a number of studies, both benzodiazepines and specific sleep medications have been prescribed in addition to an anti-depressant. It is the clinical opinion that selection of medication should depend on the cause of the sleep difficulties. Benzodiazepines appear to be most effective when sleep disturbance is the result of anxiety or hyper-arousal. Furthermore, research suggests a more rapid response to the various forms of insomnia when such medications are started at the same time as the anti-depressant. However, federal agencies recently issued warnings about the use of sleep medications other than benzodiazepines.

J. One study demonstrated positive changes in some aspects of sleep architecture in folks who recover from a depression after they complete some psychological therapy without the use of any medications. Sleep continuity and phasic REM sleep has been shown to improve or normalize along with a modest reduction in REM latency but little change occurs in the amount of slow-wave sleep.

K. Some studies have demonstrated that changing the amount and/or timing of sleep may be effective in folks with depression. For example, at one time there was excitement generated by studies showing a significant decrease in depression if folks forced themselves to stay awake for a full night each week. Recent studies indicate that the positive changes in depression level (a) occurred in 50% to 60% of folks whose depression seemed to follow a pattern of being worse in the morning and gradually improving over the course of the day but (b) only lasted until the next night's sleep. Among those for whom giving up a night's sleep was helpful, allowing them to sleep between the hours of 1700 and 2400 the next night (advanced sleep phase) before returning to normal hours of sleep within the next week extended the duration of the positive effect on depression levels.

Darkening the room significantly during the first few nights of the change and then exposure to a bright light upon awakening (as is done in the light therapy treatment of seasonal affective disorder) would likely also further phase advance the circadian rhythm. There is reason to believe that this method might alter the timing of many parts of the circadian rhythm and serve to both decrease the latency to REM and reduce total REM time.

L. One study found that depriving those with depression of REM sleep led to a decrease in depression equivalent to that found with the anti-depressant imipramine. This study was replicated and found (a) a similar anti-depressant effect with deprivation of REM sleep and (b) an even greater anti-depressant effect by depriving folks of non-REM sleep. In the replication study, both types of deprivation increased the duration of non-REM sleep, which might turn out to be the key variable.

M. Numerous studies have demonstrated abnormalities of neuroendocrine systems in people with depression. Increased HPA activity such as increased cortisol secretion is one of the most consistent findings in depression. Furthermore, there is some evidence that this abnormality, along with the finding of lowered secretion of growth hormone, corresponds to the findings of reduced slow wave brain activity in the sleep of folks with a depression.

N. The conclusion reached by various studies investigating the EEG during sleep is that the reasons for characteristics demonstrated among those with depression cannot yet be explained fully. Much more research is required.

O. Among those with depression, night time evaluations by Dr. Armitage and colleagues found (a) lower levels of coherence between left and right brain hemispheres in the theta and beta frequencies and (b) lower coherence in delta and beta frequencies within each hemisphere. This led to their hypothesis that in depression there is a breakdown in the organization of sleep EEG rhythms.

APPENDIX VIII: REFERENCES

Below are the sources I have cited in the development of this treatment manual. With reference to professional journals, sometimes I have been able to obtain the entire article and sometimes I have relied on an abstract since I do not have access to a university library. Other information sources not indicated below include annual meetings, conversations with colleagues, the insights provided by clients, and my own wonderings and trials.

Abramowitz, E. G., Barak, Y., Ben-Avi, I., & Knobler, H. Y. (2008). Hypnotherapy in the treatment of chronic combat-related PTSD patients suffering from insomnia: A randomized, zolpidem-controlled clinical trial. *International Journal of Clinical and Experimental Hypnosis, 56*(3), 270-280.

Adamou, M., Puchalska, S., Plummer, W., & Hale, A. S. (2007). Valproate in the treatment of PTSD: Systematic review and meta analysis. *Current Medical Research and Opinion, 23*(6), 1285-1291.

Adler, A. B., Litz, B. T., Castro, C. A., Suvak, M., Thomas, J. L., Burrell, L., McGurk, D., Wright, K. A., & Bliese, P. D. (2008). A group randomized trial of critical incident stress debriefing provided to U.S. peacekeepers. *Journal of Traumatic Stress Studies, 21*(3), 252-263.

Adler, A. B., Wright, K. M., Bliese, P. D., & Eckford, R. (2008). A2 diagnostic criterion for combat-related posttraumatic stress disorder. *Journal of Traumatic Stress, 2*(3), 310-308.

Afari, N., Wen, Y., Buchwald, D., Goldberg, J., & Plesh, O. (2008). Are post-traumatic stress disorder symptoms and temporomandibular pain associated? Findings from a community-based twin registry. *Journal of Orofacial Pain, 22*(1), 41-49.

Alfarez, D. N., Wiegert, O., & Krugers, H. J. (2006). Stress, corticosteroid hormones and hippocampal synaptic function. *CNS Neurological Disorders and Drug Targets, 5*(5), 521-529.

Amstrader, A. B., Acierno, R., Richardson, L. K., Kilpatrick, D. G., Gros, D. F., Gaboury, M. T., Tran, T. L., Tam, N. T., Tuan, T., Buoi, L. T., Ha, T. T., Thach, T. D., & Galea, S. (2008). Posttyphoon prevalence of posttraumatic stress disorder, major depressive disorder, panic disorder, and generalized anxiety disorder in a Vietnamese sample. *Journal of Traumatic Stress, 22*(3), 180-188.

Andreassi, J. L. (2007). *Psychophysiology: Human behavior and physiological response* (5th ed.). Mahwah, NJ: Lawrence Erlbaum Associates.

Anticevic V., & Britvic. D. (2008). Sexual functioning in war veterans with posttraumatic stress disorder. *Croatian Medical Journal, 49*(4), 499-505.

Aroche, J., Tukelija, S., & Askovic, M. (2009). Neurofeedback in work with refugee trauma: Rebuilding fragile foundations. *Biofeedback, 37*(2), 53-55.

Amstadter, A. (2007). Emotion regulation and anxiety disorders. *Journal of Anxiety Disorders, 22,* 211-221.

Austin, J. H. (1998). *Zen and the brain.* Cambridge, MA: MIT Press.

Ayers, S., Copland, C., & Dunmore, E. (2009). A preliminary study of negative appraisals and dysfunctional coping associated with post-traumatic stress disorder symptoms following myocardial infarction. *British Journal of Health Psychology, 14*(3), 459-471.

Ayers, S., Joseph, S., McKenzie-McHarg, Slade, P., & Wijma, K. (2008). Post-traumatic stress disorder following childbirth: Current issues and recommendations for future research. *Journal of Psychosomatic Obstetrics and Gynecology, 29*(4), 240-250.

Baker, M., Akrofu, K., Schiffer, R., & Boyle, M. W. (2008). EEG patterns in mild cognitive impairment patients. *Open Neuroimaging, 2,* 52-55.

Begic, D., Hotujac, L., & Jokic-Begic, N. (2001). Electroencephalographic comparison of veterans with combat-related post-traumatic stress disorder and healthy subjects. *International Journal of Psychophysiology, 40,* 167-172.

Billette, V., Guay, S., & Marchand, A. (2008). Posttraumatic stress disorder and social support in female victims of sexual assault: The impact of spousal involvement in the efficacy of cognitive behavioral therapy. *Behavior Modification, 32*(6), 876-896.

Bisson, J. I. (2008). Pharmacological treatment to prevent and treat post-traumatic stress disorder. *Torture, 18*(2), 104-106.

Bjork, M., & Sand, T. (2008). Quantitative EEG power and asymmetry increases 36 h before a migraine attack. *Cephalagia,* 960- 968. doi:10.1111/j.1468-2982.2008.01638.x

Boggio, P. S., Rigonatti, S. P., Ribero, R. B., Myczkowski, M. L., Nitsche, M. A., Pascual-Leone, A., & Hregni, F. (2008). A randomized, double-blind, clinical trial on the efficacy of cortical direct current stimulation for the treatment of major depression. *International Journal of Neuropsychopharmacology, 11*(2), 249-254.

Bonne, O., Vythillingham, M., Inagaki, M., Wood, S., Neumeister, A., Nugent, A. C., Snow, J., Luckenbaugh, D. A., Bain, E. E., Dvets, W. C., & Charney, D. S. (2008). Reduced posterior hippocampal volume in posttraumatic stress disorder. *Clinical Psychiatry, 69*(7), 1087-1091.

Bostanov, V., Hautzinger, M., & Kotchoubey, B. (2008). *Event related potentials validate attentional changes after mindfulness training in chronic depression.* Paper presented at the Society for Psychophysiological Research Annual Conference.

Bradley, M. M., Miccoli, L., Escrig, M., & Lang, P. J. (2008). The pupil as a measure of emotional arousal and autonomic activation. *Psychophysiology, 45,* 602-607.

Bremner, J. D. (2006). The relationship between cognitive and brain changes in posttraumatic stress disorder. *Annals of the New York Academy of Science, 1071,* 80-86.

Bremner, J. D. (2002). *Does stress damage the brain: Understanding trauma-related disorders from a mind-body perspective.* New York: Norton.

Bremner. J. D., Elzinga, B., Schmahl, C., & Vermetten, E. (2008). Structural and functional plasticity of the human brain in posttraumatic stress disorder. *Progress in Brain Research, 167,* 171-186.

Bremner, J. D., & Marmar, C. A. (Eds.). (1998). *Trauma, memory, and dissociation.* Washington, DC: American Psychiatric Press.

Bremner, J. D., Quinn, J., Quinn, Q., & Veledar, E. Surfing the net for medical information about psychological trauma: An empirical study of the quality ands accuracy of trauma-related websites. *Medical Informatics and the Internet in Medicine, 31*(3), 227-236.

Breslau, N., Davis, G. G., Peterson, E. L., & Schulz, L. R. (2000). A second look at comorbidity in victims of trauma: The post-traumatic stress disorder-major depression connection. *Biological Psychiatry, 48,* 902-909.

Brewin, C. R. (2001). A cognitive neuroscience account of posttraumatic stress disorder and its treatment. *Behaviour Research and Therapy, 39,* 373-393.

Briere, J. (1997). *Psychological assessment of adult posttraumatic stress.* Washington, DC: American Psychological Association.

Briere, J. (2004). *Psychological assessment of adult posttraumatic stress: Phenomenology, diagnosis, and measurement.* Washington, DC: American Psychological Association.

Briere, J., Kaltman, S., & Green, B. (2008). Accumulated childhood trauma and symptom complexity. *Journal of Traumatic Stress, 21*(2), 223-226.

Broman-Fulks, J. J., Ruggiero, K. J., Green, B. A., Smith, D. W., Hanson, R. F., Kilpatrick, D. G., & Saunders, B. E. (2009). The latent structure of posttraumatic stress disorder among adolescents. *Journal of Traumatic Stress, 22*(2), 246-152.

Bryant, R. A. (2003). Early predictors of posttraumatic stress disorder. *Biological Psychiatry, 53*, 789-795.

Bryant, R. A. (2006). Longtitudinal psychophysiological studies of heart rate: Mediating effects and implications of treatment. *Annals of the New York Academy of Science, 1071*, 19-26.

Bryant, R. A., Creamer, M., O'Donnell M., Silove, D., & McFarlane, A. C. (2008). A multisite study of initial respiration rate and heart rate as predictors of posttraumatic stress disorder. *Journal of Clinical Psychiatry, 69*(11), 1694-1701.

Bryant, R. A., Felmingham, K. L., Kemp, A. H., Barton, A., Peduto, A. S., Rennie, C., Gordon, E., & Williams, L. M. (2005). Neural networks of information processing in posttraumatic stress disorder: A functional magnetic resonance imaging study. *Biological Psychiatry, 58*, 111-118.

Bryant, R. A., Felmingham, K., Whitford, T. J., Kemp, A., Hughes, G., Peduto, A., & Williams, L. M. (2008). Rostral anterior cingulate volume predicts treatment response to cognitive behavioural therapy for posttraumatic stress disorder. *Psychiatry Neuroscience, 33*(2), 142-146.

Bryant, R. A., & Harvey A. G. (2000). *Acute stress disorder: A handbook of theory, assessment, and treatment.* Washington, DC: American Psychological Association.

Bryant, R. A., Mounds, M. L., Guthrie, R. M., Dang, S. T., Mastrodomenico, J., Nixon, R. D., Felmingham, K. L., Hopwood, S., & Creamer, M. (2008). A randomized controlled trial of exposure therapy and cognitive restructuring for posttraumatic stress disorder. *Journal of Clinical and Consulting Psychology, 76*(4), 695-703.

Bryant, R. A., Moulds, M. L., & Nixon, R. V. D. (2003). Cognitive behaviour therapy of acute stress disorder: A four-year follow-up. *Behaviour Research and Therapy, 41*, 489-494.

Calhoun, L. G., & Tedeschi, R. G. (2006). *Handbook of post-traumatic growth: Research and practice.* New York: Lawrence Erlbaum Associates.

Calhoun, P. S., Wiley, M., Dennis, M., & Beckham, J. C. (2009). Self-reported and physician diagnosed illness in women with posttraumatic stress disorder and major depressive disorder. *Journal of Traumatic Stress, 22*(2), 122-130.

Caspi, Y., Saroff, O., Suleimani, N., & Klein, E. (2008). Trauma exposure and posttraumatic reactions in a community sample of Bedouin members of the Israel Defense Forces. *Depression and Anxiety, 25*(8), 700-707.

Celle, S., Peyron, R., Faillenot, I., Pichot, V., Alabdullah, N., Gaspoz, J. M., Laurent, B., Barthelemy, J. C., & Roche, F. (2009). Undiagnosed sleep-related breathing disorders are associated with focal brainstem atrophy in the elderly. *Human Brain Mapping, 30*(7), 2090-2097.

Chen, X., Magnotta, V., Duff, K., Boles Ponto, L., & Schultz, S. (2006). Donepizil effects on cerebral blood flow in older adilts with mild cognitive impairment. *Journal of Neuropsychiatry and Clinical Neurosciences, 18*(2), 178-182.

Clark, C. R., Galletly, C. A., Ach, D. J., Moores, K. A., Penrose, R. A., & McFarlane, A. C. (2009). Evidence based medicine evaluation of electrophysiological studies of anxiety disorders. *Clinical EEG and Neuroscience, 40*(2), 84-112.

Clark, C. R., McFarlane, A. C., Morris, P., Weber, D. L., Sonkkilla, C., Shaw, M., Marcina, J., Tochon-Danguy, H. J., & Egan, G. F. (3003). Cerebral function in posttraumatic stress disorder during verbal working memory updating: A positron emission tomography study. *Biological Psychiatry, 53*, 474-481.

Congedo, M. (2009, Summer). Interview with E. Roy John. *NeuroConnections*, 15.

Cook, F., Ciorciari, J., Varker, T., & Devilly, G. J. (2009). Changes in long term neural connectivity following psychological trauma. *Clinical Neurophysiology, 120*(2), 309-14.

Cordova, M. J., Cunningham, L. L. C., Carlson, C. R., & Andrykowski, M. A. (2001). Posttraumatic growth following breast cancer: A controlled comparison study. *Health Psychology, 20*, 176-185.

Courtney, R. (2008). Strengths, weakness, and possibilities of the Buteyko breathing method. *Biofeedback, 32*(6), 59-63.

Courtois, C. A., & Gold, S. N. (2009). The need for inclusion of psychological trauma in the professional curriculum: A call to action. *Psychological Trauma: Theory, Research, Practice, and Policy, 1*(1), 3-23.

Creamer, M., & Parslow, R. (2008). Trauma exposure and posttraumatic exposure in the elderly: A community prevalence study. *American Journal of Geriatric Psychiatry, 16*(10), 853-6.

Dalai Lama (Tenzin Gyatso). (1996). *The good heart.* London: Medio Media.

Daley. A. (2008). Exercise and depression: A review of reviews. *Journal of Clinical Psychology in Medical Settings, 15*(2), 140-147.

Daly, E. S., Gulliver, S. B., Zimering, R. T., Knight, J., Kamholz, B.W., & Morisette, S. B. (2008). Disaster mental health workers responding to

Ground Zero: One year later. *Journal of Traumatic Stress, 21*(2), 227-230.

Darves-Bornoz, J., Alonso, J., de Girolamo, G., de Graaf, R., Haro, J. M., Kovess-Masfety, V., Lepine, J. P., Nachbaur, G., Negre-Pages, L., Vilagut, G., Gasquet, I., & ESEMeD/MHEDEA 2000 Investigators. (2008). Main traumatic events in Europe: PTSD in the European study of the epidemiology of mental disorders survey. *Journal of Traumatic Stress, 21*(5), 455-462.

David, A. S., Farrin, L., Hull, L., Unwin, C., Wesseley, S., & Wykes, T. (2002). Cognitive functioning and disturbances of mood in UK veterans of the Persian Gulf War: A comparative study. *Psychological Medicine, 32*, 1357-1370.

Davidson, J. R. T. (2004). Long-term treatment and prevention of posttraumatic stress disorder. *Journal of Clinical Psychiatry, 65*(Suppl. 1), 44-48.

Davidson, R. J. (2004). What does the prefrontal cortex "do" in affect: Perspectives on frontal EEG asymmetry research. *Biological Psychiatry, 67*, 219-233.

Davidson, J., Baldwin, D., Stein, D. J., Kuper, E., Benattia, I., Ahmed, S., Pedersen, R., & Musgnung, J. (2006). Treatment of posttraumatic stress disorder with venlafxine extended release: A 6-month randomized control trial. *Archives of General Psychiatry, 63*(10), 1158-1165.

Davidson, J., Rothbaum, B. O., Tucker, P., Asnia, G., Benattia, I., & Musgnung, J. J. (2006). Venlafaxine extended release in posttraumatic stress disorderA a sertraline- and placebo-controlled study. *Journal of Clinical Psychopharmacology, 26*(3), 259-267.

Davis, L. L., Davidson, J. R., Ward, L. C., Bartolucci, A., Bowden, C. L., & Petty, F. (2008). Divalproex in the treatment of posttraumatic stress disorder: A randomized, double-blind, placebo-controlled trial in a veteran population. *Journal of Clinical Psychopharmacology, 28*(1), 84-88.

Davis, L. L., English, B. A., Ambrose, S. M., & Petty, D. (2001). Psychotherapy for post-traumatic stress disorder: A comprehensive review. *Expert Opinion on Pharmacotherapy, 2*(10), 1-13.

Davis, L. L., Frazier, E. C., Williford, R. E., & Newell, J. M. (2006). Long-term pharmacotherapy for post-traumatic stress disorder. *CNS Drugs, 20*(6), 465-476.

Davis, L. L., Jewell, M. E., Ambrose, S., Farley, J., English, B., Bartolucci, A., & Petty, F. (2004). A placebo-controlled study of nefazodone for the treatment of chronic posttraumatic stress disorder. *Journal of Clinical Psychopharmacology, 24*(3), 291-297.

Davis, R. C., Lurigio, A. J., & Skogan, W. G. (Eds.). (1997). *Victims of crime* (2nd ed.). Thousand Oaks, CA: Sage Publications.

Davydow, D. S., Gifford, J. M., Desai, S. V., Needham, D. M., & Bienvenu, O. J. (2008). Posttraumatic stress disorder in general intensive care unit survivors: A systematic review. *General Hospital Psychiatry, 30*(5), 421-434.

Defrin, R., Ginzbury, K., Solomon, Z., Polad, E., Bloch, M., Govezensky, M., & Schreiber, S. (2008). Quantitative testing of pain perception in subjects with PTSD–implications for the mechanism of the coexistence between PTSD and chronic pain. *Pain, 138*(2), 450-459.

Demakis, G. J., Gervais, R. O., & Rohling, M. L. (2008). The effect of failure on cognitive and psychological symptom validity tests in litigants with symptoms of post-traumatic stress disorder. *Clinical Neuropsychology, 22*(5), 879-895.

Dement, W. C., & Vaughan, C. (1999). *The promise of sleep.* New York: Dell.

Dickie, E. W., Brunet, A., Akerib, V., & Armony, J. L. (2008). An fMRI investigation of memory encoding in PTSD: Influence of symptom severity. *Neuropsychologia, 46,* 1522-1531.

DiGrande, L., Perrin, M. A., Thorpe, L. E., Thalji, L., Murphy, J., Wu, D., Farfel, M., & Brackbill, R. M. (2008). Posttraumatic stress symptoms, PTSD, and risk factors among Lower Manhattan residents 2-3 years after the September 11, 2001 terrorist attacks. *Journal of Traumatic Stress Studies, 21*(2), 264-273.

Dileo, J. F., Brewer, W. J., Hopwood, M., Anderson, V., & Creamer, M. (2008). Olfactory identification dysfunction, aggression, and impulsivity in war veterans with post-traumatic stress disorder. *Psychological Medicine, 38*(4), 523-531.

Drevets, W. C., Savitz, J., & Trimble, M. (2008). The subgenual anterior cingulate cortex in mood disorders. *CNS Spectrum, 13*(8), 663-681.

Dyster-Aas, J., Willebrand, M., Wikehult, B., Gerdin, B., & Ekselius, L. (2008). Major depression and posttraumatic stress disorder symptoms following severe burn injury in relation to lifetime psychiatric morbidity. *Journal of Trauma, 64*(5), 1349-1356.

Dzubur, K. A., Kucukalic, A., & Malec, D. (2008). Changes in plasma lipid concentrations and risk of coronary artery disease among army veterans suffering from chronic posttraumatic stress disorder. *Croatia Medical Journal, 49*(4), 506-514.

Eskelinen, V., & Himanen, S. (2007). CPAP treatment of obstructive sleep apnea increases slow wave sleep in prefrontal EEG. *Clinical EEG and Neuroscience, 38*(3), 148-154.

Evans, J. T. (Ed.). (2007). *Handbook of neurofeedback: Dynamics and clinical applications.* New York: Haworth Medical Press.

Evans, J. R., & Abarbanel, A. (Eds.). (1999). *Introduction to quantitative EEG and neurofeedback.* San Diego, CA: Academic Press.

Falconer, E. M., Felmingham, K. L., Allen, A., Clark, C. R., McFarlane, A. C., Williams, L. M., & Bryant, R. A. (2008). Developing an integrated brain behavior and biological response profile in posttraumatic stress disorder (PTSD). *Journal of Integrative Neuroscience, 7*(3), 439-456.

Farrar, J., Ryan, J., Oliver, E., & Gillespie, M. B. (2008). Radiofrequency ablation for the treatment of obstructive sleep apnea: A metaanalysis. *Laryngoscope, 118*(10), 1878-83.

Feinstein, A., & Botes, M. (2009). The psychological health of contractors working in war zones. *Journal of Traumatic Stress, 22*(2), 102-105.

Feldner, M. T., Monson, C. M., & Friedman, M. J. A critical analysis of approaches to targeted PTSD prevention: Current status and theoretically derived directions. *Behavior Modification, 31*(1), 80-116.

Feldner, M. T., Smith, R. C., Babson, K. A., Sachs-Ericcson, N., Schmidt, N., & Zvolensky, M. J. (2009). Test of the role of nicotine dependence in the relationship between posttraumatic stress disorder and panic spectrum problems. *Journal of Traumatic Stress, 22*(1), 36-44.

Figley, C. R., & Nash, W. P. (Eds.). (2007). *Combat stress injury: Theory, research, and management.* New York: Routledge.

Foa, E. B., Keane, T. M., & Friedman, M. J. (Eds.). (2000). *Effective treatments for PTSD: Practice guidelines from the International Society for Traumatic Stress Studies.* New York: Guilford.

Foa, E. B., & Rothbaum, B. O. (1998). *Treating the trauma of rape: Cognitive behavioral therapy for PTSD.* New York: Guilford.

Fontana, A., & Rosenbeck, R. (2008). Treatment-seeking veterans of Iraq and Afghanistan: Comparison with veterans of previous wars. *Journal of Nervous and Mental Disease, 196*(7), 513-521.

Forbes, D., Parslow, R., Creamer, M., Allen, N. McHugh, T., & Hopwood, M. (2008), Mechanisms of anger and treatment outcome in combat veterans with posttraumatric stress disorder. *Journal of Traumatic Stress, 21*, 142-149.

Ford, E., & Ayers, S. (2008). Stressful events and support during birth: The effect on anxiety, mood and perceived control. *Journal of Anxiety Disorders, 23*(2), 260-268.

Foster, M., & Spiegel, D. P. (2008). Use of donepezil in the treatment of cognitive impairments of moderate traumatic brain injury. *Journal of neuropsychiatry and clinical neuroscience, 20*(1), 106.

Freeman, T. W., Hart, J., Kimbrell, T., & Ross, E. D. (2008). Comprehension of affective prosody with chronic posttraumatic stress disorder. *Journal of Neuropsychiatry and Clinical Neuroscience, 21*(1), 52-58.

Frewen, P. A., Dozois, D. J. A., Neufeld, R. W. J., & Lanius R. A. (2008a). Meta-analysis of alexithymia in posttraumatic stress disorder. *Journal of Traumatic Stress, 21*(2), 243-246.

Frewen, P.A., Lane, R. D., Neufeld, R. W. J., Densmore, M., Stevens, T., & Lanius, R. (2008). Neural correlates of levels of emotional awareness during trauma script-imagery in posttraumatic stress disorder. *Psychosomatic Medicine, 70*, 27-31.

Frewen, P. A., & Lanius, R. A. (2006). Toward a psychobiology of posttraumatic self-dysregulation: Re-experiencing, hyperarousal, dissociation, and emotional numbing. *Annals of the New York Academy of Science, 1071*, 110-124.

Frewen P. A., Lanius, R. A., Pain, C., Hopper, J. W., Densmore, M., & Hopper, J. W. (2008). Clinical and neural correlates of alexithymia in posttraumatic stress disorder. *Journal of Abnormal Psychology, 117*(1), 171-181.

Friedman, M. J., Donnelly, C. L., & Mellman, T. A. (2003). Psychopharmacology for PTSD. *Psychiatric Annals, 33*(1), 57-62

Friedman, M. J., Keane, T. M., & Resnik, P. (Eds.). (2007*). Handbook of PTSD: Science and practice.* New York: Guilford Press.

Fujita, G., & Nishida, Y. (2008). Association of objective measures of trauma exposure from motor vehicle accidents and posttraumatic stress symptoms. *Journal of Traumatic Stress, 21*(4), 425-429.

Gale, C.R., Deary, I. J., Boyle, S. H., Barefoot, J., Mortensen, L. H., & Batty, G. D. (2008). Cognitive ability in early adulthood and risk of 5 specific psychiatric disorders in middle age: The Vietnam experience study. *Archives of General Psychiatry, 65*(12), 1410-1418.

Galea, S., Tracy, M., Norris, F., & Coffey, S. F. (2008). Financial and social circumstances and the incidence of and course of PTSD in Mississippi during the first two years after Hurricane Katrina. *Journal of Traumatic Stress, 21*(4), 357-368.

Galletly, C. A., McFarlane, A. C., & Clark, R. (2008). Differentiating cortical patterns of cognitive dysfunction in schizophrenia and posttraumatic stress disorder. *Psychiatry Research, 159,* 196-206.

Gao, K., Muzina, D., Gajwani, P., & Calabrese, J. R., (2006). Efficacy of typical and atypical antipsychotics for primary and comorbid anxiety symptoms of disorders: A review. *Journal of Clinical Psychiatry, 67*(9), 1327-1340.

Gaylord, K. M., Cooper, D. B., Kennedy, J. E., Yoder, L. H., & Holcomb, J. B. (2008). Incidence of posttraumatic stress disorder and mild traumatic brain injury in burned service members: Preliminary report. *Journal of Trauma, 62*(Suppl. 2), S206-6.

Gerardi, M., Rothbaum, B. O., Ressler, K., Heekin, M., & Rizzo, A. (2008). Virtual reality exposure therapy using a virtual Iraq: Case report. *Journal of Traumatic Stress Studies, 21*(2), 209-213.

Germain, A., Hall, M., Shear, M. K., Nofzinger, E. A., & Buysse, D. J. (2006). Ecological study of sleep disruption in PTSD. *Annals of the New York Academy of Science, 1071,* 438-441.

Geuze, E. Vermetten, E., deKloet, C. S., Hijman, R., & Westenberg, H. G. (2009). Neuropsychological performance is related to current social and occupational functioning in veterans with posttraumatic stress disorder. *Depression and Anxiety 26*(1),7-15.

Geuze, E., Westenberg, H. G., Heinecke, A., de Kloet, C. S., Goebel, R., & Vermetten, E. (2008). Thinner prefrontal cortex in veterans with posttraumatic stress disorder. *Neuroimage, 41*(3), 675-681.

Gismondi, M. (2009, Fall). Interview with Dr. Robert Thatcher on the evolution of his 19 channel live z-score and LORETA training system. *NeuroConnections,* 37-39.

Gismondi, M. (2009, Fall). Tech talk. *Neuroconnections,* 39-40.

Goenjian, A. K., Noble, E. O., Walling, E. P., Goenjian, H. A., Karayan, I. S., Ritchie, T., & Bailey, J. N. (2008). Heritabilities of symptoms of posttraumatic stress disorder, anxiety, and depression in earthquake exposed Armenian families. *Psychiatric Genetics, 18,* 261-266.

Grant, D. M., Beck, J. G., Palyo, S. A., & Clapp, J. D. (2008). The structure of distress following trauma: Posttraumatic stress disorder, major depressive disorder, and generalized anxiety disorder. *Journal of Abnormal Psychology, 117*(3), 662-672.

Griffin, M. G. (2008). A prospective assessment of auditory startle alterations in rape and physical assault survivors. *Journal of Traumatic Stress Studies, 21*(1), 91-99.

Gross, J. J. (Ed.). (2007). *Handbook of emotion regulation*. New York: Guilford.

Hammond, D. F., Walker, J., Hoffman, D., Lubar, J., Trudeau, D., Gurnee, R., & Horvat, J. (2004). Standards for the use of quantitative electroencephalography (qeeg) in neurofeedack: A position paper of the International Society for Neuronal Regulation. *Journal of Neurotherapy, 8*(1). Retrieved from http://www.flexiblebrain.com/Hammond%202004%20JNT%208(1)%205-29.pdf

Hara, E., Matsuoka, Y., Hakamata, Y., Nagamine, M., Inagaki, M., Imoto, S., Murakami, K., Kim, Y., & Uchitomi, Y. (2008). Hippocampal and amygdalar volumes in breast cancer survivors with posttraumatic stress disorder. *Journal of Neuropsychiatry and Clinical Neuroscience, 20*(3), 302-308.

Harris, I. A., Young, J. M., Rae, H., Jalaludin, B. B., & Solomon, M. J. (2008). Predictors of post-traumatic stress disorder following major trauma. *Australia New Zealand Journal of Surgery, 78*(7), 583-587.

Hart, J., Kimbrell, T., Fauver, P., Cherry, B. J., Pitcock, J., Booe, L. Q., Tillman, G., & Freeman, T. W. (2008). Cognitive dysfunctions associated with PTSD: Evidence from WWII prisoners of war. *Journal of Neuropsychiatry and Clinical Neuroscience, 20*(3), 309-316.

Hedges. D. W., Thatcher, G. W., Bennett, P. J., Sood, S., Paulson, D., Creem-Regehr, S., Brown, B. L., Allen, S., Johnson, J., Froelich, B., & Bigler, E. D. (2007). Brain integrity and cerebral atrophy in Vietnam combat veterans with and without posttraumatic stress disorder. *Neurocase, 13*(5), 401-410.

Hepp. U., Moergeli, H., Buchi, S., Bruchhaus-Steinert, H., Kraemer, B., Sensky, T., & Schnyder, U. (2008). Posttraumatic stress disorder in serious accidental injury: 3-year follow-up study. *British Journal of Psychiatry, 192*(5), 276-383.

Hiskey, S., Luckie, M., Davies, S., & Brewin, C. R. (2008). The activation of posttraumatic distress in later life: A review. *Journal of Geriatric Psychiatry and Neurology, 21*(4), 232-241. DOI: 10.1177/0891988708324937

Ipser, J., Seedat, S., & Stein, D. J. (2006). Pharmacotherapy for post-traumatic stress disorder: A systematic review and meta-analysis. *South African Medical Journal, 96*(10), 1088-1092.

Jacobson, I. G., Ryam, M. A. K., Hooper, T. I., Smith, T. C., Amoroso, P. J., Boyko, E. J., Gackstetter, G., D., Wells, T. S., & Bell, N. S. (2008). Alcohol use and alcohol-related problems before and after military combat deployment. *Journal of the American Medical Association, 300*, 663-675.

Joffe, D. (2009). Connectivity assessment and training: A partial directed coherence approach. *Journal of Neurotherapy, 12*(2,3), 111-122.

Jokic-Begic & Begic. (2003). Quantitative electroencephalogram (qEEG).in combat veterans without-traumatic stress disorder (PTSD). (2003). *Nordic Journal of Psychiatry, 57*(5), 351-355.

Jones, E., Hodgins, R., McCartney, V. H., Beech, C., Palmer, I., Hymas, K., & Wessely, S. (2003). Flashbacks and post-traumatic stress disorder: The genesis of a 20th century diagnosis. *British Journal of Psychiatry, 182,* 158-163.

Karl, A., Malta, L. S., & Maercher, A. (2005). Meta-analytic review of event-related potential studies in post-traumatic stress disorder. *Biological Psychiatry, 71,* 123-147.

Karl, A., Schaefer, M., Malta, L. S., Dorfel, D., Rohleder, N., & Werner, A. (2006). A meta-analysis of structural brain abnormalities in PTSD. *Neuroscience and Biobehavioral Reviews, 30,* 1004-1031.

Kasai, K., Yamasue, H., Gilbertson, M. W., Shenton, M. E., Rauch, S. L., & Pitman, R. K. (2008). Evidence for acquired pregenual anterior cingulate gray matter loss from a twin study of combat-related posttraumatic stress disorder. *Biological Psychiatry, 63*(6), 550-556.

Keane, T. M., Caddell, J. M., & Taylor, K. L. (1988). Mississippi scale for combat-related posttraumatic stress disorder: Three studies in reliability and validity. *Journal of Consulting and Clinical Psychology, 56*(1), 85-90.

Kendall-Tackett K. (2009). Psychological trauma and physical health: A psychoneuroimmunology approach to etiology of negative health effects and possible interventions. *Psychological Trauma: Theory, Research, Practice, and Policy, 1*(1), 35-48.

Kennedy, C. H., & Zillmer, E. A. (Eds.). (2006). *Military psychology: Clinical and operational applications.* New York: Guildford.

Kern, S., Oakes, T. R., Stone, C. K., McAuliff, E. M., Kirschbaum, C., & Davidson, R. J. (2008). Glucose metabolic changes in the prefrontal cortex are associated with HPA axis response to a psychosocial stressor. *Psychoneuroendocrinology, 33,* 517-529.

Kilpatrick, D. G., Resnick, H. S., & Acierno, R. (2009). Should PTSD Criterion A be retained. *Journal of Traumatic Stress, 22*(6), 374-383.

Kimerling, R., Ouimette, P., & Wolfe, J. (Eds.). (2002). *Gender and PTSD.* New York: Guilford.

Kiss, M., & Eimer, M. (2008). ERPs reveal subliminal processing of fearful faces. *Psychophysiology, 45*, 318-326.

Klawe, J.J., Laudencka, A., Miskoweic, I., & Tafil-Klawe, M. (2005). Occurrence of obstructive sleep apnea in a group of shift worked police officers. *Journal of Physiology and Pharmacology, 56*(4), 1115-1117.

Koenigs, K., Huey, E. D., Raymont, V., Cheon, B., Solomon, J., Wassermann, E. M., & Grafman, J. (2008). Focal brain damage protects against post-traumatic stress disorder in combat veterans. *Nature Neuroscience, 11*(2), 232-237.

Kongsted, A., Bendix, T., Querama, E., Kasch, H., Bach, F. W., Korsholm, L., & Jensen, T. S. (2008). *European Journal of Pain, 12*(4), 455-463.

Krakow, B., Lowry, G., Germain, A., Gaddy, L., Hollifield, K., Moss, M., Tandberg, D., Johnston, L., & Melendrez. (2000). A retrospective study on improvements in nightmares and posttraumatic stress disorder following treatment for co-morbid sleep-disordered breathing. *Journal of Psychosomatic Medicine, 49*(5), 291-298.

Krakow, B., Melendrez, D., Pederson, B., Johnston, L., Hollifield, M., Germain, A., Koss, M., Warner, T. D., & Schrader, R. (2001). Complex insomnia: Insomnia with sleep-disordered breathing in a consecutive series of crime victims with nightmares and PTSD. *Biological Psychiatry, 49*(11), 948-953.

Krakow, B., Melendrez, D., Warner, T. D., & Clark, J. O. (2006). Signs and symptoms of sleep-disordered breathing in trauma survivors: A matched comparison with classic sleep apnea patients. *Journal of Nervous and Mental Diseases, 194*(6), 433-439.

Krakow, B., & Zadra. A. (2006). Clinical management of chronic nightmares: Imagery rehearsal therapy. *Behavioral Sleep Medicine, 4*(1), 45-70.

Kwok, F. Y., Lee., T. M., Leung, C. H., Poon, W. S. (2008). Changes of cognitive functioning following mild traumatic brain injury over a 3-month period. *Brain Injury, 22*(10), 740-752.

Ladwig, K. H., Baumert, J., Marten-Mittag, B., Kolb, C., Zrenner, B., & Schmitt, C. (2008). Posttraumatic stress symptoms and predicted mortality in patients with implantable cardioverter defibrillators: Results from the prospective living with an implanted cardioverter defibrillator study. *Archives of General Psychiatry, 65*(11), 1324-30.

Lanius, R. A., Frewen, P. A., Girotti, M., Neufeld, R. W., Stevens, T. K., & Densmore, M. (2007). Neural correlates of trauma script-imagery in posttraumatic stress disorder with and without comorbid major depression: A functional MRI investigation. *Psychiatry Research, 155*(1), 45-56.

Laudencka, A., Klawe, J. J., Tafil-Klawe, M., & Ztomanczuk, P. Does night-shift work induce apnea events in obstructive sleep apnea patients? *Journal of Physiology and Pharmacology, 55*(5), 345-347).

Lagos, L., Vaschillo, E., Vaschillo, B., Lehrer, P., Bates, M., & Pandina, R. (2008). Heart rate variability feedback as a strategy for dealing with competitive anxiety: A case study. *Biofeedback, 36*(3), 109-115.

Lazarus, A. A., Lazarus, C. N., & Fay, A. (1993). *Don't believe it for a minute: Forty toxic ideas that are driving you crazy.* Atascadero, CA: Impact Publishers,

Lehrer, P. M., Woolfolk, R. L., & Sime, W. E. (2007). *Principles and practice of stress management* (3rd ed.). New York: Guildford.

Leuchter, A. F., Cook, I. A., & Korb, A. (2009). Use of clinical neurophysiology for the selection of medication in the treatment of major depressive disorder: The state of the evidence. *Clinical EEG and Neuroscience, 40*(2), 78 ff.

Levine, S. Z., Laufer, A., Hamama-Raz, Y., Stein, E., & Solomon, Z. (2008). Posttraumatic growth in adolescence: Examining its components and relationship with PTSD. *Journal of Traumatic Stress, 21*(5), 492-496.

Liberzon, I., & Sripada, C. S. (2008). The functional neuroanatomy of PTSD: A critical review. *Progress in Brain Research, 167,* 151-169.

Liberzon, I., King, A. P., Britton, J. A., Phan, K. L., Abelson, J. L., & Taylor, S. F. (2007). Paralimbic and medial prefrontal cortical involvement in neuroendocrine responses to traumatic stimuli. *American Journal of Psychiatry, 164*(8), 1250-1258.

Liberzon, I., & Martis, B. (2006). Neuroimaging studies of emotional responses in PTSD. *Annals of the Ney York Academy of Science, 1071,* 87-109.

Licht, C, M., deGeus, E. J., Zitman, F. G., Hoogendijk, W. J., van Dyck, R., & Penninx, B. W. (2008). Association between major depressive disorder and heart rate variability in the Netherlands Study of Depression and Anxiety (NESDA). *Archives of General Psychiatry, 65*(12), 1358-1367.

Linden, M., Rotter, M., Baumann, K., & Lieberei, B. (2007). *Posttraumatic embitterment disorder: Definition, evidence, diagnosis, treatment.* Cambridge, MA: Hogrefe.

Liston, C., McEwen, B.S., & Casey, B. J. (2009). Psychosocial stress reversibly disrupts prefrontal processing and attentional control. *Proceedings of the National Academy of Science, 106*(3), 912-917.

Malta, L.S., Wyka, K.E., Giosan, C., Jayasinghe, N., & Difede, J. (2008). Numbing symptoms as predictors of unremitting posttraumatic stress disorder. *Journal of Anxiety Disorders, 23*(2), 223-229.

Marks, D. M., Park, M. H., Ham, C., Patkar, A. A., Masand, P. S., & Pae, C. U. (2008). Paroxetine: Safety and tolerability issues. *Expert Opinion and Drug Safety, 7*(6), 783-794.

Masmas, T. N. et al. (2008). Asylum seekers in Denmark. *Torture, 18*(2), 77-86.

Matsakis, A. (1998). *Trust after trauma: A guide to relationships for survivors and those who love them.* Oakland, CA: New Harbinger Publications.

Mayorga, M. A., & Riechers, R. (2008). Science of blast-induced injury and proposed mechanisms of brain injury. *Revista Espanola de Neuropsicologia, 10*(1), 109-115.

McDonald, S. D., Beckham, J. C., Morey, R., Marx, C., Tupler, L. A., & Calhoun, P. S. (2008). Factorial invariance of posttraumatic stress disorder symptoms across three veteran samples. *Journal of Traumatic Stress, 21*(3), 309-317.

Metzger, L. J., Clark, C. R., McFarlane, A. C., Veltmeyer, M. D., Lasko, N. B., Paige, S. R., Pitman, R, K., & Orr, S. P. (2009). Event-related potentials to auditory stimuli in monozygotic twins discordant for combat: Association with PTSD. *Psychophysiology, 46,* 172-178.

Metzger, L. J., Paige, S. R., Carson, M. A., Lasko, N. B., Paulus, L. A., Pitman, R. K., & Orr, S. P. (2004). PTSD arousal and depression symptoms associated with increased right-sided parietal EEG asymmetry. *Journal of Abnormal Psychology, 113*(2), 324-329.

Metzger, L. J., Pitman, R. K., Miller, G. A., Paige, S. R., & Orr, S. P. (2008). Intensity dependence of auditory P2 in monozygotic twins discordant for Vietnam combat; associations with posttraumatic stress disorder. *Journal of Rehabilitation Research and Development, 45*(3), 437-450.

Meyerson, M., & Asher, L. (1999). *Six keys to creating the life you desire: Stop pursuing the unattainable and find the fulfilment you truly need.* Oakland, CA: New Harbinger.

Miller, B. L., & Cummings, J. L. (Eds.). (2007). *The human frontal lobes: Functions and disorders.* New York: Guilford.

Miller, L. A., Collins, R. L., & Kent, T. A. Language and the modulation of impulsive aggression. (2008). *Journal of Neuropsychiatry and Clinical Neurosciences, 20*(3), 261-273.

Miller, T. W. (Ed.). (1997). *Clinical disorders and stressful life events.* Madison, CT: International Universities Press.

Mohamed, S., & Rosenheck, R. A. (2008). Pharmacotherapy of PTSD in the U.S. Department of Veterans Affairs: Diagnostic- and symptom-guided drug selection. *Journal of Clinical Psychiatry, 69*(6), 959-965.

Mohamed, S., & Rosenheck, R. (2008). Pharmacotherapy for older veterans diagnosed with posttraumatic stress disorder in Veterans Administration. *American Journal of Geriatric Psychiatry, 16*(10), 804-812.

Monti, A., Cogiamanian, F., Marceglia, S., Ferrucci, R. Mameli, F., Mrakic-Sposta, S., Vergari, M., Zago, S., & Priori, A. (2008). Improved naming after transcranial direct current stimulation in aphasia. *Journal of Neurology, Neurosurgery, and Psychiatry, 79*(4), 451-453

Monti, D. A., Stoner, M. E., Zivin, G., & Schlesinger, M. (2007). Short-term correlates of neuro emotional technique for cancer-related traumatic stress syndromes: A pilot case series. *Journal of Cancer Survivorship: Research and Practice, 1*, 161-166.

Moores, K. A., Clark, C. R., McFarlane, A. C., Brown, G. C., Puce, A., & Taylor, D. J. (2006). Abnormal recruitment of working memory updating networks in PTSD. *Clinical EEG and Neuroscience, 37*(2), 162.

Moreau, C., & Zisook, S. (2002). Rationale for posttraumatic stress spectrum disorder. *Psychiatric Clinics of North America, 25*, 775-790.

Morey, R. A., Petty, C. M., Cooper, D. A., Labar, K. S., & McCarthy, G. (2008). Neural symptoms for executive and emotional processing are modulated by symptoms of posttraumatic stress disorder in Iraq war veterans. *Psychiatric Research, 162*(1), 59-72.

Morina N & Ford, J. D. (2008). Complex sequelae of psychological trauma among Kosovar civilian war victims. *International Journal of Social Psychiatry, 54*(5), 425-436.

Morrell, M. J., & Twigg, G. (2006). Neural consequences of sleep disordered breathing: The role of intermittent hypoxia. In R. C. Roach (Ed.), *Hypoxia and Exercise* (chapter 8). New York: Springer.

Mundy, E., & Baum, A. (2004). Medical disorders as a cause of psychological trauma and posttraumatic stress disorder. *Current Opinion in Psychiatry, 17*, 123-127.

Nagai, N., Kishi, K., & Kato, S. (2007). Insular cortex and neuropsychiatric disorders: A review of recent literature. *European Psychiatry, 22*(6), 387-394.

Nagata, S., Funakosi, A., Amae, S., Yosida, S., Ambo, H., Kudo, A., Yakota, A., Ueno, T., Matsuoka, H., & Hayashi, Y. (2008). Posttraumatic stress disorder in mothers of children who have undergone surgery for congenital disease at a pediatric surgery department. *Journal of Pediatric Surgery, 43*(8), 1480-1486.

Nampiaparampil, S. E. Prevalence of chronic pain after traumatic brain injury: A systematic review. *Journal of the American Medical Association, 300*(6), 711-719.

Nash, E. (2009, Summer). It's all neurofeedback. *NeuroConnections*, 22-23.

Nikolova, R., Aleksiev, L., & Vukov, M. (2007). Psychophysiological assessment of stress and screening of health risk in peacekeeping operations. *Military Medicine, 172*, 44-48.

Noble, A. J., Baisch, S., Mendelow, A. D., Allen, L., Kane, P., & Schenk, T. (2008). Posttraumatic stress disorder explains reduced quality of life in subarachanoid hemorrage patients in both short and long term. *Neurosurgery, 63*(6), 1095-1104.

Noble, A. J., & Schenk, T., (2008). Posttraumatic stress disorder in in the family and friends of patients who have suffered spontaneous subarachanoid hemorrage. *Journal of Neurosurgery, 109*(6), 1027-1033.

Norris, A. E., & Aroian, K. J. (2008). Avoidance symptoms and assessment of posttraumatic stress disorder in Arab immigrant women. *Journal of Traumatic Stress, 21*(5), 471-478.

Norris, F. H., VanLandingham, M. J., & Vu, L. (2009). PTSD in Vietnamese Americans following Hurricane Katrine: Prevalence, patterns, and predictions. *Journal of Traumatic Stress, 22*(2), 91-101.

Nuwer, M. R., Hovda, D. A., Schrader, L. M., and Vespa, P. M. (2005). Routine and quantitative EEG in mind traumatic brain injury. *Clinical Neurophysiology, 116*, 2001-2025.

Nuwer, M. R., Hovda, D. A., Schrader. L. M., & Vespa, P. M. (2005). Routine and quantitative EEG in mind traumatic brain injury. *Clinical Neurophysiology, 116*, 2001-2025.

Owens, G. P., Dashevesky, B., Chard, K. L., Mohamed, S., Haji, U., Heppner, P. S., & Baker, D. G. (2009). The relationship between childhood trauma, combat exposure, and posttraumatic stress disorder in male veterans. *Military Psychology, 21*, 114-125.

Pae, C. U., Lim, H. K. Peindl, K., Ajwani, N., Serretti, A., Patkar, A. A., & Lee, C. (2008). The atypical antipsychotics olanzapine and respiridone in the treatment of posttraumatic stress disorder: A meta-analysis of randomized, double-blind, placebo-controlled clinical trials. *International Clinical Psychopharmacology, 23*(1), 1-8.

Pae, C. U., & Patkar, A. A. (2007). Paroxetine: Current status in psychiatry. *Expert Reviews of Neurotherapeutics, 7*(2), 107-120.

Palmieri, P. A., Canetti-Nisim, D., Galea, S., Johnson, R. J., & Hobfoll, S. E. (2008). The psychological impact of the Israel-Hezbollah War on Jews and Arabs in Israel. *Social Science and Medicine, 67*(8), 1208-1216.

Palyo, S. A., Clapp, J. D., Beck, J. G., & DeMOnd, M. G. (2008). Unpacking the relationship between posttraumatic numbing and hyperarousal in a sample of help-seeking motor vehicle accident survivors: Replication and extension. *Journal of Traumatic Stress, 21*(2), 235-238.

Park, D. H., Shin, C. J., Hong, S. C., Yu, J., Kim, E. J., Shin, H. B., & Shin, B. H. (2008). Correlation between the severity of obstructive sleep apnea and heart rate variability indices. *Journal of Korean Medical* Science, *23*(2), 226-231.

Paton, D., & Violanti, J. M. (1996). *Traumatic stress in critical occupations: Recognition, consequences, and treatment.* Springfield, IL: Charles C Thomas.

Paton D., Violanti, J. M., Dunning, C., & Smith L. M. (2004). *Managing traumatic stress risk: A proactive approach.* Springfield, IL: Charles C Thomas.

Pawels, E. K & Volterrani, D. (2008). Fatty acid facts, Part 1: Essential fatty acids as treatment for depression, or food for mood? *Drug News Perspectives, 21*(8), 446-451.

Payne, J. M., & Gevirtz, R. (2009). Psychophysiologic assessment and combat post traumatic stress disorder. *Biofeedback, 37*(1), 18-23.

Peres, J. F., McFarlane, A., Nasello, A.G., & Moores, K. A. (2008). Traumatic memories: Bridging the gap between functional neuroimaging and psychotherapy. *Australian & New Zealand Journal of Psychiatry, 42*(6), 478-488.

Piefke, M., Pestinger, M., Arin, T., Kphl, B., Kastrau, F., Schnitker, R., Vohn, R., Weber, J., Ohnhaus, M., Erli, H. J., Perlitz, V., Paar, O., Petzold, E. R., & Flatten, G. (2007). The neurofunctional mechanisms of traumatic and non-traumatic memory in patients with acute PTSD following accident trauma. *Neurocase, 13*(5), 342-357.

Pole, N. (2007). The psychophysiology of posttraumatic stress disorder: A meta-analysis. *Psychological Bulletin, 133*(5), 725-746.

Pole, N. (2006). Moderators of PTSD-related psychophysiological effect sizes: Results from a meta-analysis. *Annals of the New York Academy of Science, 1071*, 422-424

Pole, N., Neylan, T. C., Best, S. R., Orr, S. P., & Marmar, C. R. (2003). Feat-potentiated startle and posttraumatic stress symptoms in urban police officers. *Journal of Traumatic Stress, 16*(5), 471-479.

Pole, N., Neylan, T. C., Otte, C., Henn-Hasse, C., Metzler, T. J., & Marmar, C. R. (2008). Prospective prediction of posttraumatic stress disorder symptoms using fear potentiated auditory startle response. *Biological Psychiatry, 65*(3), 235-240.

Poulin, M. J., Silver, R. C., Gil-Rivas, V., Holman, E. A., & McIntosh, D. N. (2009). Finding social benefits after a collective trauma: Perceiving societal changes and well-being following 9/11. *Journal of Traumatic Stress, 22*(2), 81-90.

Powers, M., & Emmelkamp, P. (2009). Virtual reality exposure therapy for anxiety disorders: A meta-analysis. *Journal of Anxiety, 22*(3), 561-569.

Rabe, S., Beauducel, A., Zollner, T., Maercker, A., & Karl, A. (2006). Regional brain electrical activity in posttraumatic stress disorder after motor vehicle accident. *Journal of Abnormal Psychology, 115*(4), 687-698.

Rabe, S., Dorfel, D., Zollner, T., Maercker, A., & Karl, A. (2006). Cardiovascular correlates of motor vehicle accident related posttraumatic stress disorder and its successful treatment. *Apoplied Psychophysiology and Biofeedback, 31*(4), 315-330.

Rabe, S., Zoellner, T., Beauducel, A., Maercker, A., & Karl, A. (2008). Changes in brain electrical activity after cognitive behavioural therapy for posttraumatic stress disorder in patients injured in motor vehicle accidents. *Psychosomatic Medicine, 70*(1), 13-19.

Rainforth, M.V., Schneider, R. H., Nidich, S. I., Gaylord-King, C., Salerno, J. W., & Anderson, J. W. (2007). Stress reduction programs in patients with elevated blood pressure: A systematic review and meta-analysis. *Current Hypertension Reports, 9*(6), 520-528.

Rauch, S. A. M., Defever, E., Favorite, T., Duroe, A., Garrity, C., Martis, B., & Liberzon, I. (2009). Prolonged exposure for PTSD in a Veterans Health Administration PTSD clinic. *Journal of Traumatic Stress, 22*(1), 28-35.

Rauch, S. A., Grunfeld, T. E., Yadin, E., Cahill, S. P., Hembree, E., & Foa, E. B. (2008). Changes in reported physical health symptoms and social function with prolonged exposure therapy for chronic posttraumatic stress disorder. *Depression and Anxiety, 26*(8), 732-738.

Ready, D. J., Thomas, K. R., Worley, V., Backscheider, A. G., Harvey, L. A., Baltzell, D., & Rothbaum, B. O. (2008). A field test of group based exposure therapy with 102 veterans with was related posttraumatic stress disorder. *Journal of Traumatic Stress, 21*(2), 150-157.

Resnick. P. A., & Miller, M. W. (2009). Posttraumatic stress disorder: Anxiety of trauma stress disorder. *Journal of Traumatic Stress, 22*(5), 384-390.

Richardson, J. D., Long, M. E., Pedlar, D., & Elhai, J. D. (2008). Posttraumatic stress disorder and health-related quality of life among a sample of treatment-seeking deployed Canadian forces peacekeeping veterans. *Canadian Journal of Psychiatry, 53*(9), 594-600.

Robertson, M., Humphreys, L., & Ray, R. (2004). Psychological treatments for posttraumatic stress disorder: Recommendation for the clinician based on a review of the literature. *Journal of Psychiatric Research, 10*(2), 106-118.

Rosenthal, N. E. (1998). *Winter blues: Seasonal affective disorder: What it is and how to overcome it.* New York: Guilford Press.

Rothbaum, B. O., Davidson, J. R., Stein, D. J., Musgnung, J., Tian, X. W., Ahmed, S., & Baldwin, D. S. (2008). A pooled analysis of gender and trauma-type effects on responsiveness to treatment of PTSD with venlafaxine extended release or placebo. *Journal of Clinical Psychiatry, 69*(10), 1529-1539.

Rottenberg, J., & Johnson, S. L. (2007). *Emotion and psychopathology: Bridging affective and clinical science.* Washington, DC: American Psychological Association.

Rudofossi, D. (2007). *Working with traumatized police officer-patients.* Amityville, NY: Baywood Publishing Company.

Sack, M., Hopper, J. W., & Lamprecht, F. (2004). Low respiratory sinus arrhythmia and prolonged psycho-physiological arousal in posttraumatic stress disorder: Heart rate dynamics and individual differences in arousal regulation. *Biological Psychiatry, 55*, 284-290.

Sailer, U., Robinson, S., Fischmeister, F. P., Konig, D., Oppenauer, C., Leuger-Schuster, B., Moser, E., Kryspin-Exner, I., & Bauer, H. (2008). Altered reward processing in the nucleus accumbens and mesial prefrontal cortex of patients with posttraumatic stress disorder. *Neuropsychologia, 46*(11), 2836-2844.

Salsman, J. M., Segerstrom, S. C., Brechting, E. H., Carlson, C. R., & Andrykowski, M. S. (2008). Posttraumatic growth and PTSD symptomatology among colorectal cancer survivors: A 3-month longitudinal examination of cognitive processing. *Psychooncology, 18*(1), 30-41.

Salzberg, S. (1997). *A heart as wide as the world.* Boston: Shamballa.

Sareen, J., Belik, S., Afifi, T. O., Asmundson, G. J. G., Vox B., J., & Stein, M. B. (2008). Canadian military personnel's population attributable fractions of mental disorders and mental health service use associated with combat and peacekeeping operations. *American Journal of Public Health, 98*(12), 2191-2198.

Sarlo, M., Buodo, G., Munafo, M., Stegagno, L., & Palomba, D. (2008). Cardiovascular dynamics in blood phobia: Evidence for a key role of sympathetic activity in vulnerability to syncope. *Psychophysiology, 45*, 1038-1045.

Sayer, N. A., Chiros, C. E., Sigford, B., Scott, S., Clothier, B., Pickett, T., & Lew, H. L. (2008). Characteristics and rehabilitation outcomes among patients with black and other injuries sustained during the global war on terror. *Archives of Physical Medicine and Rehabilitation, 89*, 163-170.

Schardt, W. H., Nusser, D., Noethen, C., Rietschel, M., Hoffman, P., Cichon S., Goschke, T., & Erk, B. (2009). Neural correlated and genetic modulation of affect regulation. *Clinical EEG and Neuroscience, 49*(1), 57.

Schuff, N., Neylan, T. C., Fox-Bosetti, S., Lenoci, M., Samuelson, K. W., Studholme, C., Kornak, J., Marmar, C. R., & Weiner, M. W. (2008). Abnormal N-acetyrasparate in hippocampus and anterior cingulate in posttraumatic stress disorder. *Psychiatry Research, 162*(2), 147-157.

Schnyer, D. M., Zeithamova, D., & Williams, V. (2009). Decision-making under conditions of sleep deprivation: Cognitive and neural consequences. *Military Psychology, 21*(Suppl.), S36-S45.

Schottenbauer, M. A., Glass, C. R., Arnkoff, D. B., Tendick, V., & Gray, S. H. (2008). Nonresponse and dropout rates in outcome studies on PTSD: Review and methodological considerations. *Psychiatry, 71*(2), 134-168.

Schwartz, M. S., & Andrasik, F. (Eds.). (2003). *Biofeedback: A practitioner's guide* (3rd ed.). New York: Guilford.

Schuff, N., Neylan, T. C., Fox-Bosetti, S., Lenoci, M., Samuelson, K. W., Studholme, C., Kornak, J., Marmar, C. R., & Weiner, M. W. (2008). Abnormal N-acetylsasparate in hippocampus and anterior cingulate in posttraumatic stress disorder. *Psychiatry Research: Neuroimaging, 162*(2), 147-57.

Scofield, H., Roth, T., & Drake, C. (2008). Periodic limb movements during sleep: Population prevalence, clinical correlates, and racial differences. *Sleep, 31*(9), 1221-1227.

Shalev, A. Y., Tuval-Masiach, R., & Hadar, H. (2004). Posttraumatic stress disorder as a result of mass trauma. *Journal of Clinical Psychiatry, 65*(Suppl. 1), 4-10

Shankman, S. A., Silverstein, S. M., Williams, L. M., Hopkinson, P. J., Kemp, A. H., Delmingham, K. L., Bryant, R. A., McFarlane, A., & Clark, C. R.

(2008). Resting electroencephalogram asymmetry and post traumatic stress disorder. *Journal of Traumatic Stress, 21*(2), 190-198.

Shelby, R. A., Golden-Kreutz, D. M.,, & Andersen, B. L. (2008). PTSD diagnoses, subsyndromal symptoms, and comorbities contribute to impairments for breast cancer survivors. *Journal of traumatic stress, 21*(2), *165-172.*

Shin, L. M., Rauch, S. L., & Pitman, R. K. (2006). Amygdala, medial prefrontal cortex, and hippocampal function in PTSD. *Annals of the New York Academy of Science, 1071,* 67-79.

Shucard, J. L., McCabe, D. C., & Szymanski, H. (2008). An event-related potential study of attention deficits in posttraumatic stress disorder during auditory and visual go/no go continuous performance tasks. *Biological Psychology, 79*(2), 223-233.

Siepmann, M., Aykac, V., Unterdorfer, J., Petrowski, K., & Mueck-Weymann, M. (2008). A pilot study on the effects of heart rate variability biofeedback in patients with depression and in healthy subjects. *Applied Psychophysiology and Biofeedback, 33,* 195-201.

Simmons, A. N., Paulus, M. P., Thorp, S. R., Matthews, S. C., Norman, S. B., & Stein, M. B. (2008). Functional activation and the neural networks in women with posttraumatic stress disorder related to intimate partner violence. *Biological Psychiatry, 64*(8), 681-690.

Simon, N. M., & Associates. (2008). Paroxetine CR augmentation for posttraumatic stress disorder refractory to prolonged exposure therapy. *Journal of Clinical Psychiatry, 69*(3), 400-405.

Sledjeski, E. M., Speisman, B., & Dierker, L. C. (2008). Does number of lifetime traumas explain the relationship between PTSD and chronic medical conditions? answers from the National Comorbidity Survey Replication (NCS-R). *Journal of Behavioral Medicine, 31*(4), 341-349.

Smith, T. C., Wingard, D. L., Ryan, M. A., Kritz-Silverman, D., Slymen, D. J., & Sallis, J. F., for the Millennium Cohort Study Team. (2008). Prior assault and posttraumatic stress disorder after combat deployment. *Epidemiology, 19,* 505-512.

Spetalen, S., Sandvik, L., Blomhoff, S., & Jacobsen, M. B. (2008). Autonomic function at rest and in response to emotional and rectal stimuli in women with irritable bowel syndrome. *Digestive Diseases and Sciences, 53*(6), 1652-1659.

Spitzer, C., Barnow, S., Volzke, H., John, U., Freyberger, H. J., & Grabe, H. J. (2008). Trauma and posttraumatic stress disorder in the elderly: Findings from a German community study. *Journal of Clinical Psychiatry, 69*(5), 693-700.

Stahl, L. A., Begg, D. P., Weisinger, R. S., & Sinclair, A. J. (2008). The role of omega-3 fatty acids in mood disorders. *Current Opinion in Investigational Drugs, 9*(1), 57-64.

Stellman, J. M., Smith, R. P., Katz, C. L., Sharma, V., Charney, D. S., Herbert, R., Moline, J., Luft, B. J., Markowitz, S., Udasin I., Harrison, D., Barln, D., Langrigan, P. J., Levin, S. M., & Southwick, S. (2008). Enduring mental health morbidity and social function impairment in World Trade Center rescue, recovery, and clean-up workers: The psychological dimension of an environmental health disaster. *Environmental Health Perspectives, 116*(9), 1248-1253.

Striefel, S. (2009). Ethical treatment of traumatic brain injury. *Biofeedback, 37*(3), 88-91.

Suvorov. N. (2006). Psycho-physiological training of operators in adaptive biofeedback cardiorhythm control. *Spanish Journal of Psychology, 9*(2), 193-200.

Swanson, K. S., Gevirtz, R. N., Brown, M., Spira, J., Guarneri, E., & Stoletnity, L. (2009). The effect of biofeedback on function in patients with heart failure. *Applied Psychophysiology and Biofeedback, 34*, 71-91.

Tafil-Klawe, M., Klawe, J. J., Zlomaczuk, P., Szczepanska, B., Sikorski, W., & Smiettanowski, M. (2007). Daily changes in cardiac and vascular blood pressure components during breath holding episodes in obstructive sleep apnea patients after day-shift and night-shift work. *Journal of Physiology and Pharmcology, 58*(Suppl. 5, part 2), 685-690.

Taylor, F. B. (2003). Tiagabine for posttraumatic stress disorder: A case series of 7 women. *Journal of Clinical Psychiatry, 64*(12), 1421-1425.

Taylor M. K., et al. (2009). Behavioral predictors of acute stress symptoms during intense m ilitary training. *Journal of Traumatic Stress, 22*(3), 212-217.

Taylor, S. (2006). *Clinician's guide to PTSD: A cognitive behavioral approach.* New York: Guilford Press.

Tedeschi, R. G., Park, C. L., & Calhoun, L. G. (Eds.). (1998). *Posttraumatic growth: Positive changes in the aftermath of crises.* Mahwah, NJ: Lawrence Erlbaum.

Teng, E. J., Bailey, S. D., Chaison, A. D., Petersen, N. J., Hamilton, J. D., & Dunn, N. J. (2008).Treating comorbid panic disorder in veterans with posttraumatic stress disorder. *Journal of Clinical and Consulting Psychology, 76*(4), 704-710.

Thompson, W. W., & Gottesman, I. I. (2008). Challenging the conclusion that lower preinduction cognitive ability increases risk for combat-related post-traumatic stress disorder in 2,375 combat-exposed Vietnam War veterans. *Military Medicine, 173*(6), 576-582.

Thornton, K. E. and Carmody, D. P. (2008). Efficacy of traumatic brain injury rehabilitation: Interventions of QEEG-guided biofeedback, computers, strategies, and medications. *Applied Psychophysiology and Biofeedback, 33,* 101-124.

Thought Technology. (n.d.). Thought Technology web site: www.thoughttechnoogy.com

Van der Kolk, B. A. (2006). Clinical implications of neuroscience research in PTSD. *Annals of the New York Academy of Science, 1071,* 277-293.

Van der Kolk, B. S., McFarlane, A. C., & Weisaeth, L. (Eds.). (1996). *Traumatic stress: The effects of overwhelming experience on mind, body, and society.* New York: Guilford.

Van Emmerik, A. A., P., Kamhuis, K. H., & Emmerkamp, P. M. G. (2008). Prevalence and prediction of re-experiencing and avoidance after elective surgical abortion: A prospective study. *Clinical Psychology and Psychotherapy, 15,* 378-385.

Van Loey, N. E., van Son, M. J., van der Heijden, P. G., & Ellis. I. M. (2008). PTSD in persons with burns: An explorative study examining relationships with attributed responsibility, negative and positive emotional states. *Burns, 34*(8), 1082-1089.

Vasterling, J. J., & Brewin, C. R. (Eds.). (2005). *Neuropsychology of PTSD: Biological, cognitive, and clinical perspectives.* New York: Guilford Press.

Vasterling, J. J., Schumm, J., Proctor, S. P., Gentry, E., King, D. W., & King, L. A. (2008). Posttraumatic stress disorder and health functioning in a non-treatment-seeking sample of Iraq war veterans: A prospective analysis. *Journal of Rehabilitation Research and Development, 45*(3), 347-358.

Veltmeyer, M. D., McFarlane, A. C., Bryant, R. A., Mayo, T., Gordon, E., & Clark, C. R. (2006). Integrative assessment of brain function in PTSD: Brain stability and working memory. *Journal of Integrative Neuroscience, 5*(1), 123-138.

Viedma-del Jesus, M. I., Delgado-Pastor, L. C., Mocaiber, I, Martinez, M., Carmona, M., & Vila, J. (2008). *Affective modulation of the startle reflex in specific phobia.* Paper presented at the Society for Psychophysiological Research Annual Conference.

Violanti, J. M., & Gehrke, A. (2004). Police trauma encounters: Precursors of compassion fatigue. *International Journal of Emergency Mental Health, 6*(2), 75-80.

Violanti, J. M., & Paton, D. (1999). *Police trauma: Psychological aftermath of civilian combat.* Springfield: Charles C Thomas.

Waelde, L. C., Uddo, M., Marquett, R., Ropelata, M., Freightman, S., Pardo, A., & Salazar, J. (2008). A pilot study of meditation for mental health workers following Hurricane Katrina. *Journal of Traumatic Stress, 21*(5), 497-500.

Walker, J. (2009). Current status of QEEG in the courtroom. *NeuroConnections,* January, 4-6.

Walker, J. (2009). Anxiety associated with post traumatic stress disorder-the role of quantitative electro-encephalograph in diagnosing and guiding neurofeedback training to remediate the anxiety. *Biofeedback, 37*(2), 67-70.

Wallen, K., Chaboyer, W., Thalib, L., & Creeedy, D. K. (2008). Symptoms of acute posttraumatic stress disorder after intensive care. *American Journal of Critical Care, 17*(6), 534-543/

Ware, J., & Morin, C. (2000). *Understanding sleep.* Washington, DC: American Psychological Association.

Wang, S. (2006). Traumatic stress and thyroid function. *Child Abuse and Neglect, 30,* 585-588.

Werner, N.S., Meindl, T., Engel, R. R., Reiser, M., & Fast, K. (2008). *Psychiatric Research.*

Whalley, M. G., Rugg, M. D., Smith, A. P., Dolan, R. J., & Brewin, C. R. (2008). Incidental retrieval of emotional contexts in post-traumatic stress disorder and depression: An fMRI study. *Brain and Cognition, 69*(1), 98-107.

Wilcox, H. C., Storr, C. L., & Breslau, N. (2009). Posttraumatic stress disorder and suicide attempts in a community sample of urban American young adults. *Archives of General Psychiatry, 66*(3), 305-311.

Wikman, A., Bhattacharyya, M, Perkins-Porras, L., & Steptoe, A. (2008). Persistence of posttraumatic stress symptoms 12 and 36 months after acute coronary syndrome. *Psychosomatic Medicine, 70*(7), 764-772.

Wild, J., & Gur, R. C. (2008). Verbal memory and treatment response in post-traumatic stress disorder. *British Journal of Psychiatry, 193*(3), 254-255.

Wilson, J. P., & Keane, T. M. (Eds.). (1997). *Assessing psychological trauma and PTSD.* New York: Guilford.

Wilson, J. P., & Keane, T. M. (Eds.). (2004). *Assessing psychological trauma and PTSD* (2nd ed.). New York: Guilford.

Williams, M., Teasdale, J., Segal, Z., & Kabat-Zinn, J. (2007). *The mindful way through depression: Freeing yourself from chronic unhappiness.* New York: Guildford.

Wittman, L., Schredl, M., & Kramer, M. (2007). Dreaming in posttraumatic stress disorder: A critical review of phenomenology, psychophysiology, and treatment. *Psychotherapy and Psychosomatics, 76,* 25-39.

Wolf, G. K., Reinhard, M., Cozolino, L. J., Caldwell, A., & Asamen, J. K. (2009). Neuropsychiatric symptoms of complex posttraumatic stress disorder: A preliminary Minnesota Multiphasic Personality Disorder Inventory Scale to identify adult survivors of childhood abuse. *Psychological Trauma: Theory, Research, Practice, and Policy, 1*(1), 49-64.

Woodward, S. (2004, Fall). PTSD sleep research: An update. *PTSD Research Quarterly,* 1-3.

Woodward, S. H., Kaloupek, D. G., Streeter, C. C., Martinez, C., Schaer, M., & Eliez, S. (2005). Decreased anterior cingulate volume in combat-related PTSD. *Biological Psychiatry, 59*(7), 582-587.

Woodward, S. H., Kaloupek, D. G., Schaer, M., Martinez, C., & Eliez, S. (2008). Right anterior cingulate cortical volume covaries with respiratory sinus arrhythmia magnitude in combat veterans. *Journal of Rehabilitation Research and Development, 45*(3), 451-464.

Woon, F. L & Hedges, D. W. (2009). Amygdala volume in adults with posttraumatic stress: A meta-analysis. *Journal of Neuropsychiatry and Clinical Neuroscience, 21*(1), 5-12.

World Health Organization. (2007). *International statistical classification of diseases and related health problems.* Retrieved from http://apps.wh0.int/classifications/apps/icd/icd10online

Yehuda, R., Schmeidler, J., Labinsky, E., Bell, A., Morris, A., Zemelman, S., & Grossman, R. A. (2008). Ten-year follow-up study of PTSD diagnosis, symptom severity and psychosocial indicies in aging holocaust survivors. *Acta Psychiatrica Scandanavica, 119*(1), 25-34.

Zatzick, D., Jurkovich, G. J., Wang, J., Fan, M. Y., Joesch, J., & Mackenzie, E. (2008). *Annals of Surgery, 248*(3), 429-437.

APPENDIX IX: ADDITIONAL READINGS

Adessky, R. S., & Freedman, S. A. (2005). Treating survivors of terrorism while adversity continues. *Digital Object Identifier, 10*, 443-454.

Anderson, S. L., Tomada, A., Vincow, E. S., Valente, E., Polari, A., & Teicher, M. H. (2008). Preliminary evidence for sensitive periods in the effects of childhood sexual abuse on regional brain development. *Journal of Neuropsychiatry and Clinical Neuroscience, 20*(3), 292-301.

Andre-Obadia, N., Mertens, P., Gueguen, A., Peyron, R., & Garcia-Larrea, L. (2008). Pain relief by rTMS: Differential effect of current flow but no specific action on pain subtypes. *Neurology, 71*(11), 833-840.

Arditi-Babchuk, H., Feldman, R, & Gilbia-Scherhtman, E. (2009). Parasympathetic reactivity to recalled traumatic and pleasant events in trauma-exposed individuals. *Journal of Traumatic Stress, 22*(3), 254-257.

Arikan, K., Boutros, N. N., Bozhuyuk, E., Poyraz, B. C., Saurun, B. M., Bayar, R., Gunduz, A., Karaali-Sauvrun, F., & Yaman, M. (2006). EEG correlates of startle reflex with reactivity to eye opening in psychiatric disorders: Preliminary results. *Clinical EEG and Neuroscience, 37*(3), 230-234.

Asmundson, G. J., G., Stapleton, J. A., & Taylor, S. (2004). Are avoidance and numbing distinct PTSD symptom clusters? *Journal of Traumatic Stress, 17*(6), 467-475.

Attias, J., Bleich, A., Furman, V., & Zinger, Y. (1996). Event related potentials in post-traumatic stress disorder of combat origin. *Biological Psychiatry, 40*, 373-381.

Badr, S. (2008). Central sleep apnea in patients with heart failure. *Heart Failure Review, 14*(3), 135-141.

Bailham, D., & Joseph, S. (2003). Post-traumatic stress following childbirth: A review of the emerging literature and directions for research and practice. *Psychology, Health, & Medicine, 8*(2), 159-168.

Baldwin, D. S., Anderson, I. M., Nutt, D. J., Bandelow, B., Bond, A., Davidson, J. R. T., Den Boer, J. A., Fineberg, N. A., Knapp, M., Scott, J., & Wittchen, H. U. (2005). Evidence-based guidelines for the pharmacological treatment of anxiety disorders: Recommendations from the British Association for Psychopharmacology. *Journal of Psychopharmacology, 19*(6), 567-596.

Ballenger, J. C., Davidson, J. R. T., Lecrubier, Y., Nutt, D. J., Marshall, R. D., Meeroff, C. B., Shalev., A. Y., & Yehuda, R. (2004). Consensus statement on posttraumatic stress disorder from the International Consensus Group on Depression and Anxiety. *Journal of Clinical Psychiatry, 65*(Suppl. 1), 55-62.

Becker, A. B., Zayfert, C., & Anderson, E. (2003). A survey of psychologists' attitudes towards utilization of exposure therapy for PTSD. *Behaviour Research and Therapy, 42*, 277-292.

Belanger. H. G., Vanderploeg, R. D., Curtiss, G., & Warden, D. L. (2007). Recent neuroimaging techniques in mild traumatic brain injury. *Journal of Neuropsychiatry and Clinical Neuroscience, 19*(1), 5-20.

Bellville, G., Guay, S., & Marchand, A. (2009). Impact of sleep disturbances on PTSD symptoms and perceived health. *Journal of Nervous and Mental Disorders, 197*(2), 126-132.

Birmes, P., Hatton, L., Brunet, A., & Schmitt, L. (2003). Early historical literature for post-traumatic symptomatology. *Stress and Health, 19*, 17-26.

Black, L. M., Hudspeth, W. J., Townsend, A. L., & Bodenhamer-Davis, E. (2009). EEG connectivity patterns in childhood sexual abuse: A multivariate application considering curvature of brain space. *Journal of Neurotherapy, 12*(2,3), 141-160.

Blanchard, E. B., & Hickling, E. J. (1997). *After the crash: Assessment and treatment of motor vehicle accident survivors.* Washington, DC: American Psychological Association.

Boggio, P. S., Sultani, N., Fecteau, S., Merabet, L., Mecca, T., Pascual-Leone, A., Basaglia, A., & Fregni, F. (2008). Prefrontal cortex modulation using transcranial DC stimulation reduces alcohol craving: A double-blind sham control. *Drug and Alcohol Dependency, 92*(1-3), 55-60.

Bolton, E. S., Glenn, D. M., Orsillo, S., Roemer, L., & Litz, B. T. (2003). The relationship between self-disclosure and symptoms of posttraumatic stress disorder in peacekeepers deployed to Somalia. *Journal of Traumatic Stress, 16*(3), 203-210.

Bonanno, G. A. (2004). Loss, trauma, and human resilience. *American Psychologist, 59*(1), 20-28.

Bowman, M. (1997). *Individual differences in posttraumatic response.* Mahwah, NJ: Lawrence Erlbaum.

Bray, R. L. (2006). Thought field therapy: Working through traumatic stress without the overwhelming responses. *Journal of Aggression, Maltreatment and Trauma 12*(1, 2), 103-123.

Brisson, J. I., Ehlers, A., Matthews, R., Piling, S., Richards, D., & Turner, S. (2007). Psychological treatments for chronic post-traumatic stress disorder: Systematic review and meta-analysis. *British Journal of Psychiatry, 190*, 97-104.

Britt, T. W., & Adler, A. B. (Eds.). (2003). *The psychology of the peacekeeper: Lessons from the field.* Westport, CT: Praeger.

Bronner, M. B., Peel, N., de Vries, A., Bronner, A. E., Last, B. F., & Grootenhuis, M. A. (2009). A community based study of posttraumatic stress disorder in the Netherlands. *Journal of Traumatic Stress, 22*(1), 74-78.

Brough, P. (2004). Comparing the influence of traumatic and organization stressors on the psychological health of police, fire, and ambulance officers. *International Journal of Stress Management, 11*(3), 227-244.

Buckley, T. C., Holohan, D., Greif, J. L., Bedard, M., & Suvak, M. (2004). Twenty-four ambulatory assessment of heart rate and blood pressure in chronic PTSD and non-PTSD veterans. *Journal of Traumatic Stress, 17*(2), 163-171.

Bystritsky, A., Kaplan, J. T., Feurner, J. D., Kerwin, L. E., Wadekar, M., Burock, M. Wu, A. D., & Icaboni, M. (2008). A preliminary study of fMRI-guided rTMS in the treatment of generalized anxiety disorder. *Journal of Clinical Psychiatry, 69*(7), 1092-1098.

Chahine, L. M., & Chemali, Z. N. (2006). Restless legs syndrome: A review. *CNS Spectrums, 11* (7), 511-520.

Chandrashekaria, R., Shaman, Z., & Auckley, D. (2008). Impact of upper airway surgery on compliance in difficult-to-manage obstructive sleep apnea. *Archives of Otolaryngology--Head and Neck Surgery, 134*(9), 926-930.

Christiansen, D. M., & Elkit, A. (2008). Risk factors predict post-traumatic stress disorder differently in men and women. *Archives of General Psychiatry, 7,* 24. doi:10.1186/1744-859X-7-24

Cluver, L., Fincham, D. S., & Seedat, S. (2009). Posttraumatic stress in AIDS-orphaned children exposed to high levels of trauma: The protective role of perceived social support. *Journal of Traumatic Stress, 22*(2), 106-112.

Cohen H., & Benjamin, J. (2006). Power spectrum analysis and cardiovascular morbidity in anxiety disorders. *Autonomic Neuroscience: Basic and Clinical, 128,* 1-8.

Cohen, H., Kaplan, Z., Kotler, M., Kouperman, I., Moisa, R., & Grisaru, N. Mania after transcranial magnetic stimulation in PTSD. (2004). *American Journal of Psychiatry, 161*(3), 515-524.

Collura, T. (2008). Neuronal dynamics in relation to normative electroencephalography assessment and training. *Biofeedback, 36*(4), 134-139.

Conrad, A., Isaac, L., & Roth, W. T. (2008). The psychophysiology of generalized anxiety disorder: 1. Pretreatment characteristics. *Psychophysiology, 43,*

366-376.

Cozza, S. J. (2005). Combat exposure and PTSD. *PTSD Research Quarterly*, 1-3.

Dalgleish, T. (2004). Cognitive approaches to posttraumatic stress disorder: The evolution of multi representational theorizing. *Psychological Bulletin, 130*(2), 228-260.

Danieli, Y., Rodley, N. S., & Weisaeth, L. (Eds.). (1996). *International responses to traumatic stress*. Amityville, NY: Baywood Publishing Company.

Deacon, B. J., & Abramowitz, J. S. (2004). Cognitive behavioral treatments for anxiety disorders: A review of meta-analytic findings. *Journal of Clinical Psychology, 60*(4), 429-441.

Dean, E. T. (1997). *Shook over hell: Posttraumatic stress, Vietnam, and the civil war*. Cambridge, MA: Harvard University Press.

Debriec, J., & Altemus, M. (2006, September). Toward a new treatment for traumatic memories. *The Dana Forum on Brain Science*, 1-11.

Drabant, E. M., McRae, K., Manuck, S. B., Hariri, A. R., & Gross, J. J. (2008). Individual differences in typical reappraisal use predict amygdala and prefrontal responses. *Biological Psychiatry, 65*(5), 367-73.

Ebert, A., & Dyck, M. J. (2004). The experience of mental death: The core feature of complex posttraumatic stress disorder. *Clinical Psychology Review, 24*, 617-635.

Ehlers, A., & Clark, D. M. (2003). Early psychological interventions for adult survivors of trauma: A review. *Biological Psychiatry, 53*, 817-826.

Everly, G. S. (2003). Early psychological intervention: A word of caution. *International Journal of Emergency Mental Health, 5*(4), 179-184.

Everstoine, D. S., & Everstine, L. (1993). *The trauma response: Treatment for emotional injury*. New York: Norton.

Falconer, E., Bryant, R., Felmingham, K. L., Kemp, A. H., Gordon, E., Peduto, A., Olivieri, G., & Williams, L. M. (2008). The neural networks of inhibitory control in posttraumatic stress disorder. *Psychiatry Neuroscience, 33*(5), 413-422.

Felmingham, K., Kemp, A., Williams, L., Das, L., Hughes, G., Peduto, A., & Bryant, R. (2007). Changes in anterior cingulate and amygdala after cognitive behavior therapy of posttraumatic stress disorder. *Psychological Science, 18*(2), 127-129.

Felmingham, K., Kemp, A. H., Williams, L., Falconer, E., Olivieri, G., Peduto, A.,

& Bryant, R. (2008). Dissociative responses to conscious and non-conscious fear impact underlying brain function in post-traumatic stress disorder. *Psychological Medicine, 38,* 1771-1780.

Follette, V. M., Ruzek, J. I., & Abueg, F. R. (Eds.). (1998). *Cognitive-behavioral therapies for trauma.* New York: Guilford.

Forbes, D., Carty, J., Elliott, P., Creamer, M., McHugh, T., Hopwood, M., & Chemtob, C. M. (2006). Is mixed-handedness a marker of treatment response in posttraumatic stress disorder?: A pilot study. *Journal of Traumatic Stress Studies, 19*(6), 961-966.

Foy, D. W. (Ed.). (1992). *Treating PTSD: Cognitive behavioral strategies.* New York: Guilford.

Freedy, J. R., & Hobfoll, S. E. (1995). *Traumatic stress: From theory to practice.* New York: Plenum Press.

Fullerton, C. S., & Ursano, R. J. (1997). *Posttraumatic stress disorder: Acute and long-term responses to trauma and disaster.* Washington, DC: American Psychiatric Press.

Geracioti, T. D., Carpenter, L. L., Owens, M. J., Baker, D. G., Ekhator, N. N., Horn, P. S., Strawn, J. R., Sanacora, G., Kinkead, B., Price, L. H., & Nemeroff, C. B. (2006). Elevated cerebrospinal fluid substance P concentrations in posttraumatic stress disorder and depression. *American Journal of Psychiatry, 163*(4), 637-643.

Gilboa, A., Shalev, A. Y., Laor, L., Lester, H., Louzoun, Chisin, R., & Bonne, O. (2004). Functional connectivity of prefrontal cortex and the amygdala in posttraumatic stress disorder. *Biological Psychiatry, 55,* 263-272.

Glannon, W. (2007). Psychopharmacology and memory. *Journal of Medical Ethics, 32,* 74-78.

Gray, M. J., & Litz, B. T. (2005). Behavioral interventions for recent trauma: Empirically informed practice guidelines. *Behavior Modification, 29*(1), 189-215.

Green, C. H. (2009). The sky is NOT falling! Building on force strength. *The Military Psychologist.*

Guriel, J., & Fremouw, W. (2003). Assessing malingered posttraumatic stress disorder: A critical review. *Clinical Psychology Review, 23,* 881-904.

Haher, M. J., Rego, S. A., & Asnis, G. M. (2006). Sleep disturbances in patients with post-traumatic stress disorder: Epidemiology, impact and approaches to management. *CNS Drugs, 20*(7), 567-590.

Hammer, M. B., Robert, S., & Frueh, B. C. (2004). Treatment-resistant posttraumatic stress disorder: Strategies for intervention. *CNS Spectrums, 9*(10), 740-752.

Harpaz-Rotem, J., Rosenheck, R. A., Mohamed, S., & Desai, R. A. (2008). Pharmacologic treatment of posttraumatic stress disorder among privately injured Americans. *Psychiatric Services, 59*(10), 1184-1190.

Harvey, A. G., Bryant, R. A., & Tarrier, N. (2003). Cognitive behaviour for posttraumatic stress disorder. *Clinical Psychology Review, 23*, 510-522.

Harvey, A. G., Brewin, C. R., Jones, C., & Kopelman, M. D. (2003). Coexistence of posttraumatic stress disorder and traumatic brain injury: Towards a resolution of the paradox. *Journal of the International Neuropsychological Society, 9*, 663-676.

Harvey, A. G., Jones, C., & Schmidt, D. A. (2003). Sleep and posttraumatic stress disorder: A review. *Clinical Psychology Review, 23*, 377-407.

Hembree, E. A., Foa, E. B., Dorfan, N. M., Street, G. P., Kowalski, J., & Tu, X. (2003). Do patients drop out prematurely from exposure therapy for PTSD? *Journal of Traumatic Stress, 16*(6), 555-562.

Herrera R. W., DeJesus, M. J., Baxter, A. S., Ines, Q. M., & Pacheco De Toledo, F. M. (2008). Prevalence of mental disorder and associated factors in civilian Guatemalans with disabilities caused by internal armed conflict. *International Journal of Social Psychiatry, 54*(5), 414-424.

Hofman, S. G., Moscovitch, D. A., Pizzagalli, D. A., Litz, B. T., Kim, H., & Davis, L. L. (2005). The worried mind: Autonomic and prefrontal activity during worrying. *Emotion, 5*(4), 464-475.

Hooper, J. W., Spinazzola, J., Simpson, W. B., van der Kolk, B. A. (2006). Preliminary evidence of parasympathetic influence on basas heart rate in posttraumatic stress disorder. *Journal of Psychosomatic Research, 60*(1), 83-90.

Isaac, C. L., Cushway, D, & Jones, G. V. (2006). Is posttraumatic stress disorder associated with specific deficits in episodic memory? *Clinical Psychology Review, 26*, 939-955.

James, L. C., & Folen, R. A. (2004). EEG biofeedback as a treatment for chronic fatigue syndrome: A controlled case report. *Behavioral Medicine, 22*(2), 1-7.

Janoff-Bulman, R. (1992). *Shattered assumptions: Towards a new psychology of trauma.* New York: The Free Press.

Jaranson, J. M., & Popkin, M. K. (Eds.). (1998). *Caring for victims of torture.* Washington, DC: American Psychiatric Press.

Jelinek, L, Jacobsen, D., Kellner, M., & Larbig, F. (2006).Verbal and nonverbal memory functioning in posttraumatic stress disorder (PTSD). *Journal of Clinical and Experimental Neuropsychology, 28*(6), 940-948.

Jonsson P., & Hansson-Sandsten, M. (2008). Respiratory sinus arrhythmia in response to fear-relevant and fear-irrelevant stimuli. *Scandinavian Journal of Psychology, 49*, 123-131.

Jovanovic, T., Norrholm, S. D., Sakoman, A. J., Esterajher, S., & Kozaric-Kovacic, D. Altered resting psychophysiology and startle responses in Croatian combat veterans with PTSD. *International Journal of Psychophysiology*, November 5.

Karam, E & Ghosn, M. B. (2003). Psychosocial consequences of war among civilian populations. *Current Opinions in Psychiatry, 16*(4), 413-419.

Kedr, E. M., Abo-Elfotoh, N., & Rothwell, J. C. (2008). Treatment of post-stroke dysphagia with repetitive transcranial magnetic stimulation. *Acta Neurologica Scandinavica, 119*(3), 155-61.

Khayat, R., Patt, B., & Hayes, D. (2008). Obstructive sleep apnea: The new cardiovascular disease. Part 1: Obstructive sleep apnea and the pathogenesis of vascular disease. *Heart Fail Review, 4*(3), 143-53.

Kim, M. J., Chey, K. J., Chung, A., Bae, S., Khang, H., Ham, B., Yoon, S. J., Jeong, D. U., & Loo, I. K. (2008). Diminished rostral anterior cingulate activity in threat-related events in posttraumatic stress disorder. *Journal of Psychiatric Research, 42*(4), 268-277.

Kimble, M., & Kaufman, M. (2004). Clinical correlates of neurological change in posttraumatic stress disorder: An overview of critical systems. *Psychiatric Clinics of North America, 27*, 49-65.

Kito, S., Fujita, K., & Koga, Y. (2008). Changes in regional blood flow after repetitive transcranial magnetic stimulation of the left dorsal lateral prefrontal cortex in treatment-resistant depression. *Neuropsychiatry and Clinical Neuroscience, 20*(1), 74-80.

Kalric, M., Franiskovic, T.,, Klaric, M., Grkovic, J., Lisica, I. D., & Stevanovic, A. (2008). Social support and PTSD symptoms in war-traumatized women in Bosnia and Herzegovina. *Psychiatria Danubina, 20*(4), 466-473.

Kobayashi, I., Boart, J. M., & Delahanty, D. L. (2007). Polysomnographically measured sleep abnormalities in PTSD: A meta-analysis. *Psychophysiology, 44*, 660-669.

Koren, D., Hemel, D., & Klein, E. (2006). Injury increases the risk for PTSD: Examination of potential neurobiological and psychological mediators. *CNS Spectrums, 11*(8), 616-624.

Koronor, H., Winje, D., Ekeberg, O., Weisaeth, L., Kirkehei, I., Johansen, K., & Steiro, A. (2008). Early trauma-focused cognitive behavior therapy to prevent post-traumatic stress disorder and related symptoms: A systematic review and meta-analysis. *BMC Psychiatry, 8,* 81. doi:10.1186/1471-244X-8-81

Kross, E. K., Gries, C. J., & Curtis, J. R. (2008). Posttraumatic stress disorder following critical illness. *Critical Care Clinics, 24*(4), 875-87.

Krotopov, J. D. (2009). *Quantitative EEG, event-related potentials and neurotherapy.* San Diego, CA: Academic Press.

Kumar, R. (2008). Approved and investigational uses of modafinil: An evidence-based review. *Drugs, 68*(13), 1803-1839.

Laffaye, C., Cavella, S., Drescher, K., & Rosen, C. (2008). Relationships among PTSD symptoms, social support, and support source in veterans with chronic PTSD. *Journal of Traumatic Stress, 21*(4), 394-401.

Larson, C, L., Schaefewr, H. S., Siegle, G. J., Jackson, C. A. B., Anderle, M. J., & Davidson, R. J. (2006). Fear is fast in phobic individuals: Amygdala activation in response to fear relevant stimuli. *Biological Psychiatry, 60,* 410-417.

Lee-Chiong, T. L. (Ed.). (2006). *Sleep: A comprehensive handbook.* Hoboken, NJ: Wiley.

Linley, P. A. (2003). Positive adaptation to trauma: Wisdom as both process and outcome. *Journal of Traumatic Stress, 16*(6), 601-610.

Linley, P. A., & Joseph, S. (2004). Positive change following trauma and adversity: A review. *Journal of Traumatic Stress, 17*(1), 11-21.

MacNair, R. M. (2002). *Perpetration-induced traumatic stress: The psychological consequences of killing.* Westport, CT: Praeger.

Mallon, L., Broman, J. E., & Hetta, J. (2008). Restless leg symptoms with sleepiness in relation to mortality: 20 year follow-up study of a middle-aged Swedish population. *Psychiatry and clinical neuroscience, 62*(4), 457-463.

Marsella, A. J., Friedman, M. J., Gerrity, E. T., & Scurfield, R. M. (1996). *Ethnocultural aspects of posttraumatic stress disorder: Issues, research, and clinical applications.* Washington, DC: American Psychological Association.

McCleery, J. M., & Harvey, Al. G., (2004). Integration of psychological and biological approaches to trauma memory: Implications for pharmacological prevention of PTSD. *Journal of Traumatic Stress, 17*(6), 485-496.

McFarlane, A. C., Yehuda, R., & Clark, C. R. (2002). Biological models of traumatic memories and post-traumatic stress disorder: The role of neural networks. *Psychiatric Clinics of North America, 25*(2), 253-270.

McGaugh, J. L. (2002). Memory consolidation and the amygdala: A systems perspective. *Trends in Neurosciences, 25*(9), 456-461.

McKibben, E. S., Britt, T. W., Hoge, C. W., & Castro, C. A. (2009). Receipt and rated adequacy of stress management training is related to PTSD and other outcomes among Operation Iraqi Freedom veterans. *Military Psychology, 2*(Suppl. 2), S68-S81.

McNally, R. J., (2003). Psychological mechanisms in acute response to trauma. *Biological Psychiatry, 53*, 779-788.

McWilliams, L. A., Cox, B.J., & Asmundson, G. J. G. (2005). Symptom structure of posttraumatic stress disorder in a nationally representative sample. *Journal of Anxiety Disorders, 19*(6), 626-641.

Mellman, T. A., & Hipolito, M. M. S. (2006). Sleep disturbance in the aftermath of trauma and posttraumatic stress disorder. *CNS Spectrum, 11*(8), 1-5.

Mellman, T. A., Knorr, B. R., Pigeon, W. R., Leiter, J. C., & Akay, M. (2004). Heart rate variability during sleep and the early development of posttraumatic stress disorder. *Biological Psychiatry, 55*, 953-956.

Merfett, S. M., Metzler, T. J., Henn-Haase, C., McCaslin, S. , Inslicht, S., Chemtob, C., Neylan, T., & Marmar, C. R. (2008). A prospective study of trait anger and PTSD symptoms in police. *Journal of Traumatic stress, 24*(4), 410-416.

Milev, R., Abraham, G., Hasey, G., & Vabaj, J. L. (2008). Repetitive transcranial magnetic stimulation for treatment of medication-resistant depression in older adults: A case series. *Journal of ECT, 25*(1), 44-49.

Morgan, C. A., Krystal, J. H., & Southwick, S. M. (2003). Toward early pharmacological posttraumatic stress intervention. *Biological Psychiatry, 53*, 834-843.

Nestorius, Y., Martin, A., Rief, W., & Andrasik, F. (2008). Biofeedback treatment for headache disorders: A comprehensive efficacy review. *Applied Psychophysiology and Biofeedback, 33*, 125-140.

Neubauer, D. N. (2008). A review of ramelteon in the treatment of sleep disorders. *Neoropsychiatric Disease and Treatment, 4*(1), 69-79.

Neylan, T. C., Otte, C., Yehuda, R., & Marmar, C. R. (2006). Neuroendocrine regulation of sleep disturbances in PTSD. *Annals of the New York Academy of Science, 1071*, 203-215.

Ng, B. H. P., & Tsang, H. W. H. (2009). Psychophysiological outcomes of health qigong for chronic conditions: A systematic review. *Psychophysiology, 46*, 257-269.

Nutt, D. J. (2005). Overview of diagnosis and drug treatments of anxiety disorders. *CNS Spectrums, 10*(1), 49-56.

Nutt, D. J., & Malizia, A. L. (2004). *Journal of Clinical Psychiatry, 65*(Suppl. 1), 11-17.

Ohman, A. (2005). The role of the amygdala in human fear: Automatic detection of threat. *Psychoneuroendocrinology, 30,* 953-958.

Ohn, S. H., Park, C. I., Yoo, M. H., Choi, K. P., Kim, G. M., Lee, Y. T., & Kim, Y. H. (2008). Time-dependent effect of transcranial direct current stimulation on the enhancement of working memory. *Neuroreport, 19*(1), 43-47.

Opler, L. A., Grennan, M. S., & Opler, M. G., (2006). Pharmacotherapy of post-traumatic stress disorder. *Drugs of Today, 42*(12), 803-809.

Osuch, E. E., Benson, B. E., Luckenbaugh, D. A., Geraci, M., Post, R. M., & McCann, U. (2008). Repetitive TMS with exposure therapy for PTSD: A preliminary study. *Journal of Anxiety Disorders, 23*(1), 54-9.

Othmer, S., & Othmer, S. F. (2009). Post traumatic stress disorder: The neurofeedback remedy. *Biofeedback, 37*(1), 24-31.

Palm, S. L., Pires, M. L., Bittencourt, L. R., Silva, R. S., Santos, R. F., Esteves, A. M., Barreto, A. T., Tulik, S., & deMello, M. T., (2008). Sleep complaints and polysomnographic findings: A study of nuclear power plant shift workers. *Chronobiology International, 25*(2), 321-331.

Pennebaker, J. W. (Ed.). (1995). *Emotion, disclosure, and health.* Washington, DC: American Psychological Association.

Pivar, I., & Field, N. P. (2004). Unresolved grief in combat veterans with PTSD. *Anxiety Disorders, 18,* 745-755.

Pressman, M., & Orr, W. (Eds.). (1997). *Understanding sleep.* Washington, DC: American Psychological Association.

Raghuraj, P., & Telles, S. (2008). Immediate effect of specific nostril manipulating yoga breathing practices on autonomic and respiratory variables. *Applied Psychophysiology and Biofeedback, 33,* 65-75.

Ray, C. (1992). Positive and negative social support in a chronic illness. *Psychological Reports, 21,* 977-978.

Read, J. D., & Lindsay, D. S. (Eds.). (1997). *Recollections of trauma: Scientific evidence and clinical practice*. New York: Plenum.

Rober, S., Hammer, M. B., Ulmer, H. G., Lorberbaum, J. P., & Durkalski, V. L. (2006). Open-label trial of escitalopram in the treatment of posttraumatic stress disorder. *Journal of Clinical Psychiatry, 67*(10), 1522-1526.

Roizwnblatt, S., Fregni, F., Gimenez, R., Wetzel, T., Rigonatti, S. P., Tufik, S., Boggio, P. S., & Valle, A. C. (2007). Site-specific effects of transcranial direct current stimulation on sleep and pain in fibromyalgia; a randomized, sham-controlled study. *Pain Practice, 7*(4), 297-306.

Rona, R. J., Hooper, R., Jones, M., Iversen, A. C., Hull, L., Murphy, D., Hotopf, M., & Wessley, S. (2009). The contribution of prior psychological symptoms and combat exposure to post Iraq deployment mental health in the UK military. *Journal of Traumatic Stress, 22*(1), 11-19.

Rosen, G. M., & Powel, J. E. (2003). Use of a symptom validity test in the forensic assessment of posttraumatic stress disorder. *Journal of Anxiety Disorders, 17*(3), 361-367.

Rosen, J. B. (2004). A neurobiology of conditioned and unconditioned fear: A neurobehavioral system analysis of the amygdala. *Behavioral and Cognitive Neuroscience Reviews, 3*(1), 23-41.

Rosenbloom, D., & Williams, M. B. (1999). *Life after trauma: A workbook for healing*. New York: Guilford.

Saigh, P. A., & Bremner, J. D. (1999). *Posttraumatic stress disorder: A comprehensive text*. Boston: Allyn and Bacon.

Sarchielli, P., Presciutti, O, Alberti, A., Tarducci, R., Gobbi, G., Galletti, F., Costa, C., Eusebi, P., & Calabresi, P. (2008). A (1)H magnetic resonance spectroscopy study in patients with obstructive sleep apnea. *European Journal of Neurology, 15*(10), 1058-1064.

Savaas, L. S., White, D. L., Wieman, M., Daci, K., Fitzgerald, S., Laday, S. S., Tan, G., Graham, D. P., Cully, J. A., & El-Serag, H. B. (2008, September 10). Irritable bowel syndrome with dyspepsia among women veterans; prevalence and association with psychological distress. *Alimentary Pharmacology and Therapeutics*.

Schauerbroeck, J., & Ganster, D. C. (1993). Chronic demands and responsivity to challenge. *Journal of Applied Psychology, 78*(1), 73-85.

Schmahl, C. G., Elzinga, B. M., Ebner, U. W., Simms, T., Sanislow, C., Vermetten, E., McGlashan, T. H., & Bremmer, J. D. (2004). Psychophysiological reactivity to traumatic and abandonment scripts in borderline personality and posttraumatic stress disorders: A preliminary report. *Psychiatry Research, 126*, 33-42.

Schnurr, P. S., & Green, B. L. (Eds.). (2004). *Trauma and health: Physical health consequences of exposure to extreme stress*. Washington DC: American Psychological Association.

Schoenfeld, F. B., Marmar, C. R., & Neylan, T. C. (2004). Current concepts in pharmacotherapy for posttraumatic stress disorder. *Psychiatric Services, 55*(5), 519-531.

Seedat, S., Warwick, J., van Heerden, B., Hugo, C., Zungu-Dirwati, N., van Kradenberg, J., & Stein, D. J. (2004). Single photon emission computed tomography in posttraumatic stress disorder before and after treatment with a selective serotonin re-uptake inhibitor. *Journal of Affective Disorders, 80*, 45-53.

Shackman, A., Sarinopolous, I, Maxwell, J. S., Pizzagalli, D. A., Lavric, A., & Davidson, R. J. (2006). Anxiety selectively disrupts visuospatial working memory. *Emotion, 6*(1), 40-61.

Sheikh, J. I., Woodward, S. H., & Leskin, G. A. (2003). Sleep in post-traumatic stress disorder and panic: Convergence and divergence. *Depression and Anxiety, 18*, 187-197.

Shore, A. N. (2002). Dysregulation of the right brain: A fundamental mechanism of traumatic attachment and the psychopathogenesis of posttraumatic stress disorder. *Australian and New Zealand Journal of Psychiatry, 36*, 9-30.

Spronk, D., Arns, M., Bootsma, A., Can Ruth, R., & Fitzgerald, P. B. (2008). Long term effects of left frontal rTMS on EEG and ERPs in patients with depression. *Clinical EEG and Neuroscience, 39*(3), 118-124.

Stimpson, N. J., Thomas, H. V., Weightman, A. L., Dunstan, F., & Lewis, G. (2003). Psychiatric disorder in veterans of the Persian Gulf. *British Journal of Psychiatry, 182*, 391-403.

Tedstone, J. E., & Tarrier, N. (2003). Posttraumatic stress disorder following medical illness and treatment. *Clinical Psychology Review, 23*, 409-448.

Thakur, G. A., Joober, R., & Brunet, A., (2009). Development and persistence of posttraumatic stress disorder and the 5-HTTLPR polymorphism. *Journal of Traumatic Stress, 22*(3), 240-243.

Toch, H. (2002). *Stress in policing*. Washington, DC: American Psychological Association.

Tranulis, C., Sepehry, A. A., Galinowski, A., & Stip, E. (2008). Should we treat auditory hallucinations with repetitive transcranial magnetic stimulation: A metanalysis. *Canadian Journal of Psychiatry, 59*(9), 577-586.

Underhill, F. J., Konopka, J., & Hines, L. (2006). Decreased neuropsychological memory functioning and beta amplitude asymmetry in a PTSD population. *Clinical EEG and Neuroscience, 37*(3), 269.

Van Leimp, S., Vermetten, E., Geuze, E., & Westenberg, H. (2006). Psychotherapeutic treatment of nightmares and insomnia in posttraumatic stress disorder. *Annals of the New York Academy of Science, 1071*, 502-507.

Villareal, G., & King, C. (2004). Neuroimaging studies reveal brain changes in posttraumatic stress disorder. *Psychiatric Annals, 34*(11), 845-856.

Vojvoda, d., Weine, S. M., McGlashan, T., Becker, D. F., & Southwick, S. M. (2008). Posttraumatic stress disorder in Bosnian refugees 3 ½ years after resettlement. *Journal of Rehabilitation Research and Development, 45*(3), 421-426.

Walker, M. P. (2008). Sleep-dependent memory processing. *Harvard Review of Psychiatry, 16*(5), 287-298.

Walker, M. P., & Stickgold, R. (2006). Sleep, memory, and plasticity. *Annual Review of Psychology, 57*, 139-166.

Wittmann, L., Moergeli, H., Martin-Soelch, C., & Schnyder, U. (2008). Comorbidity in posttraumatic stress disorder: A structural equation modelling approach. *Comprehensive Psychiatry, 49*(5), 430-440.

Yehuda, R. (2004). Risk and resilience in posttraumatic stress disorder. *Journal of Clinical Psychiatry, 65*(Suppl. 1), 29-36.

Yehuda, R. (Ed.). (1999). *Risk factors for posttraumatic stress disorder.* Washington, DC: American Psychiatric Press.

Yehuda, R. (Ed.). (1998). *Psychological trauma.* Washington, DC: American Psychiatric Press.

Yehuda, R. (2002). Post-traumatic stress disorder. *The New England Journal of Medicine, 346*(2), 108-114.

Yehuda, R. (2003). Adult neuroendocrine aspects. *Psychiatric Annals, 33*(1), 30-36.

Yehuda, R., & McFarlane, A. C. (Eds.). (1997). *Psychobiology of posttraumatic stress disorder.* New York: New York Academy of Sciences.

Yule, W. (Ed.). (1999). *Post-traumatic stress disorders: Concepts and therapy.* New York: Wiley.

Zuker, T. L., Samuelsonj, K. W., Muench, F., Greenberg, M. A., & Gevirtz, R. N. (2009). The effects of respiratory sinus arrhythmia biofeedback on heart rate variability and posttraumatic stress disorder symptoms: A pilot study. *Applied Psychophysiology and Biofeedback, 34,* 135-143.